ASPECTS OF GREEK AND ROMAN LIFE

General Editor: H. H. Scullard

* * *

GREEK AND ROMAN TECHNOLOGY

K. D. White

Greek and Roman Technology

K·D·White

CORNELL UNIVERSITY PRESS
Ithaca, New York

To the memory of John Ward Perkins

First published 1984 by Cornell University Press

International Standard Book Number
0/8014/1439/3
Library of Congress Catalog Card Number
82-74518
Printed and bound in Hungary

Contents

Introduction   6

PART I

1 Environment of Classical Technology   9

2 Investigating Technical Development   14

3 Technical Apparatus of Classical Civilization   18

4 Innovation and Development: a Survey   27

5 Power Resources of the Classical World   49

PART II

6 Agriculture and Food   58

7 Building   73

8 Civil Engineering and Surveying   91

9 Mining and Metallurgy   113

10 Land Transport   127

11 Ships and Water Transport   141

12 Hydraulic Engineering   157

Conclusions   172

Sources of Information   174

Appendices   189

Tables   221

Bibliography   241

Notes   255

List of Illustrations   265

Index   268

Introduction

The study of any branch of the history of technology is, of necessity, an integrated process. More precisely, the analysis requires integration at two levels; the material remains must be examined in close association with the documentary evidence, while the technical processes and their development must be seen in the context of the society in which they occured. As Professor Finley so cogently put it, 'The history of technology is more than a Boys' Book of Mechanics... The evolution of pottery-making, the development of iron technology or the restricted use of the water-mill... are not just the story of new knowledge of new skills. They are as much the story of labor, transport facilities and costs, the level of demand, social pressures favoring investment in land, even politics.'[1]

The overall design and detailed content of this book reflect this conviction, treating technology as an essential part of the ancient economy. To this end, emphasis is laid, not on the processes, machines or devices *per se*, but on the developmental aspect, so that the technical process may be viewed as emerging in an historical context, in response to a specific need or demand, or to some shift in the balance of the economy, such as a critical shortage of an important raw material or manufactured product, or manpower. Again, technology is not a recent phenomenon, a child of the Industrial Revolution; it has been an integral part of human societies since Palaeolithic men made the first burins, handaxes and scrapers. The importance of the historical context may be seen by comparing the actual power sources available to the classical and mediaeval worlds respectively, and then examining the differences between the two civilizations in the extent to which those power sources were applied. Writing of mediaeval advances in technology, in the spheres of agriculture and the applied arts, J. D. Bernal[2] pointed out that improvements in these areas 'took the direction of a substitution of mechanical for human action; of animal- and water-power for manpower. There is nothing, it is true, that the mediaeval craftsman could do that could not have been done by the Greeks or Romans, but they lacked the compelling incentive, the need to do more work with fewer men.' Within the classical world itself, the context of a technical invention, innovation or development is often of vital importance, and occasionally of momentous significance; the point is clearly brought out in some recent studies of an early advance in Greek engineering, the water tunnel constructed by the engineer *164* Eupalinos for Polycrates, tyrant of Samos, in connection with the water supply to his capital city. Writing on the subject in 1965, Jane Goodfield and Stephen Toulmin[3] treated the tunnel as an isolated phenomenon, having no relation to the history of civil engineering or of hydraulics. Much more illuminating was the approach of Alfred Burns[4], who set the tunnel 'in the context of the evolution... of surveying, its technique, its instrumentation, and its peculiar body of knowledge.'

The need for an up-to-date account of the technical resources of the classical world, of the methods by which they were exploited, and of their relationship with the socio-economic environment, has long been obvious to economic and social historians. Until quite recently, technology, its achievements and its limitations, have occupied a very lowly place in the hierarchy of classical studies. The fact that great scientists like Archimedes, and first-class engineers like Hero of Alexandria were interested in the appropriate complementary disciplines, Archimedes in the practical application of science to warfare, and Hero in the principles of mechanics, made it impossible for historians of ancient science to ignore completely their contributions to practice and theory respectively, but the subject has been admitted grudgingly and half-heartedly, and with some doubt as to whether it has a proper place in the history of science, as may be clearly seen in the following paragraph, 'The extent to which technological material should enter into a source book of Greek science is another difficult question. We have included chiefly material that seems either to have been a direct application of Greek scientific theory or to have contributed to its subsequent development.'[5] The hopes entertained in the early 1950s that this unsatisfactory state of affairs would come to an end with the appearance of the first two volumes of the monumental *History of Technology*[6] were unfortunately not fulfilled; the materials and techniques described by the various contributors were handled like the specimens in the cases of old-fashioned museums, totally cut off from the realities of the cultural milieu in which they had been produced, developed or matured. Some advance was signalled by the appearance during the same decade of the first volumes in Robert J. Forbes' *Studies in Ancient Technology*.[7] Here at last was an attempt at the systematic treatment of technical processes, based on, and securely anchored in, the ancient literary sources. Unfortunately, the value of this indispensable corpus has been marred by two serious defects: first, the author's all too evident dependence on translations of the Greek and Roman writers, not on direct acquaintance with the original texts; secondly, his failure to make use of the growing volume of important evidence provided by excavation.[8] Among more recent contributions to the subject, Henry Hodges' *Technology in the Ancient World*[9] suffers from the opposite defect; the author, who is an archaeological technologist of great experience, has no difficulty in handling the archaeological materi-

al. but either ignores the evidence provided by documents (as in his chapters on Egyptian and Babylonian technology), or, (as in the classical chapters) 'betrays ignorance of both the ancient writings and the best modern research'.[10]

The considerations set out above have dictated both the arrangement of the contents and the method of handling the material. The book is divided into two parts. Part I contains some general observations on the relationship between history and technology, on the transmission of technical processes, and on methods of investigating technical development (Chs 1 and 2). Chapter 3 deals with the technical apparatus of classical civilization, and discusses the interrelations of technology and the physical environment.

Chapter 4 contains a survey of areas of innovation and development, and Chapter 5 rounds off this part of the book with a review of the available sources of power for meeting the basic requirements, which include food, water, buildings, transport by sea, river and land, and defence. Part II is concerned with the major areas of activity in which the more important technological developments took place, from building construction to the processing of grain. The appendices, which cover a wide range of technical problems, have been designed either to take care of highly controversial questions such as the development of the water-mill (Appendix 6) or to give the detailed attention required in connection with the design and construction of merchant ships (Appendix 11) when discussion of the essential details would have disturbed the flow of the narrative. Finally, in order to maintain the emphasis laid on innovation and development in Part I, a series of tables offers the reader a convenient conspectus of technical development in a wide area of operations. The result is that while treatment *in extenso* is limited to the seven topics covered by the chapters of Part II, up-to-date information is provided on other subjects and there are few aspects that do not, in the end, find a place either in the chapters, the appendices or the tables.

The Mediterranean world in classical times was much affected by influences from beyond its borders, not least in the technical sphere, and consideration is given to the contributions made to its development by barbarian communities and peoples. Attention is also given to the rate of technical change, and to time-lags in the application of new or improved techniques or sources of power, as well as to areas of technical stagnation or recession. As for the scale of treatment of particular machines and processes, the recent appearance of a number of monographs, such as those of Eric Marsden, J. F. Healy, Paul Vigneron and the author of this book[11] in the military, mining, land transport and agricultural areas respectively, offering more accurate interpretations of the ancient documents and archaeological remains than that provided by earlier writers, have made it possible to focus attention on the part played by technology in the historical process, rather than on the details of particular machines or processes. Specially significant in this connection was the publication in 1978 of John Landels' *Engineering in the Ancient World*.[12] This monograph, which spells out in precise detail both the engineering construction and the operational technique of many important technical devices, as well as attempting estimates of output and efficiency, is complementary to the present study, and frequent reference will be made to the author's findings in the pages that follow.

Acknowledgments

A book such as this, which attempts to cover such numerous and diverse fields of study, owes so much to so many that it would be impossible for the author to acknowledge all his obligations — to historians, engineers, archaeologists and scientists.

I should like, however, to make special mention of three friends to whom a special debt is owing: Mark Hassall, an hour of whose company has invariably yielded a rich harvest; John Crook, who has 'castigated' the early drafts of several chapters, and produced order out of chaos; and Rod Tolley, of Penang, who has guided my faltering steps towards some understanding of the mysteries of metallurgy.

Among the many institutions which have given generous help, I wish to thank Anne Healy and her colleagues in the Library of the Institute of Classical Studies; the Council of the British School at Rome which elected me to the Balsdon Senior Research Fellowship and the Libarian and her colleagues, and the Leverhulme Trustees for the opportunities provided by the award of an Emeritus Fellowship. Thanks are also due to Mrs Lysbeth Merrifield for preparing the index, and to the Science Museum for giving permission to reproduce the passage on p. 210 with a quotation from G. S. Laird Clowes, *Sailing Ships,* Part I, Historical Notes, 3rd edition, published 1932. I am particularly indebted to the following for valuable assistance on specific topics: Lionel Casson, Don Emanuele and Eugen Maroti; above all, to my wife for her continuing forbearance and support.

Kenneth D. White

1 Environment of Classical Technology

Technology and history

The history of technology, unlike that of science, has been much neglected by historians, and for two interconnected reasons: first, because until very recently education has had a bias towards aristocratic and intellectual goals, while the practical arts and crafts have been disdained as inferior ('only the duds attend the woodwork class'!); secondly, because the so-called liberal education enjoyed by the majority of professional historians has effectively cut them off from contact with work in the factory, in the forge, or on the lathe. The principles on which some of the power-driven machines used in classical times depended are by no means easy for laymen to comprehend, and the reconstructions which are essential to the making of any credible estimate of their efficiency require the collaboration of scholars familiar with the ancient texts and engineers familiar with the processes they describe. Joint seminars and projects such as those mentioned in Landels' book (1978, 107 ff.) have had the effect, among a number of interesting results, of exposing clearly, and sometimes painfully, the necessity of relating technical processes closely to their historical environment, and of underlining the contention made earlier that technology is woven into the fabric of a society, and that it cannot be profitably studied in isolation.

Techniques and the environment

The earliest phases carry us back to man's first beginnings as a social animal, to his inherited capacity to co-ordinate hand and eye, first as tool-user, then as tool-maker, and beyond that stage of development to his more inventive role as machine-maker.[13] The development of new techniques has invariably led to two further results: first, an improvement in man's control of physical enviroment; secondly, an alteration in the environment itself, which in turn has thrown up new problems that could only be solved by more technical change. Thus fur clothing, windproof huts and mastery over fire enabled early men to leave their warmer habitats and spread into colder climates; but the destruction of the natural vegetational cover, by 'slash-and-burn' agriculture, and by uncontrolled felling for industrial or domestic fuel, has so altered, and continues to alter, the delicate balance of the environment, in marginal areas of low rainfall, as to cause widespread erosion turning grassland into scrub, and scrub into desert, and threatening the existence of entire populations. It is, therefore, essential that the physical environment of the Mediterranean region and its neighbouring lands, the patterns of settlement which arose within its borders, the methods of tillage and of animal husbandry which it favoured, and the incidence and accessibility of the fuels and raw materials required by its industries, should each receive its full share of attention.[14] The influences, whether favourable or constraining, may be simple or complex, and where the pattern is complex, the threads which are woven into it may have their origins in any part of the environment, natural or man-made, geological, climatic, social or economic. A few examples, taken from the history of agricultural technology and that of iron and steel-making respectively, may serve to demonstrate the need for close yet wide-ranging analysis. In agriculture, labour-saving machines in the shape of steam-ploughs or mechanical harvesters have made their appearance in areas where large open fields have predominated, the former in the wheatlands of East Anglia, the latter on the American prairie. The earliest known form of mechanical harvester, an animal-powered heading machine for grain, appeared in the first century AD in the

47 open plains of north-eastern France; in looking for signs of its appearance elsewhere, one would not expect to find it in peninsular Italy, where the inter-cultivation of sown and planted crops was the rule.[15] Turning to a very different area, we observe first that iron in its pure state is too soft for the making of tools or weapons; to produce the necessary toughness and strength the iron must be combined with very small quantities of carbon; the result of this alloying process is steel. The product of a Greek or Roman furnace was a spongy mass of slag, ash and other impurities, with globules of

24 iron embedded in it. Releasing the metal involved a series of reheatings and repeating hammerings. No wonder the product is known as 'wrought iron'! To turn this material into steel required more heating and hammering, followed by quenching and tempering. In the Roman province of Noricum two factors occurring together (one technical, the other environmental) made it possible to simplify methods and produce by direct process a high grade steel known everywhere from its place of origin. What did the trick was a combination of an improved furnace and a local ore rich in carbon.[16] Many centuries later the denudation of the forests led to a shortage of charcoal. Early attempts to use coal, which was both abundant and easy to win from near the surface, were unsuccessful (coal must first be converted into coke), but the success achieved by the English ironmaster Abraham Darby at Coalbrookdale in Shropshire was in part due to the low sulphur content of the local coal, which had been a major obstacle to its industrial use. Outside England coke was not used with success even by English experts specially brought over to continental Europe, and large-scale production there was delayed for half a century.[17]

10

Transmission of technical processes

Anyone who has attempted to trace the origin and transmission of a particular invention will have discovered that the evidence is in many cases both scarce and difficult to interpret. Was the invention in question invented only once, and transmitted from a single centre, or was it invented in several places? These difficulties are mainly of concern to the student of pre-classical history, since the Greeks and Romans inherited from earlier civilizations not only such inventions as fire, writing and the wheel, but four out of the five fundamental technical devices for the redirection of muscular effort, namely the lever, the wedge, the windlass and the pulley; the fifth, the screw, was, as far as we know, an innovation of the third century before Christ.[18] We shall, therefore, be mainly concerned with innovation and development from these basic inventions. In examining the transmission of technical inventions and processes, we can identify five methods: by migration of the possessor of a new technique; by trade; by war and conquest; by publication of discoveries and improved techniques, and by interaction between scientists and technologists. An attempt will be made in what follows to estimate the extent to which these categories apply to the transmission of techniques to and within the classical world.

Transmission by migration of the inventor or the craftsman

The notion of an original inventor who carried his invention far and wide, and thus attained virtual immortality (if he was not already an immortal!), is deeply embedded in the legends of ancient Greece, Prometheus, who stole the gift of fire from heaven, and brought its precious secret down to men, being the first of a long line of benefactors of mankind. Among those crafts which arose from the taming of the power of combustion and its application to the service of the human race, none has stood higher in esteem than that of the smith,[19] the master of the processes by which the elements of copper and tin are fused to make bronze, and of the much more numerous and difficult stages by which iron ore is made to give up its many impurities and is 'wrought' into useful tools or lethal weapons of iron or steel. The itinerant smith figures strongly in the legends of many peoples, a fact which may be adduced in support of the widely held opinion that the simplified process of making iron attributed to the Chalybes of the Caucasus region which ushered in the Iron Age in the Near East was a 'one-off' invention, which spread fairly rapidly in spite of attempts to keep it a secret.[20] Transfer of ideas by migration is, of course, well attested for scientists and philosophers (many distinguished migrants in fifth-century Greece were both), who have, during many historical periods, been men of itinerant habit. The Sophists, recently dubbed 'the first internationalists',[21] covered between them a very broad spectrum of enquiry, including some areas of applied science. Smiths have always been travellers, and so have engineers and masons.

Transmission by trade

From early times the interchange of raw materials and manufactured goods, and the movement of the people involved in such exchanges, have provided a stimulus to the transmission of techniques. The fact that the itinerant trader from a society enjoying a more advanced culture than that of his customers is in a position to communicate his experience of new ideas and unfamiliar methods makes him a potential purveyor of much more than the range of trade goods he has to sell. Although information on the spread of particular inventions or processes is lacking for the Graeco-Roman world, transmission through trade will have been an important factor in the rapid spread of so many technical processes and products. When Pliny, for example, reports a new device or technique as 'recently discovered' in this or that provincial area, we may reasonably infer that one of the most likely sources of his information will have been traders from the area in question.

1 Greek black-figured vase depicting a blacksmith's forge.

Transmission by war and conquest

Warfare has been from early times a notable means by which technical processes and skills have been transferred from one area to another. The transfer could take several forms, depending on the relationships between rivals, and the methods by which the victor consolidated his conquests. The all-out struggle for scientific and technical supremacy that was waged with increasing intensity throughout World War II to reach its climax in the indecisive rocket and the successful atomic bomb can be seen as a recent and highly dramatic stage in a competition that can be traced back to the third millennium BC, when Sargon created the first large-scale land empire. The wholesale transfers of conquered peoples within the territories of the Assyrian Empire represent a very different type of transmission; in this instance we have a conquering power importing skilled craftsmen from among its subjects, leaving only the peasantry to cultivate the land for their new masters. A well-known example of this policy is recorded in the story of the final overthrow of the kingdom of Judah by Nebuchadnezzar of Babylon.[22] In the classical period, the Roman conquest of the eastern Mediterranean led rapidly to a counter-conquest by 'captive Greece' of her 'fierce conqueror', in science, medicine, engineering and architecture, as well as in the liberal arts; and the later conquest of

11

what is now France and Belgium brought their Roman masters into fruitful contact with the superior skills of the Celtic barbarians, notably in metallurgy and road vehicle building.[23] It should be noticed, however, that reactions to conquest have varied greatly both on the part of the conqueror and on that of the conquered; the slowing down of the economy of the Roman Empire, and the failure of its rulers to ward off decline by technical innovation, to which attention will be directed in a later chapter, appear to illlustrate what Andreski called the 'parasitic appropriation of surplus', which in his view constitutes one of the most powerful among the factors that inhibit technical progress.[24]

Transmission by publication

It has been customary to assume that, before the invention of printing, communication of scientific and technical discoveries was sporadic and ineffective, but the available evidence does not support this notion. A major source of information on technical developments in the Roman period is the *Historia Naturalis* of Pliny. When he tells his readers about some technical improvement or innovation in agriculture, such as the Gallic whetstone (*HN* 18.261) or the wheeled plough (*HN* 18.172), it is surely reasonable to infer that he had means of communication with distant provinces, and perhaps foreign correspondents who supplied him with information.[25] As it happens, agriculture is of all technical subjects the best documented, at least for the Roman period, and it is surely significant that a practical handbook like Palladius' *Opus Agriculturae* was widely read. Why may we not assume that other technical subjects possessed a literature now lost to us? Dr Landels, in his chapter on power and energy sources, calls attention to the fact that we know very little about either the methods used or the standards reached in the manufacture of that highly important fuel, charcoal, 'it was a craft industry, like so many others, and it is most unlikely that any technical manuals were ever written' (1978, 32). The inference to be drawn from the above

statement appears to be that all crafts were matters of traditional techniques passed on within a closed circle of practitioners, and that nothing was committed to writing. Is this because the craftsmen could neither read nor write? The notion is *a priori* difficult to accept, and recent work in this area suggests that text-books and practical manuals did exist for many practical subjects.[26] Pliny, we are told, was an inveterate note-taker and abstracter of published material. Is it not likely that a great deal of his reading consisted of this type of information?

Development through co-operation between science and technology

The invention of new manufacturing processes, and the introduction of a wide range of gadgets and devices designed to cheapen the unit cost of articles intended for mass consumption, have been due in large measure to the close association of technologists with scientists working in their laboratories. The development of many industrial processes has owed much to a fruitful interchange of ideas between the bench and the laboratory and in both directions. In the ancient world, however, there was a gulf separating scientist from craftsman that tended to widen as time went on. The effect of this attitude of contempt for 'banausic' occupations was extremely restrictive: 'the fertile interplay between science and technology was almost totally absent'.[27]

The Greeks had no word to describe the scientist: their philosophers attached little importance to manual labour, or to advanced technology; even doctors, whose manipulative skills reached high levels, were just *technitai*. The principal motive animating Greek scientists was the intellectual satisfaction that arises from speculating about man and nature – an attitude of mind so congenial to certain types of Victorian don that one of the latter declared that to him the Greeks were 'Fellows of another College'.[28] As for the Romans, it is easy enough, with the aid of a few carefully selected passages from the writings of Cicero (who would no doubt have found the donnish attitude to science very much to his liking), to dis-

2 Wall painting, tomb of Trebius Iustus, Via Latina, Rome, depicting bricklaying in progress.

miss the unimaginative Romans as hostile to science and scientists. But the matter is not so simple as that. Pliny, writing in the 70s of the first century AD, has some scathing comments to make on the contemporary neglect of scientific as well as literary research, 'In spite of generous official patronage,' he writes, 'no addition whatever is being made to knowledge by means of original research, and in fact even the discoveries of our predecessors are not being thoroughly studied.' To this he adds a further comment that those concerned with commerce are so blinded by avarice that they do not understand that 'knowledge is a more reliable means even of making a profit' (2.117–18). Pliny's point is not without its relevance to contemporary research: today, in an age of incredibly swift technological advance, it is still not uncommon to hear of enterprises, both public and private, which do not allocate sufficient funds to research and development. Pliny's comment reveals no hostility to science; what he is concerned about is the slowing down of a process from a time when research was

flourishing to a time when stagnation has begun to set in. To see genuine hostility to applied science one must turn to the well-known passage in Plutarch's *Marcellus,* written two generations later, in which the scientist and inventor Archimedes is represented as despising the practical devices he invented to ward off the Roman attack on his native city of Syracuse as unworthy of his lofty scientific spirit: 'although his inventions had won for him a name and fame for superhuman sagacity, he would not consent to leave behind him any treatise on this subject, but regarding the work of the engineer and every art that ministers to the needs of life as ignoble and vulgar, he devoted his earnest efforts only to those studies whose charm and subtlety are not affected by the claims of necessity... And although he made many excellent discoveries, he is said to have asked his kinsmen and friends to place over his grave a cylinder enclosing a sphere, with an inscription giving the proportion by which the containing solid exceeds the contained' (Plut. *Marcellus* 17).[29]

2 Investigating Technical Development

Invention, innovation and application

Progress in the study of the technical infrastructure of the ancient world has been hampered by the fact that many of those who have written about it have failed to grasp what the history of technology is about, and have erroneously interpreted it as the study of the history of inventions.

Since very few major technical discoveries were made in classical times, *ergo,* so the argument runs, the classical world can be largely ignored.[30] The Greeks, it is argued, were theoreticians, the Romans practical engineers; consequently, neither Greeks nor Romans invented anything of importance. The argument, however, is founded on a misunderstanding; invention, in the sense of an entirely new source of power, such as steam, or atomic fission, or a revolution in the means of communication, such as wireless telegraphy or television, has been exceedingly rare. On the other hand, some of the most significant advances in man's control of his environment have come about, not as a result of an invention, but via one of its applications. The effective exploitation of an invention depends on many different factors, which may be either purely technical (for example, the presence or absence of certain materials or skills which are needed to translate the original idea into a machine, or a process), or they may be economic: an invention may not for a long period get beyond the drawing-board stage because of the high cost factor involved in turning out the product (see Ch. 5, pp. 55 ff. on water-mills). In other cases, again, the inhibiting factor may be a social one: where, for example, the society in which the invention was made lacked the organizational capacity to sustain the practical application of the idea. Often, especially in recent history, it is the economic factor that has been uppermost. Thus in aviation, jet propulsion was a British invention, but partly as a result of the weakness of the British economy, and partly through lack of a large internal market, the exploitation of this great invention has fallen almost exclusively into the hands of the great aircraft corporations of the USA. The complex nature of the problems that arise in this area of enquiry, the variety of possible contributing factors, and the consequent need for caution against drawing firm conclusions from incomplete data can best be gauged by looking at one or two examples from the ancient world. The first one, the pulley, was already in use in Assyria before the classical period in Greece; it makes its first appearance on Assyrian monuments as early as the eight century BC, over two thousand years later than the first appearance of the wheel, of which it is a development.[31] How can one account for the time-lag? Should it be regarded as an instance of a technical development being held back by the social structure of a Bronze Age society, or is it perhaps due to the fact that pulleys could not be produced cheaply before the Iron Age? The compound pulley is essential to the operation of a crane, whether powered by tread-wheel, or by steam or by electric winch; at what point, and in response to what simulus, did cranes first appear? In the present state of knowledge no satisfactory answer can be given to either of these questions. On the other hand, Vitruvius' detailed account of the subject (*De Arch.* X. 2. 1–10) provides valuable information on the types of crane in use in his time, and some clear indications of developments in their design (details

72–7

and illustrations in Landels 1978, 84 ff). The
famous bas-relief on the family tomb of the
Haterii of a century later than Vitruvius shows
a quintuple pulley system on the hoisting cable,
and a tread-wheel, with a crew of nine, five to
provide the motive power, two probably to
control the movement, and to communicate
with the construction workers and two to
control the ropes at the top of the jib.

3

The next example is the screw. There is a
strong tradition that the screw was a Greek invention of the third century BC (above, p. 10).
One of its earliest applications, a cylinder enclosing an endless screw, used for raising water
for irrigation, is associated with Archimedes of
Syracuse, and bears his name. There is no record
of any other applications for more than two centuries, but the prolific inventor Hero of Alexandria included in the third book of his treatise
Mechanike a detailed description of a double-
screw press, and of the screw-cutting machine
tool required for making the female screw-
thread which is essential to the construction of
the press.[32] It has been argued that the screw-
cutter, and with it the screw-nut that combines
with the externally threaded bolt to provide
engineers and builders with a simple uniting
device of universal application, were not invented before about AD 50, but this cannot be
proved. The questions raised here are of the
greatest importance to the furtherance of our
understanding of the world of classical antiquity. We need to scrutinize all the available
sources of information on particular processes,
literary, epigraphic and monumental, with the
aim of uncovering evidence of innovation and
development.

12

Methods of investigating technical processes

We must first examine the sources of information available to us (whether primary, in the
form of surviving tools, machines, manufacturing equipment and products of a particular
manufacturing process, or secondary, in the

3 Monument of the Haterii, Musei Vaticani, Rome.
Tread-wheel crane. The relief, *c.* 2nd century AD,
despite its misleading perspectives, contains much
valuable detail. Top, two men fasten a rope; below
them, three lugs fasten the stay ropes to the derrick.
Five men work the wheel, assisted by two others
standing underneath.

form of descriptions of the machine, equipment
or process concerned). Then, where this is possible with the quantity of information thus
made available, we must recreate the conditions
under which the item was made or the process
carried on in classical times. All too often in the
past we have tended to confine our attention to
the literary references, with scarcely more than
a cursory glance at the monuments on which
the object or the process has been depicted. *153*
How many of those who make confident asser- *154*
tions about the limited sailing capacities of
Roman merchantmen have looked at more than *155*
a handful of the extant representations—and
there are more than 600 (see p. 144).[33] With each

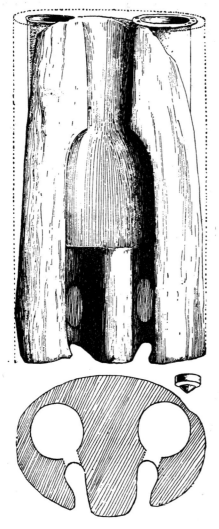

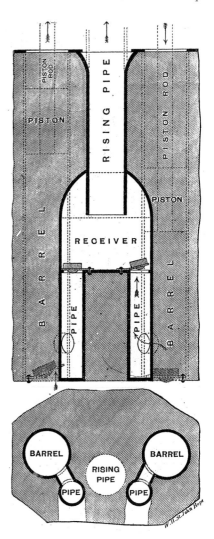

4 Ctesibian water-pump from Silchester, Britain. Above, oak 'cylinder' head; below, cross-section of cylinder head.

5 Detailed diagram of ill. *4;* above, half-section; below, cross-section.

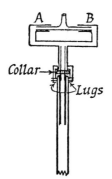

6 Diagram of nozzle attached to the Ctesibian pump for fire-fighting; it could be tilted up and down, and swivelled in any direction. The joints at A and B are 'sleeved' and a collar is fixed around the outer pipe. L-shaped lugs are attached to the inner one to engage with it.

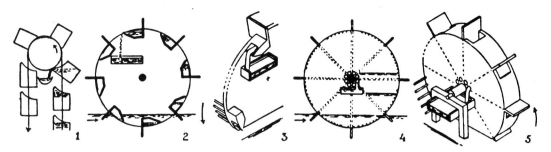

7 Three types of water-lifting machine described by Vitruvius: *1*, bucket-chain hoist; *2, 3*, scoop-wheels; *4, 5*, compartment wheel with paddles as for undershot water-mill.

kind of evidence the call is for greater precision. In dealing with literary texts we need more accurate texts, we need up-to-date commentaries,[34] and we need to free ourselves of misconceptions regarding the aims of those who wrote them; Hero of Alexandria was a competent engineer, not a mere dilettante inventor of mechanical toys, and the Elder Pliny deserves more than a disparaging 'brief mention'.[35] A good example of the kind of problem to be solved occurs in that writer's list of the methods in use in his time for raising irrigation water for the kitchen-garden. They include the *163* simple bucket-hoist, the swing-beam or *shaduf*, and the *organum pneumaticum*. I am convinced that this last item refers to the suction-pump *4* known as the 'Ctesibian' device' *(Ctesibica machina)*, so named after the Hellenistic engineer Ktesibios who is said to have invented it. The device is mentioned several times in Hero's *Pneumatika* (1.28, etc.), and the design and specifications are given by Vitruvius (*De Arch*. X. 7). Numerous devices answering to the description have been found, yet Landels (1978, 75 ff.) has provided the first detailed discussion of its operation, noting the important evidence for the development of the invention over the two and a half centuries that divide Ktesibios from Vitruvius.[36]

Reconstruction of the device or process with the greatest possible degree of conformity to the known conditions of antiquity is then necessary. This type of research, which calls for close collaboration between the archaeologist, the historian and the technologist, is a fairly recent development, in which pre-classical archaeologists have taken the lead. Some very important work has already been carried out on the manufacture of iron and steel, and on the construction of catapults.[37] The results so far published have both pointed the way to further experiments and underlined the need for close interdisciplinary co-operation; the history of technology cannot be satisfactorily investigated either by engineers, however well qualified, who cannot study the ancient writers at first hand, but are dependent on translations or references picked up from modern writers, or by historians who are unfamiliar with the techniques referred to by the ancient writers.

3 Technical Apparatus of Classical Civilization

The next task is to attempt an assessment of the technical demands occurring in those sectors of the economy which were involved in providing food, shelter, clothing and defence, together with other basic domestic needs, and with the means of transport needed to convey raw materials and manufactured products from place to place. This is indeed a formidable task, with source material both thin on the ground, and unevenly distributed in time as well as in department; thus for Athenian silver-mining surviving documentary evidence enables us to form a picture of the industry during the fourth century BC, but of its evolution during the crucial fifth century we know only what can be pieced together from scattered literary allusions.[38] Again, for agricultural technology there are many surviving handbooks covering the Roman period, whereas documents on land transport are extremely rare.[39] The researcher can learn much from workers in later and better documented periods of history; thus the student of sea transport, wrestling with the tangle of evidence on problems of ship-construction, of cargoes and their distribution, and so on, now emerging from the sea-floor of the Mediterranean, will find much to help him in the profusely documented pages of Fernand Braudel's *magnum opus*.[40]

In the absence of statistical information an adequate model of the economy must be constructed, which leads straight into an area of controversy, with two focal points that will now engage our attention: the first concerns the nature of the ancient classical economy, the second the role of industry within that economy. For reasons which will become obvious in the course of the ensuing discussion, it is proposed to detach the economy of the classical Greek *polis* from those of the Hellenistic monarchies and Rome. In classical Greece economic activity was regulated by a variety of non-economic considerations and concepts. First and most important of these was an attitude towards work standing in sharp contrast to that still prevailing in modern industrialized communities, 'The condition of the free man', writes Aristotle, 'is that he does not live for the benefit of another.' Free citizens of a Greek *polis* who found themselves obliged to work for a living, endeavoured to work on their own account, and not for anyone else ('the hireling flees because he is an hireling').[41] The second, of equal importance, was the belief that agriculture was the foundation of civilized life, with a concomitant down-grading of commercial and industrial occupations. Both concepts had considerable influence over the way in which the Greeks organized their industries. Over and above these attitudes must be taken into account the absence of anything resembling a 'work ethic', still less a productivity motive. 'In many Greek writers there are ideas which may superficially recall the modern theory of the division of labour, but looking at them more closely one realises that they have nothing to do with division of labour in the modern sense, i.e. they are concerned not with an increase in production, but with an improvement in the quality of the goods produced.'[42]

In his book *The Ancient Economy,* M. I. Finley sought to apply this model to the economy of the Roman Empire, leaving out of consideration the large units of political organization which developed after the decline of the independent *polis*. Not only, he argues, was agriculture the dominant economic activity, but the powerful men who owned most of the landed property exhibited all the behaviour patterns we have come to recognize as typical of their class. In particular, they displayed a negative attitude towards all forms of economic activity that were not connected with the land; their goal was leisure *(otium)*, not business *(negotium)* ; those, on the other hand, who occupied themselves in business, industry or trade, suffered politically and socially.[43] The distinction is a valid one, but can easily be pressed too hard. The evidence from late Republican times does not support a rigid division between landed aristocrat and man of business; even as early as the second century BC, if we are to believe Plutarch's account of his business connections,[44] the Elder Cato found time for other lucrative enterprises, apart from farming and politics. It is, of course, possible, though unlikely that, like a modern shareholder, he took no part whatever in the business activities in which he had invested his capital. With the coming of the Empire, the edges of the distinction became progressively blurred. As for the thesis that those engaged in industry suffered politically and socially, it does not seem to fit the evidence from Pompeii; a recent study of the wool industry in that city[45] shows that successful 'finishers' like Lucius Veranius Hypsaeus stood just as high in public esteem as the woollen manufacturer Baronius Sura of Aquinum, who became a member of the local senate and eventually reached the highest office in the local community, that of *duumvir*.

To return for a moment to the Elder Cato and his business ventures, it is of some importance to notice that all of them (see n. 44) belong to the same category, namely that of private contracting. This characteristically Roman form of economic activity played a major role in the economy, while at the same time, through the widespread use of business agents, leading men were able, without loss of prestige, to engage in these lucrative ventures.[46]

What then was the nature of the 'ancient economy'? It will be evident from the foregoing argument that I do not believe that the question as framed can be answered in any

meaningful sense. The 'primitive' system propounded by Finley can be seen to be true of many of the smaller city-states of Greece, but the economies of the Hellenistic monarchies that followed the break-up of Alexander's world empire, together with that of Rome after her conquest of the Mediterranean, seem to belong to a different order of things; a single illustration, taken from a fairly well-documented period, that of Rome's rapid expansion during the first half of the second century BC, will show the inadequacy of the 'primitive' system of economic relations. It comes from a high demand area—that of supplying military clothing and horses for the Roman army operating against the Macedonians. The job was let out to private contractors, and it called for the supply of 6,000 togas, 30,000 tunics, and 200 Numidian horses—no small order even in modern terms.[47] How could such a contract have been speedily executed within the confines of a 'non-market' economy? In view of the heavy demands made on its resources by the army, it is not surprising to find that the textile industry at Pompeii had developed features which are recognizably those of a market economy. These include factory production with male weavers, bulk sales of finished cloth, and a rationally organized system of production and distribution.[48] Nor were such advances confined to those areas of the Roman economy which were exposed to the stimulus of high demand from the authorities responsible for imperial defence, and which might, therefore, be regarded as untypical, and outside the play of normal economic forces. In a recent study of the quarrying industry John Ward Perkins has drawn attention to some remarkable improvemens effected by the Romans, both at the technical and the organizational level, including the stock-piling of 'standard' ready-cut stone blocks, and an imperially controlled accounting system. In sharp contrast, the records for the pre-Roman period reveal primitive conditions in the industry, the supply of stone being carried out on a hand-to-mouth basis.[49] The details of the system are discussed at length in Chapter 7, but it is clear that

8 Bas-relief from Linares, a major mining centre in Spain, showing a gang of miners on their way to work, wearing protective leather aprons, and carrying picks and lamps.

swing there may have been many thousands of slaves at work in the galleries, yet for this vital period the sources of information are minimal. The result of this part of the enquiry is the conclusion that the evidence does not justify the notion of a 'classical economy' of a non-market type; what it does appear to point to is there being at one and the same time differing levels of economic organization, with primitive 'reciprocal' and advanced 'market' economics existing side by side.[50] Thus, if we look closely at the economy of classical Athens, it can be readily agreed that it generally satisfies the definition of a 'primitive' economy as originally defined by Hasebroek;[51] there was no such thing as an economic policy in the modern sense of the word, for the very good reason that the greater part of economic activity in commerce and industry was carried on by outsiders (resident aliens and slaves), who had no access to political decision-making. The economic activities of the state were, in fact, restricted to ensuring that those commodities which were regarded as essential to the life of the people, and had to be imported, were brought in without obstruction—hence the attention given to controlling the routes by which the corn reached Athens, and the operations of the corn-dealers who handled the cargoes on arrival (see below, p. 25f). A society of this type, the economic interest of

any model which does not take account of sophisticated operations such as those mentioned above, cannot stand unmodified. As a further precaution against undue dogmatizing, it is important to remember how thin and badly distributed is the evidence for many sectors of the economy. For the Athenians, the exploitation of their rich resources of silver in the Laurion deposits was, at the zenith of their imperial power in the fifth century, of capital economic importance; when mining was in full

9 Mosaic from a large vaulted subterranean tomb at Sousse, Tunisia, *c.* 250 AD, depicting a coaster unloading in the surf. Bardo Museum, Tunis.

whose members was confined to consumption, and in which productive industry and commerce were in the hands of outsiders, will scarcely have provided a favourable climate for technical invention or innovation. The Roman economy, on the other hand, contained elements which do not fit this framework, and belong to a different economic milieu.

The next task is to examine the areas of high demand in the classical world, and to try to find out how far appropriate technical innovation and development came in by way of satisfying those demands. We shall first take a look at the elements that make for technological innovation, using well-documented examples from more recent history, and then proceed to review the fundamental technical requirements of the various sectors of the classical world.

In a paper on the history of agriculture, J. T. Schlebeker[52] has identified four elements that make for technological innovation: *i* accumulated knowledge; *ii* evident need; *iii* economic possibility; *iv* cultural and social acceptability.

Under his first heading we may observe that innovation does not come about *in vacuo*; the first phase of an inventive process has often been no more than a matter of aroused curiosity, giving rise to philosophical investigation of an inexplicable natural effect. This can be illustrated very clearly in the history of electricity, where research into magnetism and electricity was carried on in this way for a long time, without any practical applications in view. The key period of advance, which led to the construction of the first electrical generators, occurred at a time when there was evident need for two things: (1) mobility and transfer in the location of power sources, and (2) flexible and rapid controllable conversion. The evident need or demand may take many different forms: it may be for a labour-saving device or method to meet a shortage of labour, or it may be for a new fuel or source of energy to take the place of one that is becoming exhausted, as in the example of charcoal mentioned earlier. Categories three and four, if taken together, may often throw light on the well-known problems of

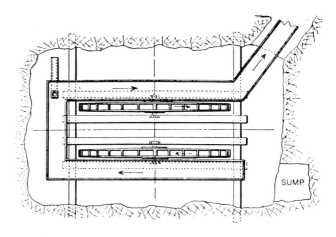

10 A pair of multiple drainage wheels, viewed from above, showing launders and opposed direction of flow. Rio Tinto, Spain.

time-lag between the first appearance of an invention and its general use. The application of water power via the water-wheel was known in the first century BC, but apparently was not used widely for several centuries (see Ch. 5, p. 56f). Failure of an invention or innovation to spread may be ascribed to the fourth category cancelling out the positive advantages included in the third.[53] We should at the same time remind ourselves that considerations of economic feasibility have frequently been set aside in favour of overriding political or military requirements. Nor should Schlebeker's list be regarded as exhaustive; the history of technology over the last two centuries provides numerous examples of developments being promoted or hindered by personal factors, more particularly the relationship between the ideas of scientists and their translation into machines or processes by engineers and technologists. To illustrate this category we may cite one of the phases in the development of the steam engine. The invention in 1797 of a screw-cutting lathe, a precision tool-making device, made it possible to turn out a variety of machine parts to a standard of accuracy not hitherto attainable. It is on record that the inventor of the lathe, Henry Maudslay, produced it in order to help the engineer Brunel, who was a close friend, realize his idea of a hydraulic press.[54]

11 Roman freighter under full sail, 3rd century AD. Notable are the vast mainsail and topsail, the small artemon, the swan awkwardly squeezed into the frame, and the deck house in front of the steersman. Relief found at Portus harbour. Museo Torlonia, Rome.

Inter-relation of technical development and the physical environment

The establishment and extension over the last 150 years of a vast network of communications spanning entire continents and stretching across the oceans that divided them in the past, have engendered fundamental changes in the relationship of natural resources and their exploitation by industrial processes. From remote times right up to and including the first Industrial Revolution there has been a close connection between the place of manufacture or processing and the place of origin of the raw material employed. The reduction and eventual exhaustion of the local supply, with which the industry had been associated from its inception, have led to an ever widening search for the necessary raw material, such as stone and timber for building, copper, lead and other basic metals. Changing levels of demand, both for the raw material, and for the transport required to carry it to the place of manufacture, have had an important bearing on the development of communications by land or by water,

and in this way on the design of ships, the extension of the inland waterways, and so on.

Turning from the sphere of industrial manufacture to that of food production, we find that the physical environment has been of capital importance in the form and location of land settlements, as well as in the spread of implements and techniques. The point comes out very strongly in a recent study of the origins and early development of agriculture in China: in a carefully documented study[55] Dr Ping-ti-Ho shows that the key factor which enabled the Chinese to anticipate the earliest farmers in the Fertile Crescent was the vast deposit of volcanic dust — the famous *loess*. This soil bonanza made possible the development of a system of cultivation based on limited rainfall, unleached soil, and drought-resistant food plants; this combination meant that they avoided regular fallowing, and the risk of increasing salinity in the soil through continuous recourse to irrigation. Chinese agriculture remained entirely distinct from the systems evolved in the Middle East, where the basic crops were domesticated wheat and barley, both of which required flooding or organized irrigation for success.

Climate and physical geography of the Mediterranean region

The general character of the physical environment of the Mediterranean region is well known, and up-to-date surveys are readily accessible.[56] Our concern here is to point to those areas where the shortcomings of the environment from the point of view of human settlement or its further development have given a stimulus to innovation. The first and most important feature of the climate is the seasonal distribution of the rainfall, the second the variation in the annual precipitation over the region. The fact that the region receives most of its rain in the winter months means that autumn-sown crops reach the final stages in their growth to maturity during the increasing heat of late spring and early summer. This in turn means that the food-grain crops must be harvested early; the onset of summer drought forbids delay; postponement results in rapid drying and shattering, with loss of grain. We shall, therefore, look to the processes of harvesting, threshing and winnowing (all distinct operations in ancient farming) as promising for technical advance.

Closely connected with the first feature is the fact that the aridity in the climate increases from north-west to south-east, with northern Spain at one end of the precipitation chart and the Nile valley at the other; the pattern is further modified by the nature of the terrain; the great backbone of the Pindus mountains not only makes communications in northern Greece difficult from west to east, but it has also the effect of forcing the storm-clouds to deposit most of their contents on the western side, so that Epirus gets 60 ins (1500 mm) per annum, and Athens a mere 16 ins (400 mm).

Characteristic of the region as a whole is the tendency for the bulk of the rainfall to be divided into two periods, towards the beginning and the end of winter respectively; these are the 'former' and the 'latter' rains of the Bible; when the former rains fail, the planting season is delayed, with the prospect of reduced yields at the other end of the cycle; failure of the 'latter' rains may well mean total disaster. In consequence, conservation of the limited water supply is essential, and technical developments may be looked for here, especially in the more arid areas of the south and south-east. The extreme example is Egypt, with virtually zero rainfall and elaborate irrigation. The areas that lay beyond the reach of the floodwaters were entirely dependent on mechanical water-raising devices, and it is significant that the standard term for both of the commonest of them, the compartment wheel or 'drum', and the 'Archimedean' screw-pump, was the undifferentiated word *mechane*—'machine', or 'device'.[57]

Many of these water-lifting devices may also be applied with success to the process of getting rid of surplus water. Well-drained soil is essen-

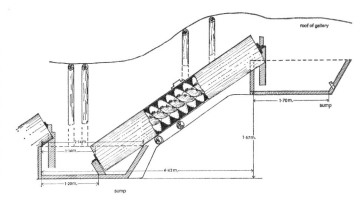

12 Archimedean screw from Centenillo, Spain. Diagram showing in section one of a series of water-screws used for mine drainage.

tial for raising sown crops such as wheat, and planted crops like grape-vines. Impervious clays which occur in the strata of many parts of the region cause waterlogging in the valley bottoms; the networks of underground drainage channels, such as those which have been traced in southern Etruria, show how Roman hydraulic engineers improved on the work of their Etruscan predecessors, to be later excelled by their successors in North Africa.[58]

Turning now to the relationship of the Mediterranean terrain to communication problems, we may note how, in mainland Greece, the complex pattern of relief and the remarkably indented coastline made land communications difficult. Ease of access to the sea reduced these to a very secondary role; the great trunk road, the 'Royal Road', that linked Sardis to the Persian capital of Susa, made a deep impression on Greek visitors to Asia, for they had nothing like it in their homeland; and it is significant that the first such road constructed in mainland Greece was the Via Egnatia, the Roman military highway that linked the Adriatic port of Dyrrachium with Byzantium, providing one of the major strands in the network of imperial roads, totalling some 56,000 miles or 90,000 km.

Human geography and technical development

Some of the points that call for attention in this part of our enquiry are obvious enough: concentrations of population in urban centres pose problems of organization of their food supplies and of the associated transport needs. The classic case is that of Rome, which suffered from intermittent food shortages, some of them serious, long before she became a power of any importance in the Mediterranean world. Helped out in this earlier phase by neighbouring communities with a surplus of wheat, she later became heavily dependent on the resources of the island of Sicily, and by the first century AD on those of the recently conquered kingdom of Egypt. The total of grain imported from Egypt was probably of the order of 100,000 tons, and the total shipping required was around 250 standard shiploads per annum. An undertaking of this magnitude might be expected to promote improvement in the design and operation of the cargo ships, and of the port installations required to handle the traffic.[19] No other city in the classical world presented technical problems of this magnitude; Syracuse, the largest urban centre in the Greek world, had some of the best wheatlands in the Mediterranean on her doorstep, and the large manufacturing cities of the East were

able, in normal years, to satisfy their grain requirements from the resources of their own hinterland. Under abnormal conditions, when supplies were reduced or interrupted by drought, pestilence or warfare, the inadequacy of land communication made it virtually impossible for neighbouring centres with a surplus to afford relief, unless both supplier and recipient were on or very near the coast. This factor also put a constraint on the growth of urban communities: 'towns could not safely outgrow the food production of their own immediate hinterlands unless they had direct access to waterways'.[60] In Greece, the case of Athens turns out to be, in this, as in many other respects, the exception to the rule. Here we have a city with a permanent food supply problem, which became more and more insistent with the growth in its population during the fifth century, the result being a discernible economic policy directed to the purpose of securing the vital supplies of imported corn from the Black Sea region, by taking and maintaining control of the Straits. The problems were, however, organizational, not technical, and no technical developments took

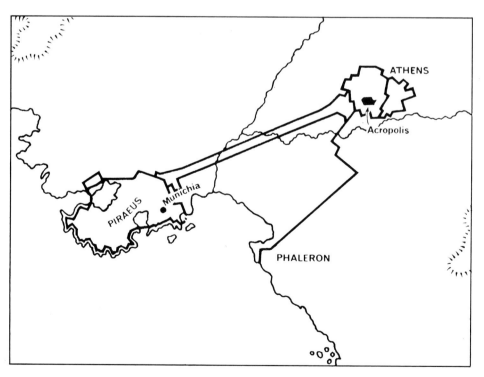

13 Plan showing the site of Athens and the harbours of her new port of Piraeus.

place. Athens was also exceptional in another way, and one that is more to our purpose. The attitude of contempt for manual occupations noted above as characteristic of ancient Greek thinking was notably absent in this city; citizens had, from the time of Solon, been required to teach their sons a trade, and technical skills were encouraged. In the course of the fifth century, aided by the splendid handling facilities created by the new harbour of Piraeus, Athens became a great centre of international trade. In this development as a clearing-house, she had one special advantage: her most valuable export commodity, silver, was available as a substantial source of profit for those importers 'who did not to take out return cargoes'.[61] Centres for the interchange of commodities often become centres for the exchange of ideas and techniques; it is no accident that many of the outstanding scientists of the ancient world worked in important entrepots of trade — Thales in sixth-century Miletus, and Archimedes in third-century Syracuse are among instances that come to mind—and this section concludes with observations on some of the more important of these commercial centres, beginning with Corinth.

Corinth On the outstanding importance of this city, whose commercial strength began to be felt from the middle of the seventh century BC, when the tyrant Cypselus overthrew the oligarchic rule of the Bacchiads, we may cite the description given many centuries later by the geographer Strabo, who visited the city during the reign of the emperor Augustus:

Corinth is said to be 'wealthy' because of its trading port. It lies on the Isthmus and controls two harbours, one of which looks straight towards Asia, the other towards Italy. The exchange of cargoes sailing in both directions is thus made easy for those who live at such a great distance from each other. And just as in former days the Sicilian Straits were dangerous for mariners, so too were the high seas, and especially the sea beyond Cape Malea, because of the contrary winds. Hence the proverb: 'When you double Cape Malea, forget about the way home.' It was therefore a welcome alternative to traders both from Italy and

from Asia to be able to unload their cargoes here. Similarly the excise duties on goods exported by land from the Peloponnese and imported into it were under the control of those who held the keys to the Isthmus. This state of things has persisted ever since ... Strabo VIII. 6. 20

Four centuries before Strabo the historian Thucydides had seen in Corinth the precursor of Athens as mistress of the seas, attributing her superiority to an important invention, that of the fighting trireme (for detailed discussion see Landels 1978, 140–151). According to Thucydides, the commercial strength of Corinth went back to a time long before the growth of seaborne commerce:

Planted on an Isthmus, Corinth had been a commercial centre from time immemorial; in olden days almost all communication between Greeks inside and outside the Peloponnese was maintained overland rather than by sea, and the highway by which it went passed through Corinthian territory. As a result of this, Corinth possessed great monetary resources, as indicated by the epithet 'wealthy' attached to the place by the old poets. When traffic by sea became more frequent among the Greeks, the Corinthians acquired a navy and suppressed piracy, and having established a market on both sides of the isthmus, they gained for their city all the power which a large revenue provides. Thucydides I. 13. 2–5

Athens At the height of her prosperity, fifth-century Athens had become both a great imperial power and a highly important entrepot of sea-borne trade, importing and exporting on her own account, and deriving substantial revenues from the transit trade through the port of Piraeus and the tribute paid by her dependent allies. The following is a passage, from a political pamphlet written around 430 BC by a shrewd opponent of democracy as practised there. The writer has a firm grasp of the realities of power; he also contributes some useful information on the range of raw materials which the Athenians had to import for the building and upkeep of their navy:

The Athenians are the only nation among the Greeks and barbarians that can secure wealth; if any state is rich in ship timber, where are they going to dispose of it unless they gain the favour of the rulers of the sea? Or if any state has plenty of iron, bronze or flax, where are they going to dispose of it, unless they get the consent of the rulers of the sea? It is, however, from these very materials that our ships are constructed; from one city comes timber, from another iron, from another bronze, from another hemp, from another wax ... I, without working, get all these benefits out of the land by means of the sea; and no other city has any two of these materials; the same city doesn't have both timber and flax, since, where flax is plentiful, the ground is level and bare of trees, nor do bronze and iron come from the same city ... (Pseud.) Xenophon, *Constitution of Athens,* II. 11 f.

The writer is here referring exclusively to materials needed for building warships; a full list of Athenian imports would include two items not mentioned here, namely grain and fish, the former of which was, as we have seen (above, p. 24), of crucial importance to the security of the state. These items, along with wool and hides, were the most important imports; exports included olive oil, wine and some manufactured articles, especially pottery containers; and last, but by no means least in importance, silver, that valuable, readily exchangeable, and easily transportable product.[62]

Trade centres of the Hellenistic East

The world conquests of Alexander the Great were short-lived in the strictly political sense; within little more than a single generation after his early death his leading marshals had ended a series of bloody struggles over the carcass of his empire by splitting it into three. Culturally, on the other hand, the changes wrought by his invading army were unifying rather than divisive, and the cultural centre of gravity shifted to the east:

As Alexander penetrated eastward, he dropped off contingents of his soldiers to found settlements; each became, as it were, an injection of the Greek way of life into the body of the ancient east ... Alongside the dwellings and places of worship of age-old oriental type rose Greek temples and theatres and porticoes and the rest of the apparatus of a Greek city-

state. Robed and turbanned locals now shared the streets with Greeks in their light tunics, and gradually got used to the sight of Greek youths working out in the nude in the newly built gymnasia, at Greek elders shouting in violent debate at town council sessions in the newly built meeting chambers, at Greeks of all ages chattering like magpies in the newly built market-places, the agoras. Lionel Casson, *Travel in the Ancient World,* London 1974, 115–16

The Mediterranean world, already steeped in Greek culture, was rapidly enlarged to incorporate a newly Hellenized East. United as never before by the common tongue, the *koine* Greek that quickly outstripped the older dialects as an easy medium of communication, by expanding trade, and by the spread of a recognizably similar way of life, the Greeks of the new 'Hellenistic' age developed a genuinely cosmopolitan culture. The great seaport that after nearly twenty-three centuries still bears the conqueror's name, was founded by him in 331 BC, and rapidly became, not merely an expanding entrepot of trade, through which the luxury products of the Orient passed to the West, not just an important manufacturing centre, specializing in glassware and jewellery, but an international meeting place where scholars and scientists from many lands could exchange ideas. Among the eminent men of science who came to work here was Eratosthenes of Cyrene, the leading geographer of his day, who employed a new method for calculating the circumference of the earth, and is reported to have come very close to the correct measurement.[63] Another distinguished immigrant was Hipparchus of Nicaea, who refined Eratosthenes' methods to a still higher standard, proposing a narrower and more carefully drawn map of latitudes and longitudes. It is said that, in order to secure greater accuracy in the determination of latitudes, he planned a co-operative venture in which groups of observers throughout the Mediterranean lands were to carry out measurements of the sun's shadow at the zenith; but nothing came of the plan—by no means the last instance of a promising scheme failing for lack of international co-operation.

4 Innovation and Development: a Survey

Some general considerations

In Chapter 2 we discussed in general terms the relationship between inventions and their application, examining the possible factors which may have inhibited or delayed the application of an original idea. This was followed up by drawing attention to particular areas where 'the shortcomings of the environment ... have given a stimulus to innovation' (above, p. 22). Our next task is to take a closer look at these 'promising' areas, describing at some length the particular innovation or development, and evaluating its importance to the economy. In this review of different areas of economic activity we shall notice that technical development follows no predictable or uniform path. Thus in some areas traditional implements and processes, many of them inherited by Greeks or Romans from earlier societies, continued to be used in traditional ways without significant change.[64] In others, again, an innovation may be seen to occur, and to be followed by a phase of rapid development; in some, by contrast, an important invention may remain undeveloped for a long period of time. The approach should be one of great caution: the appearance of fresh evidence may show that the 'gap' was fortuitous, the result of the loss of important evidence or of failure on the part of the investigator to recognize a piece of evidence, or to place it in its proper sequence.[65] A careful cataloguing of the material, together with a listing of promising areas may encourage the investigator to be on the lookout for missing evidence; 'problems', as Professor Hopkins has recently reminded us, 'do not present themselves, they need to be conceived'.[66] Getting at the archaeological evidence is, alas, no easy task; to avoid drawing unwarranted conclusions from the often limited array of facts, we may find it useful to follow one or two guidelines. These include: (1) not to attach importance to the search for origins—we could do worse than follow the Greeks here; they just slapped on a name (god, hero or man) as inventor, and went on from there; (2) conservative forces are strong: the new or improved machine does not immediately supplant the old and well tried;[67] time-lags are common; (3) failure of an invention to spread may be due to any one of a variety of causes; the hold-up may turn out to have been due to an unsuspected factor in the environment; (4) 'Every invention', a distinguished historian of technology has pointed out, 'is born into an uncongenial society, has few friends and many enemies, and only the hardiest and luckiest survive.'[68]

We are now in a position to review the areas of the economy in which technical innovation and development can be identified. The information, which inevitably covers a wide variety of material, literary, epigraphical and archaeological, including the supply areas of raw materials (see map, p. 28), the main transport routes for carrying them and the manufactured products, together with other necessary data, is assembled and classified in a series of tables, with separate headings for the basic needs, the operations required to satisfy those needs, the raw materials, the instruments or machines needed for processing them, and the products. The reader who may want to explore any given *14* area in detail is referred to the relevant chapter in Part II.

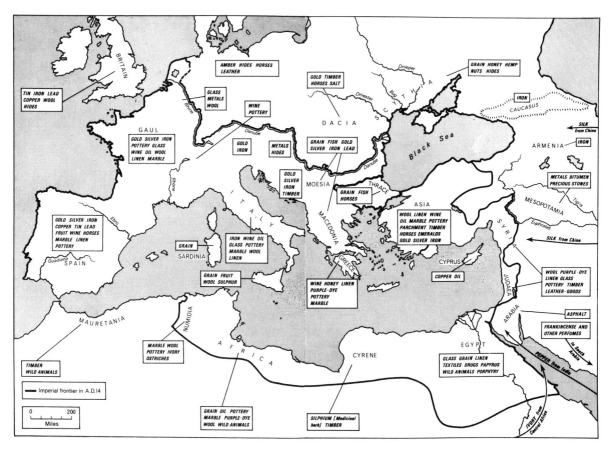

14 Supply areas of raw materials in the Roman world.

Areas of innovation and development

In agriculture

From its earliest beginnings in the Neolithic Age right up to the introduction of mechanization in the early part of last century, agriculture has been a labour-intensive occupation. Innovation and development have occurred, both in hand tools and in machines.

(a) Hand implements

In these we can identify three distinct lines of technical development.

(i) The invention of a double-ended implement to carry out two different tasks within the same process. Good examples are to be seen in the axe/adze, and a near relative belonging to the same family, the pick/mattock. In clearing woodland or bush the woodsman needs an axe to chop away roots, and an adze to claw out the stumps. In digging trenches the sharp pick breaks the surface, while the broad-bladed mattock bites into the less compacted earth below surface level. Double-ended implements are first noticed on Roman sites, and may be of Roman invention. Much more advanced is the multi-purpose vine-dresser's knife (see below, p. 58).

44

(ii) Improved versions of a traditional implement designed to make the task easier for the operator. An outstanding example is the so-called 'balanced' sickle. The action of this grain-harvesting implement is a backward pull across the stalk, which involves much tiring wristwork. The 'balanced' sickle has the angle

28

15 'Balanced' Roman sickle *(falx messoria)*, open-bladed type, from Pompeii.

between haft and blade set backwards, thus easing the strain on the wrist. An exactly similar development has recently taken place in the design of the hacksaw and the jigsaw, the backward sloping handle being much less tiring than the older vertical type.

(iii) Specialized variants of existing implements designed to deal with an increasing range of tasks involving the same basic technique. A good example is the scythe, a long-handled mowing implement consisting of a wide-arced blade fitted with a heavy backstrip to prevent 'whip'. The action of this implement is the reverse of that of the sickle (see (ii) above), consisting of a forward sweep at ground level. Improvements designed to increase output include lengthening of the blade, substitution of a suitably curved handle for the traditional straight haft, and provision of handgrips—all three being designed to increase the area enclosed by each sweep, and so speed up the operation (see White 1967, 98–103).

16

(b) Machines

45 (i) Ploughs. The symmetrical sole-ard, which appears on all surviving representations of ploughing in Greece, was well suited to the light alluvial soils which are commonly found in the relatively few areas of good arable suitable for the raising of cereal crops; no significant variations are attested. Italian soils, on the other hand, vary considerably, from the light soils of the south to the much heavier loams of Umbria and the Po Valley. Here we can trace the development of a number of variants of the

sole-ard designed to suit different soils (details in White 1967, 123–45).

(ii) Mechanical grain-harvesters. These are Gallic inventions, embodying a new principle in reaping grain. The heads were cut off, not with a sawing action, as with the hand-sickle, but by forcing them to pass between rows of close-set knife-blades, which cut off the heads; much time is saved, both in the reaping and in the subsequent treatment of the grain. Manual harvesting normally calls for three successive processes: viz. cutting, binding of the sheaves, and removal of the stalks. Two types of these machines are attested, which have evidently evolved separately from two different types of animal-powered vehicle: the *vallus* (from a form meaning a 'palisade'), with the toothed frame mounted on a single axle, and an underslung receiver for the heads of grain, and the *carpentum* (which takes its name from a form of two-wheeled passenger carriage), fitted with a tum-

16 Two panels from the Porte de Mars, Reims; lower left, haymaking (July); centre, sharpening the scythe-blade; right, long-handled Gallic scythe in operation; upper right, reaping with the *vallus* (August).

17 *Carpentum*. Restoration from Palladius' detailed description (vii.1) of the larger tumbril-type harvesting machine powered by an ox in shafts.

bril-type body into which the toothed frame is set, the body itself serving as a large container. The lighter *vallus* was pushed from behind by a mule or donkey, the *carpentum* by an ox (see White 1967, 157–73).

(iii) Threshing machines. The first recorded improvement on the age-old method of trampling out the grain with animals was the threshing-sledge *(tribulum)*, which was fitted on the underside with sharp flints, and drawn by an animal; a definite improvement on this machine was the 'Punic cart' *(plostellum Punicum)*, mentioned by Varro (late first century BC), and originating in eastern Spain, which was a Carthaginian possession before its conquest by the Romans. This was a sledge mounted on small rollers, with a seat for the driver (see White 1967, 152–56).

Food processing

(a) Grain-mills

The oldest known device for grinding grain into meal consisted of a sloping slab of stone (the 'saddle-stone') and a smaller stone (the 'grain-rubber'). The operator knelt on the ground, and rubbed the smaller stone up and down on the slab (for the position of the operator see *Exodus* 11:5 ('the maidservant that is behind the mill'), and the well-known series of painted terracotta figurines (Moritz 1958, pl. I, a–c). This is the 'up-and-down' mill referred to by Cato *(De Agri.* 10.4; 11.4) as *mola trusatilis*, or the 'pushable mill'. The main drawback of

this device (the saddle-quern, as it is usually called) is lack of continuous working; both hands were needed for the rubbing process; only a little grain could be placed between the stones at a time: hence frequent halts. That this method was not satisfactory, and called for improvement is shown by the appearance, some time in fourth-century Greece, of a pair of stones, the upper one hollowed out and fitted with a narrow slit, which enabled it to receive a quantity of grain and to admit it continuously and gradually on to the lower stone. The next improvement was to make suitable grooves in the stone, making the process of breaking the grain more efficient, and at the same time controlling the size of the particles. Devices of this type are known as 'hopper-rubbers'. Next came an improved version of the hopper-rubber in which a lever was provided to operate the upper stone, thus converting it into a machine. Note that the intake of grain, the discharge of the meal, and the size of the particles are all controlled by mechanical means, marking an important step in the development of grain-milling by reducing the human element in the process of converting grain into meal.

(b) Sifters

The conversion of meal into the various grades of flour was achieved by different combinations of the processes of milling and sifting. The limited evidence for classical Greek practice suggests that the sifting will have done little more than eliminate the lumps of bran, together with unground or partially ground grains; what passed for purified flour 'must have been very dark indeed by our standards' (Moritz 1958, 163). Technical improvements in the design of sieves in Roman times are noticed by Pliny *(HN* 18.108), who reports that fine horse-hair sieves were invented in the Gallic provinces, while linen sifters, which could produce flour of the highest grade, came from Spain.

(c) Pounders

Most food-grains can be reduced to meal by grinding, but some of the tougher kernels (e.g. emmer wheat *(triticum dicoccum)* and maize *(zea)*

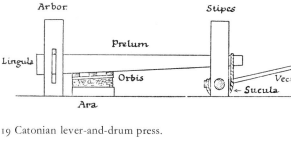

19 Catonian lever-and-drum press.

18 Greek black-figure amphora, 6th century BC, depicting two girls pounding grain with pestles. Hermitage Museum, Leningrad.

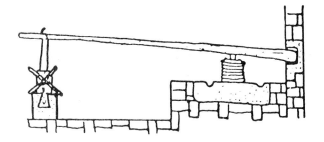

20 Modern Catonian lever-press with counter-weight and capstan, northern Algeria; note Lewis bolt suspension (see *ill. 76*).

require pounding. The former was used extensively by the Romans for making porridge *(alica)*. The cracking was done in antiquity, as it still is in many parts of Africa for pounding *18* maize, by means of a heavy wooden pestle and a narrow mortar. Pliny's account (*HN* 18.97) includes mention of an improved pestle fitted with an iron cap, which first stripped the husked grain, and then cracked the kernel; this heavy, slogging task was assigned to chain-gang slaves (Pliny, *loc. cit.*). Modern 'cracked wheat' is produced by pounding machines, which employ the same principle, but use power-driven machinery.

Presses for grapes and olives

In this department we can identify at least one
61 invention, the direct screw-press, mentioned both by Hero and Pliny, which enormously reduced both the press-room area, and the quantity of high-grade timber required for the
21 press-beams and fulcrum beams of the Catonian press, and several examples of development,
19 e.g. the substitution of the Catonian lever-press

for the primitive stone-weighted tree-trunk used in early Greece, or the later substitution of the more efficient drum-and-pulley method of pulling down the press-beam for the simple rope or thong. The gaps in the sequence of development may well be filled, and the numerous technical problems clarified by the results of excavation of farm sites now going on at an accelerating pace. In his pioneering study (*Ancient Oil-mills and Presses* 1932), A.G. Drachmann examined the evidence for technical development in this area, and demonstrated in detail the progress towards greater efficiency in the process of juice-extraction, from the primitive stone-weighted tree-trunk of early Greece to the Catonian lever-press, and on to the later varieties of screw-press. He also sought to associate the important innovation by which the lever-press was equipped with a screw-down mechanism, with the screw-cutting device (see. *ill. 61*) described by Hero. According to Drachmann, it was this invention which made possible the production of internally threaded nuts to fit externally threaded

31

21 Reconstruction of a single-screw press from Pompeii.

bolts. This neat and plausible correlation cannot be accepted: internal threads can easily be cut without a screw-cutter. This apart, the evidence of development of the lever-press is impressive, as is that of the innovative screw-press, which extended the squeezing action well beyond the limit set by the lever-press in its most advanced form (details in White 1975, Appendix A, pp. 225–33).

Oil-mills

After pressing, olives intended for oil were treated in a mill, not for the purpose of crushing the kernels but in order to dislodge them from the fruit. Of the various devices mentioned only one (Cato's *trapetum*) is described *63* (*De Agri.* 20–22), so that while there is clear evidence of experiment, we cannot prove innovation or development. Of critical importance in the efficient operation of a mill is the provision of some means of adjusting the distance between the revolving stones and the mortar containing the fruit. Cato's device, it appears, could only be reset for larger or smaller fruit by removing the stones and inserting a deeper col-

lar on the spindle, so that it could not take account of variations in size, unless the fruit had been previously sorted for size, for which there is no evidence. Columella's list (*De Re Rustica* 12.52–6) of four milling devices includes one called the 'olive-mill' (*mola olearia*) *64* which, as he explains, is an improvement of the *trapetum*, in that the clearance between rollers and mortar can be adjusted while the machine is working: 'mills', he tells us, 'make it very easy to organise the 'crushing' process; they can be lowered or even raised to suit the size of the berries, and to prevent crushing of the kernel, which spoils the flavour of the oil' (*loc. cit.*).

Water-raising equipment

(a) Piped water-supply

This uses the 'U-bend' or 'inverted siphon' as it is sometimes called. There is some evidence of Roman improvements on earlier Greek engineering here; but note that at Pergamon the closed-pipe system installed by Greeks was replaced in Roman times by an open channel, *22* two arcades and a tunnel. For the difficulties of closed-pipe systems (excessive pressure, air-locks, etc.) see Landels 1978, 42 ff.

(b) 'Archimedean' screw (kochlias, cochlea)

For lifting irrigation water, draining mine- *12* workings, etc. We have evidence of improvements in design (e.g. number of rotor blades increased to give higher lift (see Landels 1978, 62), and in materials for increased durability; a surviving screw from a Spanish mining site has blades of 1/8 in. (3.2 mm) sheet copper attached to rotor and case by metal brackets (Landels, *loc. cit.*).

(c) Bucket-wheel

For lifting water to a higher level than (b); pre- *23* sumably developed from the primitive 'wheel of pots' still in use in many parts of the Middle East. An obvious improvement (? Roman, undated), mentioned by Vitruvius (*De Arch.* 10.4) has buckets shaped like the standard Roman corn-measure (*modius*), from which they get their *modioli*; this means that, unlike the pots, they will not have started to discharge their

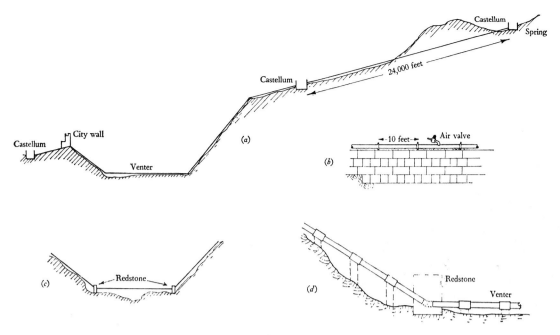

22 Vitruvius' aqueducts for lead and clay pipes; *a,* aqueduct for lead pipes with detail of venter *(b)* ; *c,* venter of aqueduct for clay pipes, with detail of 'elbow'.

contents before commencing to descend, as Vitruvius points out. Landels (1978, 67) is more cautious : 'if this was in fact achieved, it represents a big improvement . . .'.

(d) Multiple water-wheels

Arranged in opposed pairs. These mark a distinct advance on the bucket-wheel in that the wheel is provided with a hollow rim divided into compartments, the apertures for collecting and discharging the water being so arranged as to ensure 'minimum spillage before, and complete emptying of the compartments after "top dead centre"' (Landels 1978, 69). The multiple system found at Rio Tinto consisted of sixteen wheels mounted in opposite pairs (they were set to revolve in opposite directions to ensure an even flow into the launders), and capable of lifting water to a head of almost 100 ft (*c.* 28 m) at a rate of 2,400 galls (*c.* 120 hl) per hour; the power was supplied by men treading.

(e) Bucket-chain

This device for lifting water appears in two related models in Vitruvius (*De Arch.* 10.4, 10.5), the first being powered by a tread-wheel on a horizontal axle, from which were suspended two endless chains with buckets attached at intervals; the second being worked by a water-wheel with paddles, the power being applied at the bottom of the chains. The output of the tread-wheel version (calculated by Landels (1978, 74) at *c.* 350 galls (1750 l) per two-hour shift by a single treader) would have made it useless for irrigation or mine-drainage. Animal power, which was successfully applied to the corn-mill, could have been harnessed to the bucket-chain for drainage, and Roman failure

23 Vitruvius' bucket-chain hoist. The standard size of bucket was one *congius* (about 3.3 litres).

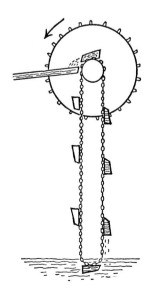

to apply it is noted by Finley (1973, 147) as an important example of a technical block. The matter is not quite so simple, however; if Landels' assumptions are reasonable in relation to the output of a tread-wheel-operated bucket-pump, the substitution of animal power would not have made an efficient drainage-pump for mines out of a device which was only capable of meeting the modest demands of a domestic water supply drawn from a well. Increasing the capacity of the buckets on such a high-lift machine would increase considerably the strain on the chains and the gearing. It is perhaps worth noting that, whereas the animal-driven wheel of pots continued in use in the Mediterranean region through the Middle Ages and right on into modern times, it was generally replaced by windmills, not by animal-powered bucket-chains.

(f) Piston-driven force-pump

4, 5 This twin-cylinder pump, correctly designated by Pliny as a 'pneumatic device', was known as

6 the *Ctesibica machina* (see above p. 17).[69] Its design and operation have been fully described and lucidly illustrated by Landels (1978, 75–83). In this likely area for innovation and development, where the descriptions given in the literary sources can be set alongside the now substantial and growing number of identified specimens, some of which are virtually complete, so that even variations in the design of the valve-systems can be assessed, the case for innovation (in valve-design), and for development (in fit and surface finish of pistons and cylinders), appears to be fairly strong; unfortunately scarcely any of the specimens can be dated within wide limits of time. Further discoveries of these widely used pumps may well lead to the establishment of a developmental sequence[70].

Mining

Both this and the following section (on metallurgy) are concerned with areas of the ancient economy in which high profit margins or state demands would be likely to promote experiment, innovation or development. Among outstanding examples may be cited the deep-level mining for silver at Laurion in response to *119* the increasing demand created by the advance of Athens as an imperial power, and the elaborate and costly installations exposed by recent excavations in the imperial Roman gold-mines of north-western Spain (see Ch. 9, for details).

(a) Prospecting

'The Greeks, in the absence of any detailed knowledge of the principles of geology, were unable to prospect in a scientific manner by modern standards' (Healy 1978, 71, who later notes that (*pace* Forbes, *Studies* VII, 110) Roman lead-working at Bottino (Etruria) shows sufficient understanding of faulting to enable the continuation of a lode over an important fault to be traced (*op. cit.* 87)).

(b) Underground mining techniques

These, including extensive galleries driven from the bases of shafts, roof-propping and ventilation techniques, were inherited from Bronze Age precursors.[71] The Laurion workings show no evidence of improvements in any of the above departments; nor, since mining stopped at sea-level, were there any drainage problems that could not be met by simple channelling and baling (Healy 1978, 84). An important development is to be noticed in the subsequent process of separating the extracted mineral from the waste; the primitive and time-consuming method of repeated crushing and sieving in water, as described by Strabo (III.2.10) being replaced in the Laurion mines by a well-designed system, which involved recycling of the water, used in large quantities, and difficult to obtain in such a rainless region. Examination of the remains of several of these ore-washeries reveals a simple but effective *25, 12.* series of operations, as Healy explains (1978, 144 f.): 'The plant consists of four main elements: (1) a large flat plastered floor or table surrounded by water channels on its four sides and provided with two round settling tanks at the two southern corners; (2) a smaller table lying beyond the northern channel, similarly plaster-faced but very slightly inclined so as to

34

drain into that channel; (3) a long water-tank next to the inclined table, having only a thin wall at the front but a massive dam-like wall at the rear, and, at the back, (4) a great round cistern or reservoir sunk into the hillside.' This ingenious combination of jet emission and gravity flow via channels and settling tanks enabled the argentiferous ore to be graded both for density and purity, and made ready for smelting. Healy *(loc. cit.)* also gives the results of recent experimental testing of the method, which yielded a high output of recovered metal.[72]

Technical development in Roman mining included the driving of connecting galleries to exploit several levels, as on Andros, at Cartagena and Rio Tinto (Healy 1978, 91), lighting by means of oil lamps attached to the miner's forehead (?),[73] and drainage by means of machines, including Archimedean screws and water-wheels. The use of water power to remove the overburden is perhaps the most spectacular technical advance attested. The techniques included the use of aqueducts and water-tunnels (often at high altitudes) to provide the necessary 'head' of water, dams and 'hushing' tanks, as well as ground-sluices (see Ch. 9, pp. 117ff.). These technical advances provide good examples of development by transfer of technology from one department of engineering to another.

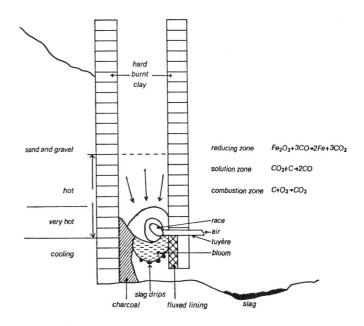

24 Diagram of the shaft furnace built by Tylecote on the basis of a 2nd-century AD furnace from Norfolk, England, in which the slag had been tapped.

(c) Health and safety of underground workers
Here Greek mining shows no evidence of any technical improvements. The labour (mainly condemned criminals and barbarians) was, as in Britain before the first Factory Acts, expendable; there is no evidence of any change in the cramped working conditions for face-workers, or in ventilation methods, based as they were

25 Washery B at Agrileza, Laurion; colonnade and roof reconstruction. The table, surrounding washing sluice and enclosing wall are still *in situ*.

on natural air circulation, aided by the use of 'punkahs' during shaft-sinking, and of pairs of parallel shafts, with fires at the junctions with galleries, to create a draught. Nor did the Roman period see any improvement: the numerous references to asphyxiation by poisonous fumes, and the actual remains of suffocated miners tell their own story.[74] But the regulations contained in a well-known inscription from Portugal, dating from the reign of Hadrian (AD 117–38), include provisions which show that the skilled men recruited for work in this remote region had to be given incentives.[75]

Metallurgy

In this department of technology, with its many varieties and sequences of technical processes appropriate to the widely differing ores and lodes in which the metals occur, we find much evidence of traditional methods and processes inherited from pre-classical times, but also solid proof of improved extraction-rates of those metals through analysis of datable artefacts, and of the contents of slag-heaps. Inventions were few; they include the introduction of the shaft-furnace around 500 BC in place of the age-old but inefficient bloomery hearth for iron-making, the vital oxygen supply being now controllable by means of blow-pipes *(tuyères)* and bellows; in bronze-casting, the invention of the delicate lost-wax process goes back to Mycenaean times but both inventions are Greek. By contrast, cast iron, which the

Chinese had been producing from as early as 500 BC was unknown, and 'it was not until the Middle Ages that the technique of handling and further working this material was properly understood in Europe'.[76] Recent experiments in the production of bloomery iron, using Roman-type furnaces and reduction methods (Cleere 1971; Aiano 1975), suggest very strongly that Roman methods were defective in two important respects: (1) very slow production (Cleere took eight hours to produce 20 lbs of metal); (2) lack of uniformity in the product, the tops of many blooms being converted into steel during their passage down the shaft. For details of the stages in the development of iron technology in classical times, the reader should refer to Healy (1978, 182 ff.). There is some evidence of improved furnace draught in the late Roman period. The earliest improvement on the primitive method of inserting a blow-pipe into the furnace wall, was an inflated animal skin fitted with a hole and a draw-string. The true bellows, with boards and a valve permitting immediate reinflation as soon as the boards have been compressed, was probably not a Hellenistic invention, as Rostovtzeff supposed (1941, Ch. VIII, n. 166), the authenticity of the two lamps supposedly depicting bellows being disputed. But Ausonius uses the comparison of a bellows in a simile describing the death agonies of a fish, strong evidence surely that the true bellows was in common use by the middle of the fourth century.[77]

Refinement of the precious metals is an area in which we should look for technical development in the search for more highly refined gold and silver for the production of jewellery and coins. Removal of the silver to obtain pure gold from the naturally occurring 'white gold' of Mediterranean deposits could be effected by cementation and cupellation. Both techniques are pre-classical, but a reference in Strabo (III.2.8) to the use of 'styptic earth' in the first of these processes suggests an important discovery by Roman refiners, who had further acquired the technique of extracting gold from other minerals such as arseno-pyrite, realgar

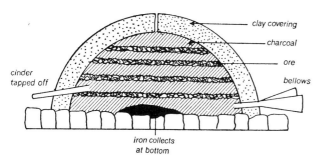

26 Reconstruction of a primitive bloomery hearth for producing wrought iron (and a lot of slag).

and stibnite (Healy 1978, 156). In addition they were responsible for the addition of brass and pewter to the artificial alloys available to the classical world.[78]

27 Black-figured Corinthian votive plaque showing a potter at work.

Pottery

The first major technical development, the potter's wheel, belongs to the late fourth millennium. The Greeks used large, heavy wheels (0.6 m/2 ft in diameter), set low to the ground; they either squatted or sat with feet wide apart. Once set in motion, the wheel had enough momentum to need only the occasional push, either from the potter or, as in some representations, from a second man sitting opposite. The older method, by which the pot was built up from the ground by winding 'snakes' of clay round and round, continued to be used for the large storage jars *(pithoi),* which are still made this way today in Greece. The various stages, some ten in all, from preparing the clay to firing the pot, are too well known to be repeated here. The outstanding technical achievement was the famous black glaze associated with black-figured Attic pottery of the sixth century BC. The method by which this astonishing permanent glaze was produced baffled researchers until 1942, when the German chemist Theodor Schuman unlocked the secret. He found that the glaze was made permanent, not by a series of firings at different temperatures, but by a single firing in three phases, the secret being the

moisture content of the clay and the fuel, and their interaction during firing. The firing began with the normal oxidizing action, but the use of green wood or wet sawdust changed this for a time to a reduction process, reverting once more to an oxidizing phase which made the glaze permanent.

Roman potters, it seems, used 'one-piece' wheels, but it is thought probable that they also used a double kind, similar to the modern wheel, consisting of a flywheel, connected by a spindle to a wheelhead, on which the potter worked at waist height. There is no evidence for the further technical improvement of a treadle-operated wheel, but the wheelhead in itself makes for easier control of the work, and facilitates the fashioning of small vessels. Roman potters made major technical advances in the application of moulded decoration, which had been invented in Hellenistic times, as in the red gloss ware widely circulating in the

28 Greek black-figure *oenochoe* (wine-decanter) depicting a shaft-furnace in operation. To left, the smith-god Hephaistos. British Museum, London.

29 Interior of a Roman pottery mould showing the impressed figures and the potter's stamp back to front. From Arezzo.

Roman world as Arretine. Roman silversmiths were well known for their use of repoussé decoration, and the similarity of design found in the two media supports the view that this is a case of transferred technology, a feature well known both to the art historian and to the student of technological history.[79] The details of the technique have been partially recovered from surviving moulds, which were made rather thick, and separately fired before use. The main problem in transferring the design will have been how to avoid trapping air in the hollow parts of the relief, and it has been suggested that the method may have been to press the clay into the mould with the fingers, which is known to have been the method used in making terracotta figurines and clay lamps. Making red gloss ware was evidently not a straightforward task; a particular type of kiln, and special firing conditions were needed (details in Strong and Brown 1976, 80). In addition, the pots were usually stacked upside down in the kiln, and

this meant rings to support the bottom pot in the pile.

The techniques of Roman glazed pottery seem to have been acquired from Anatolian or North Syrian sources, the latter area being already famous for glass manufacture. Most of the glazed ware has been found to be made of lead silicate glass, to which copper oxide has been added for the green glaze, and ferrous oxide for the brown. Workshop remains show that the pots to be glazed were set on stilts and totally enclosed to seal them from the kiln gases.

Recent experimental work on the constituents of glosses and glazes shows that Roman potters disposed of a variety of techniques in producing the various categories of pottery ware. The red gloss surface technique, which was lost after the end of the Roman period and only recently rediscovered, is based on the addition to a liquid clay of a suitable agent which will prevent some of the smallest and finest clay particles from settling when the liquid is left to stand. The glossy surface has been found to be due to the presence, among the three minerals which make up the fine suspension, of a high proportion of illite, the molecules of which 'lie flat against one another giving an overall flat and glossy surface' (Strong and Brown 1976, 83).

Textiles

The production of finished cloth from either of the commonest materials available to the classical world, namely wool and flax, involved a variety of processes, carried out by traditional methods inherited from remote times, and handed down with little change. The techniques employed in the major processes of spinning and weaving are so well known that detailed treatment is unnecessary, but there are two areas, that of scouring and other cleansing operations on the one hand, and those of dyeing and finishing on the other, which show evidence of technical improvement, and, therefore, call for more detailed discussion. The various processes are clearly set out in Table 10. Here we shall review the treatments that were applied to remove impurities from the raw wool and prepare the fibre for spinning. The wool clip as it comes off the animal's back must be thoroughly cleaned which means first removing foreign bodies, chiefly dirt and burrs, then the natural grease. Most of the routine procedures and the recipes used, including the chemicals, are already well known to us from the earlier Egyptian and Jewish sources. Many of these processes are identical with those *31* used in the fulling, bleaching and general refurbishing of dirty clothing by fullers and laundrymen, whose activities are often recorded on wall paintings or other monuments, and whose guilds have a prominent place in the surviving inscriptions that deal with trades and professions. For similar reasons the dyeing of wool *30* and of new cloth, as well as the re-dyeing of old garments, were often carried out on the same premises and under the control of the laundry proprietors.

30 Dyer (?), redrawn from a funerary monument at Arlon, Belgium.

31 Scenes from a fuller's tombstone. Musée de Sens, Yonne, France.

The first stage of cleaning was to wash the wool in hot water to get rid of surface dirt; after drying the wool was then beaten to separate the fibres. Next came the most important of these preliminary processes, that of scouring with chemicals (chiefly fullers' earth, *creta fullonica*) to remove the grease. The varieties of detergent chemicals used are described by Pliny (17.46) in connection with the cleaning and bleaching of used clothing by laundrymen. In the last two preliminary stages, the wool was

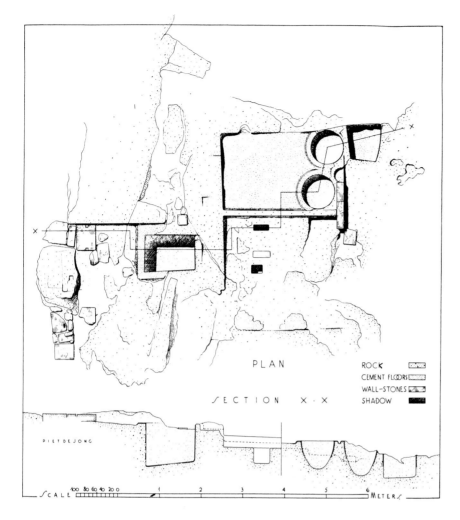

PLAN

SECTION X · X

ROCK
CEMENT FLOORS
WALL-STONES
SHADOW

PIET DE JONG

SCALE
100 80 60 40 20 0

1 2 3 4 5 6

METERS

32 Dyeing and weaving works at Isthmia, Greece. Ground plan (above) and section (below) showing a pair of circular vats for double-dyeing on a platform for working with detergents, and square vats for rinsing. 4th-3rd century BC.

plucked, and then either combed (with a special type of comb) or carded (with a natural teazle, *carduus,* giving its name to this process) or with an artificial one), the choice of treatment being determined by whether the surface was to be smooth or raised. In this very traditional trade the same distinction of treatment is observed between wool for light suitings and for overcoatings. All this by way of preparation for the first major operation, that of spinning. Combing laid the tangled fibres parallel, whereas carding set them crosswise. Forbes (*Studies* IV, 21) states that there is no evidence that these distinct technical processes were known to antiquity. Could the matter not be put to the test by examining surviving pieces of woollen fabric? Such testing has produced important conclusions in relation to the dyeing process.

Successful dyeing involves the solution of three major technical problems.

(1) How to get the required colours (Forbes (*Studies* IV, 99–127) gives a fully annotated list of the dyestuffs used in antiquity, but how many of the items in the records were actually used?).

(2) How to get fast dyes (a fundamental problem, which still presents difficulties for the dyer in spite of the enormous advances made in colour chemistry and its application during the past century).

(3) How to get the dyestuff to adhere to the fabric (a major reason for dyeing 'in the wool', and the source of the proverbial use of the expression!).

As in so many technical processes, the techniques and recipes were of considerable anti-

40

quity, and many of the traditional methods and formulae will have been 'handed down from father to son and closely guarded as trade secrets' (Forbes, *Studies* IV, 127). We do, however, have some valuable published papyri of the Hellenistic period from which it is clear that books containing recipes for dyes did exist, though none now survive. Two main lines of research have been pursued: that of K. Reinking, who examined the chemical papyri (Pap. Holmiensis and Pap. Leidensis) and sought to extract from them the basis for a full account of the dyeing techniques employed in the Hellenistic period, and that of K. Pfister, who adopted an entirely different and thoroughly practical approach, designing a method of analysis of the dyestuffs mentioned in the texts, and then testing actual samples of surviving dyed materials (Egyptian, Syrian and Coptic). He found that 'the multiplicity of dyes mentioned in the papyri was theoretical, and that in actual dyeing we can trace only a few dyes like indigo, madder, kermes and purple', the desired colours being obtained by the admixture to the basic dyestuff of slight amounts of one of the others.

31   The processes of washing, bleaching, fulling and felting are fully discussed by Forbes (*Studies* IV, 82–98) and those of dyeing in a later section of the same chapter (pp. 127–43). In an area so technically advanced, especially in Egypt, long before the Hellenistic period—the superb quality of some of the linen material found in the tombs has often been noticed—we should perhaps not expect significant developments in the classical period. Moreover, information is very defective: no extant fabrics of the Hellenistic period, and no representations to compare with either the paintings on vases of the classical period, or those from Pompeii and Herculaneum. Only two innovations seem to be attested for textiles, both of them Roman: (1) an improved type of vertical loom, and (2) an improved cloth-press (there is a specimen from Herculaneum, and a painted representation at Pompeii). An extended study in Naples might unearth some more. As for dyeing, there are two

33 Redrawn from a late black-figured cup from Thebes (now in the Ashmolean Museum, Oxford) depicting Circe offering a potion to Odysseus, while standing beside a Greek vertical warp-weighted loom.

surviving papyri (one in Leiden, the other in Uppsala—references in Forbes, *loc. cit.*) which contain extracts from Bolus' *On Dyeing,* with interesting evidence of a widening demand stimulating the search for cheap substitutes for the more expensive dyes.

Glass

The glass-workers of ancient Greece inherited three techniques; first, the sand-core method, in which a lump of sand or clay, shaped in the desired form, is put on the end of a bronze or iron rod, which is then immersed in molten glass, and rotated over a hard surface. The core

34 Small Roman glass bowl with pinched vertical ribbing and marvered white trails. Diameter 10 cm. Ashmolean Museum, Oxford.

is then removed; this method was possibly evolved from the technique of applying a glaze to earthenware pots; the second was mould-pressing, with either an exterior mould, or an inverted interior mould, over which the molten glass was poured, or by a combination of both techniques, a plunger with an interior moulded design being thrust into the molten glass inside an exterior mould; this was another case of transferred technology, a technique for clay or metal being applied to glass. The third was cold-cutting, another example of transferred technology, this time from gem- and stone-cutting. The classical period witnessed two important technical developments. The first occurred during the fourth century BC, when the first and second techniques were combined, the glass being first cast, then cut and polished. This technique reached a peak in the third century with the invention of two-layer gold-glass, in which two layers of glass were produced, the outside of the inner layer being decorated with gold leaf. Glass blowing had not yet arrived, so the glass 'must have been formed by two castings, one inner, one outer, that were later so carefully ground and polished that the outer fitted the inner one like a glove'.[80] The following three centuries saw no further technical evolution, but towards the end of the first century BC came the revolutionary invention of glass-blowing, presumably by someone who made the discovery that it was possible to obtain a more even flow of glass into a mould by blowing through a hollow tube than by pouring. The effect of the invention was dramatic, both in the speed with which it spread, and in the range of glass products which could be easily made. Both effects are very much to our purpose; by the middle of the first century AD, glass works were already established in the Po Valley and in Aquileia in North Italy (two glass jugs found at Linz in Austria are stamped with the name of the Aquileian glass-maker Sentia Secunda), by Syrian glass-workers moving there; by AD 50 glass-making had spread as far north as Cologne on the Rhine. The subsequent developments have been admirably summed up

as follows: 'Perhaps at no time during the first 27 centuries of glass history were there as many factories in production as in the 2nd, 3rd and 4th AD. Given the cohesive influence of Roman civilization, glass markets could be developed aggressively ... Consumers reacted favourably to a broadened base of distribution which was intelligently established by a versatile and production-minded industry of remarkable resourcefulness.'[81] (*Corning Museum of Glass*, 1957).

Roman imperial glass is also remarkable both for the ingenuity of the craftsmen and the variety of decoration, which included tooling and relief work, threading and painted glass. Glass, unlike earthenware, is extremely flexible, and can be shaped into almost endless varieties *34* of vessel. It is typical of the Roman attitude towards art that one of their earliest commercial successes was the manufacture of a translucent glass, closely resembling natural crystal. Their range was astonishing, from such intricate bizarreries as the enclosing of glass within metal (the *diatreta*) to the practical window-glass, a direct product, it is thought, of the invention of blowing (Harden 1968, 45). Most important of all is the economic and social impact of the Roman glass industry. By the first century AD, this material, which was still rare in Hellenistic times, 'became, during the first century AD, a common material of everyday use for all kinds of domestic purposes, whether practical or ornamental' (Harden 1968, 66).

Mosaic

This remarkable technique, which assumed a great variety of forms in the treatment of floors, walls and vaults, makes a very early appearance in the ancient East; a third-millennium temple at Erech in Mesopotamia had its walls and columns decorated with plaster-work into which finger-sized cones of coloured brick were pressed leaving the surface covered with a pattern of polka-dots. Floor treatment begins humbly with coloured pebbles, dating back to eighth-century Anatolia, and flourishing in Greece by the fourth century.

35 Wall mosaic, 1st century BC, depicting a Hellenistic galley at action stations with marines on deck. Palazzo Barberini, Palestrina.

True mosaic, consisting of cut cubes set in concrete, is essentially a Roman development with a Hellenistic ancestry. The materials used were pieces of stone, terracotta or glass, broken pottery and glassware being often put to use in making the cubes for tessellated work. Vitruvius mentions two distinct types of flooring: *pavimentum sectile,* in which the surface was composed of pieces cut *(secta)* into regular forms (triangular, hexagonal or square), which were composed into formal patterns and *pavimentum tessellatum,* in which the components were cut into regular cubes, forming a different type of formal pattern (see Vitruv. 7.1.4). A further technical development, which gave great scope for the treatment of landscapes or animal and human figures, was to make the *tesserae* follow the contours of the object represented (see Pliny, *HN* 36.15 f.) The convoluted appearance of the pieces when viewed from a distance, gave rise to the technical term by which it was designated, namely *pavimentum vermiculatum.*

The method was as follows: for flooring guidelines were scored on the uppermost layer of the foundation, the squared *tesserae* being laid with pincers on a thin mortar bed (the current method is still the same). They could be laid in one of three ways, either directly (straight on to the mortar), or indirectly (first set in a bed of sand, then covered with glued paper, reset in mortar after the glue has dried, the paper being removed later with hot water), or in reverse (glued face down on a sheet of coloured paper or cloth, the design appearing in reverse when relaid). These surfaces have proved remarkably durable, and the thousands of surviving examples exhibit great differences in skill of execution, from the masterpieces of Piazza Armerina or the Palace of the Emperors at Istanbul to the crude and repetitive 'pattern book' work found on many provincial sites.

The most interesting technical developments in the application of mosaic to walls and vaults belong to the Roman period. Natural grottoes, very common in the limestone forma-

35

43

tions of the Mediterranean, provided inviting surfaces for applied decoration, first with seashells, to be replaced later by the more sophisticated *tesserae*. These shady retreats, the abode of local nymphs (hence the standard term *nymphaeum* for these apsidal architectural forms) were later invaded by *tesserae* of glass, making patterns of dancing light on the water below. An important technical development here was the introduction of Egyptian blue, or 'blue frit', a chemical compound of silicon, malachite, calcium carbonate and *natron* (sesqui-carbonate of soda), raised to a temperature of 850 °C. When ground up it was used as a background colouring. The next development was the use of glass, first as broken pieces, later in manufactured discs and rods, which came in during the early first century AD. The method used was necessarily that of direct application to the wet plaster, traces of paint where *tesserae* have fallen out proves that the guidelines were painted on the wet plaster (Sear 1976, 236). Sear also notes (from the differentials recorded in the *Edictum de pretiis* of Diocletian) that mosaic workers who worked on vaults got a higher rate of pay than the floor workers.

Heating

The control of fire, whether for cooking or for warming dwellings, remained, for an immense period, completely innocent of technical innovation. The charcoal brazier represents a considerable advance on the open wood fire; it produces much less smoke, and can be moved about as required. An inheritance from pre-classical times, it served both purposes, its only modification being a variety of jacketed containers, heated either by fire or by water at a high temperature, after the manner of a chafing dish. The extant baking ovens in the bakeries at Pompeii have two arched apertures, the lower for the fuel; the design has remained unchanged and may still be seen today in Naples. The smelting-oven, on the other hand, underwent several modifications, in response to the need for higher temperatures and for forced draught (see above, p. 36 and below, p. 121f.).

The Roman period saw one major invention in domestic heating, that of underfloor heating of rooms, attributed to one Caius Sergius Orata (early first century BC) by Pliny, who also credits him with an ingenious method for maintaining an all the year round supply of oysters on his estate on the Bay of Baiae, near Naples. According to Pliny, Orata's motive in both activities was financial: houses were bought up, fitted with the fashionable new heating, and sold at a handsome profit.[82] The term 'hypocaust' universally associated with this Roman invention occurs first in a poem of Statius (late first century AD), written to celebrate the elaborate bath system of a villa (*Silv.* I.5.59), and Vitruvius (V.10.2) gives specifications for the 'hanging floors' *(suspensurae)* of the hot rooms *(caldaria)* of the public baths. Portions of these systems, with their arrays of short, sturdy brick piers, have survived, together with the flues (see below s.v. 'baths'). Some years ago a heating engineer (see Otto Neugebauer, *The Exact Sciences in Antiquity,* London 1932, 258) raised the question of whether all such suspended floors could have been heated by hypocaust, since, in many cases, he found no sign of carbon, ash or soot; he, therefore, claimed that those with specially thick floors were being damp-proofed, not heated. Hypocausts, with their external furnaces fed by wood, will have been expensive to maintain, and the less expensive method of heating by means of pipes laid diagonally under the floors is known from several sites (e.g. Saalburg, West Germany); its two main disadvantages were that it gave uneven distribution of heat, and allowed for no regulation of the temperature. The hypocaust, on the other hand, when applied to the public baths, quickly superseded the ancient braziers, and made rapid progress, leading to ingenious and often spectacular solutions of numerous technical problems.[83] The fully developed bath systems (when reintroduced many centuries after their disappearance they were rightly described as 'thermal establishments') required not only a series of rooms with a variety of temperatures

36

44

ranging from cold to steaming hot, but supplies of water laid on at different temperatures as well. Technical developments in the boiler-house can be examined on excavated sites from different periods, the evidence being correlated with that from literary sources. Vitruvius' requirement (*De Arch.* V.10.1) of three separate boilers is neatly illustrated at the Forum Baths at Herculaneum (Maiuri 1958, 109–11 and figs 68 and 88). Much more economical (and certainly neater!) is the solution adopted later at Exeter *(Isca)* in Britain; its 3,000 gallon reservoir was fed by an aqueduct, and its 900 gallon boiler (of lead with a bronze base-plate) matches exactly the specification of Faventinus (*De Diversis Fabricis Architectonicae,* 16). The Exeter system was fitted with stop-cocks for controlling the temperature, and a bib-cock for draining the boiler.[84]

The well-known system from villa no. 13 at Boscoreale, near Naples,[85] had a brick furnace, over which stood the main boiler for hot water, with a second boiler above the chimney that took away the furnace gases to provide tepid water. The plumbing was rather untidy, but the ingenious arrangement of faucets and stop-cocks in the boilers and the adjacent reservoir merits careful study.

Conservation of heat While the furnaces of Roman bath houses continued their prodigal consumption of fuel, signs of technical development in the distribution system are not lacking. These fall into three categories: (a) the economical placing of facilities with a view to making maximum use of the heat supply; (b) use of spent gases from the furnace to heat the hypocausts; (c) improvements in boiler design (already discussed above). So far as the siting of facilities is concerned, the clustering of kitchen, boiler room and *caldarium* is an obvious arrangement, adopted quite early as a standard procedure in the design of private houses, along with the equally obvious orientation of the *caldarium* facing south-west. Interesting developments noticed from the larger bath establishments are, (1) location of the sweat-room *(laconicum)* immediately above the boiler, and

36 Forum baths, Pompeii. The hot chamber *(caldarium)* showing in the apse, basin for ablutions and, above it, a circular opening, fitted with a lid which could be raised or lowered to control the temperature.

surrounded by other rooms so as to minimize loss of heat by radiation; (2) insertion of a lobby between the heated rooms and the exterior of the building to minimize loss of heat through external walls. The second type of conservation arrangement is well illustrated from the women's section of the Stabian Baths at Pompeii, where the required high temperature was maintained (with heat conservation) by siting the bath directly over a pillared room through which the gases from the furnace passed on their way to the chimney.

A recent study of the so-called 'Small Baths' at Hadrian's Villa at Tivoli[86] has revealed what appears to be an unique architectural complex, with a highly developed technology harnessed to its specific requirements. Nothing now remains of the hydraulic equipment or of the heating installation, but the main lines of both systems have been recovered. The water supply was fed in by gravity flow from a source on high ground to the south-east of the buildings. Seven rooms were jacketed with square pipes made of box-tiles, and three furnaces provided the various apartments with a range of temperatures. The special features included a *nymphaeum,* containing a scenic water display, and 177 there was possibly a water-organ to provide music. How much of this ingenious hydraulic system one wonders was still *in situ* when the hydraulic engineers of a much later age set about their task of providing the gardens of the neighbouring Villa D'Este with its world-famous aquatic spectacles?

The numerous references to Orata and his inventive capacities suggest that he was a rather untypical Roman; and the centre of his activities, the Bay of Naples, could very well have been the source of his invention of the 'hanging bath-house' *(balneum pensile),* for the sweat-baths of the fashionable resort of Baiae were heated by natural volcanic steam. They may also have given him the germ of the idea for the warm tanks in which he kept his fish growing during the winter months. Hypocaust heating caught on rapidly, and soon spread throughout the Empire, disappearing with the collapse of the western part of the Roman dominion, only to be reintroduced many centuries later from the East under the name of 'Turkish' baths.

Technology in medicine

Greek surgery

The sections of the Hippocratic Corpus entitled *On Fractures and Joints* contain full descriptions of the apparatus required, including bandages and splints for fractures, leverage and wooden devices for dislocations (Phillips, 1973,93). The requisite traction was provided either manually or by windlass. The famous 'Hippocratic bench' was provided with various fittings, including straps, cords and windlasses, a post for holding the patient against extension of the leg, wooden

levers set in grooves, and cross-bars. A bellows-inflated wineskin was used for inward dislocation of the femur. For surgical instruments, large numbers of which are to be found in the museums, J. S. Milne's comprehensive and profusely illustrated study (*Surgical Instruments in Greek and Roman Times,* Oxford 1907) is still indispensable. In addition to chapters on the major categories – knives, probes, forceps, etc., the work includes a classified inventory of instruments from more than thirty museums, and there are 54 plates.

Roman surgery

Searborough (1970) has little to contribute, except for an important reference to the instruments found at Colophon (K. Caton, *JHS* 24 (1914), 114–18, which point to technical improvements in surgical implements, possibly arising from important metallurgical techniques.

Military and naval development

The areas of activity represented in the seven chapters of Part II of this book differ widely in the degree to which the development of technical skills and processes can be accurately assessed. For agriculture and food processing we have a valuable collection of technical manuals, and numerous references elsewhere in the literary record, plus a good deal of archaeological material, both in artefacts and representations. In ship-building on the other hand, until the recent advances in underwater exploration, the student had only the visible remains of harbours, plus a large number of mosaics and reliefs, of which only a limited number have been available in print. One particular area of technology, that of weaponry, stands apart from all the others. There are two reasons for this situation: the first is the comparatively large bulk of the written sources, especially those which describe the catapults and ballistae; 37 the second is the fact that the development of these weapons was not subjected to the constraints which limited development in other areas of the economy. In classical antiquity, as in other periods of history, the impulse to

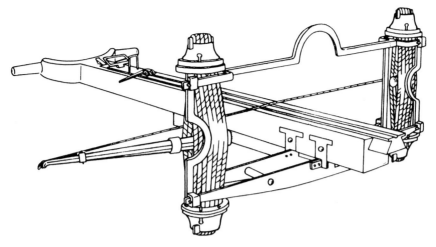

37 Reconstruction of Hero's *cheiroballistra* which was evidently a manually-operated device like the earlier Greek *gastrophetes,* and not drawn by windlass.

match, and if possible to surpass, the offensive capabilities of real or potential enemies overrode considerations of expense.

So far as concerns artillery, the department which saw the most spectacular advances, the abundant literary sources made available by the patient researches of Eric Marsden, make it possible to present a clear and well-documented account of the various types of weapons, and their development in the search for greater firepower and accuracy in reaching their targets. This task has now been accomplished by John Landels, whose monograph is readily available. His detailed and lucid account, which is written from a 'problem-solving' angle highly appropriate to the subject, has covered the ground thoroughly.[87]

Military and naval development in the Hellenistic period

The three areas concerning which we are well informed are those of: (a) missiles; (b) defensive devices in siege warfare; (c) warship design and armament.

(a) Missiles

Landels (1978, 123 f.) has identified six developments in the design of catapults, four of which are of Greek and two of Roman origin. The first of these is concerned with firing speed, and the man responsible for the improved version, an engineer named Dionysius from Alexandria, was able to automate three of the six steps involved in the firing process, and simplify two of the remainder, making it possible for a single soldier with little or no training to operate it. The second improvement was a new method of tensioning the springs of the torsion catapult, which allowed for a greater length of rope, and increased leverage. The third, which consisted of the substitution of bronze springs for sinewropes as the source of power, probably never left the drawing board, while the fourth, a compressed air type, in which the bronze plates of No. 3 were replaced by pistons and cylinders, never got beyond the experimental stage.

(b) Defensive devices in siege warfare

Polybius' account of the siege of Syracuse (VIII. 4–6) mentions the setting up by Archimedes of a battery of stone-throwers and arrow-shooters of varying trajectories, enabling the defenders to cripple the advancing vessels and reduce the numbers of the assault marines at different stages of their advance.

The devices employed to destroy their scaling ladders and the ships from which they were mounted called for no new inventions but only

37

for slight, yet highly effective, variations of the pivoting crane (Landels 1978, 94–95 and fig. 31), which was used for two quite different purposes: (a) dropping heavy stones or lumps of lead on to the decks; (b) clawing up the bow section by means of a grab.

(c) Warship design and armament

Hellenistic naval development seems to have been confined to a great competition between the opposing powers to build bigger and bigger warships. Soon after the arrival of the quinquereme or fiver, which probably had the same arrangement of rowers as the highly successful trireme (see p. 142), but with more men to the oar in the two upper tiers, seveners, thirteeners and sixteeners make their appearance, and at the height of a building race reminiscent of the 'Dreadnought' competition which led up to World War I, two thirtiers were built, but we have no information about the disposition of the oars. The last of these monsters, a 'fortier',

was described by a writer, Callixenos of Rhodes, some fifty years after its launching. If we are to believe the specifications (details in Landels 1978, 153) she was over 130 m (425 ft) in length, with a beam of 17.6 m (nearly 60 ft). Her four steering oars were 13.8 m (45.5 ft) long, and the top bank sweeps 17.5 m (57.3 ft)! It would be difficult to find a more appropriate conclusion to a chapter concerned with technical innovation and development than to cite the final paragraph of Landels' discussion of this remarkable vessel:

Given the personnel on deck to man a huge array of catapults and missile-droppers, and to put big boarding-parties on to any captured vessel, such a warship was without doubt unapproachable, unsinkable and altogether invincible. There was just one slight problem—it was also practically immovable. According to Plutarch, it was just a showpiece which stayed at its moorings and never went into action, and although here (as so often) he is complaining about the extravagance of the rich, he may be right.

5 Power Resources of the Classical World

In an age when the complicated process of harnessing atomic energy has begun and directed its fearsome potentialities into peaceful channels, it is scarcely necessary to underline the comparatively meagre nature and range of the power and energy sources available to the Graeco–Roman world. The fact that we still employ in our day-to-day calculations the old-fashioned base of one horse-power enables us to see the contrast in all its starkness:

One gallon of petrol . . . if used in an ordinary engine of average efficiency . . . will do the equivalent work of about 90 men, or of nine horses of the smallish size used in the ancient world, for one hour.

Landels (1978,9)

Of the power resources available to advanced countries in the nineteenth century, namely muscle (human and animal), water, wind, hot air and steam, the possibilities of the last three were known, but applied only minimally and mainly for amusement. Water power was employed, but its use is not attested before the first century BC, and its extension was apparently rather slow; manpower and animal power dominated the working scene throughout classical antiquity.

We shall now examine these power resources in turn, and try to see with what degree of efficiency they carried out the tasks they were called upon to perform. In dealing with human and animal muscle power, we shall look closely at the mechanical devices and techniques which were introduced to direct, convert or extend it, and, as was forecast in an earlier chapter, we shall take special note of any important innovations or developments of basic techniques.

Human muscle power

Before going into details we should take careful note of two important features common to pre-industrial societies. The first is the dominant role of human muscle power, which can persist long beyond the introduction of equines (horses, mules, asses) with their much greater load-bearing capacity. The second point is that where human muscle power is the major, or the dominant, source of energy, increased demands on that resource (for instance for the shifting of heavy loads) is normally met by multiplying the units of burden or traction, as the case may be. This second point is well illustrated in a passage in his treatise *De Architectura,* in which the Roman writer Vitruvius is discussing the problem of leverage (*De Arch.* 3.7):

Again, when very heavy loads are carried by four or six porters, they are balanced exactly in the middle of the carrying-poles, so that the undivided solid weight is shared in fixed proportions, and each workman carries an equal portion of the load. For the central portions of the poles, on which the straps of the porter are fastened, are fitted with pegs; these prevent the straps from sliding into one part of the pole.

Harnessing and redirecting human muscular power

Hand tools, as extensions of that ingeniously fashioned, and highly adaptable piece of equipment, the human hand, have vastly increased, and continue through fresh inventions to increase, his capacity to accomplish an astonishing variety of tasks under the most diverse working conditions, from sea-bed to outer space. Among the important tools invented or developed to some new purpose in the classical period are the Roman balanced sickle and the *15*

British long scythe in agriculture, and in carpentry, the brace and the frame saw. In addition, the employers of labour could avail themselves of a number of mechanical devices for harnessing and redirecting human effort.

Table of muscular movements

(a) Arms

1   pressing down (working bellows)
2   pressing up and down (working force-pump/fire appliance)
    (= *Ctesibica machina, organum pneumaticum.* Pliny)
3   swinging   (a)   backwards and forwards (breaking ground with hoe)
                    (b)   sideways (scything hay/grass)
                    (c)   downwards (excavating ore in mine with pick; forging metal with sledge-hammer)
4   pulling towards one (sickling grain, sawing wood)
5   reaching up and pulling down
                    (a)   drawing water with swing-beam/swipe
                          (= *keloneion* (Gr.); *tolleno* (Lat.))
                    (b)   lifting weight through pulley-blocks
6   hauling up (bucket from well, baskets of earth from trench, etc.)
7   kneeling with up-and-down movement (grinding grain with saddle-quern) (= *mola trusatilis* 'pushing mill')
8   rotating arm (grinding grain with hand-mill) *(mola)*

(b) Legs

1   HAULAGE walking with load (hauled on sledge); on river barge head-loaded; back-loaded; shoulder-loaded; pushed in front
2   PORTERAGE 'treading' (grapes, olives, in vat)
    water-screw (*cochlias,* Archimedes' screw)
                    (a)   on cylinder
                    (b)   on tread-wheel, see below
    tread-wheel   (on crane, on compartment wheel ('drum', *tympanum*))
                    (on treadmill)

(c) Arms and legs combined

1   digging, trenching with spade
2   rowing

Among the most important of these was the circular capstan fitted with removable handgrips (the 'handspakes' or 'handspikes'), which could be set horizontally or vertically to winch up from below (e.g. an anchor from the sea floor), or to haul up over pulley-blocks (e.g. a yard-arm, or any heavy tackle on shipboard or a-shore). Contrary to a popular view, it seems that work of this kind was not adversely affected by lack of that mediaeval invention, the crank.[88] The standard well-head installation of a wooden roller on which the rope and its bucket were wound seems unthinkable without a crank handle; yet the restriction on the thrust to be applied if the handle is not to fly backwards means that the substitution of a capstan by a crank 'would have lowered the handling capacity [of a crane] by some 20–30%' (Landels 1978, 11). The only device which would have gained considerably from the use of a cranking mechanism was the repeater catapult, which could have been reloaded very much faster (Landels, *loc. cit.* for details).

Of equal importance to classical technology was the tread-wheel. Whereas the capstan produced an increased power output and secured a continuous work-flow by multiplying the units of energy and co-ordinating them into a team, the tread-wheel used the very efficient muscular combination required for the steady climbing of a slope, the action resembling that of riding a bicycle, except that the muscular drive was largely confined to the legs. The tread-wheel is so called because its rim is fitted with wooden treads, on which the operator treads continuously to set the wheel going and maintain it at the required speed. To obtain maximum torque he should maintain his treading position on a level with the axle of the wheel. Both design and size of tread-wheels, as well as the number of operators will have varied according to the nature of the job, and the power output required. Unfortunately our knowledge is far from complete. None of the standard works contains a systematic treatment of the various problems. The only extant representation, the bas-relief on the family tomb of the Haterii,

50

38 Detail of bas-relief from Avezzano, now in the Museo Torlonia, Rome, depicting two men turning a capstan, to which ropes are attached for winching, 1st century AD.

3 represents not a wheel with outside treads, but a winding drum for a crane, with the treaders inside the drum. The most spectacular example of the use of tread-wheels is the system of eight superimposed pairs of wheels, with hollow rims, installed by Roman engineers to drain mine-installed by Roman engineers to drain mine-shafts on the Rio Tinto in Spain. The team of sixteen men will have been able to lift about 2,400 gallons per hour to a height of almost 100 feet.

Animal muscle power

The working animals used in classical antiquity comprised, in the bovine category, the ox and the cow, and among the equines the donkey and the two products of its crossing with the horse, the mule and the hinny. There is some evidence for the use of entire bulls by both Greek and Roman farmers for drawing the plough, and of mares for the ancient method of threshing the grain by trampling, but apart from the use of worn-out horses in the endless 39 drudgery of the rotary grain-mill, this noble animal was almost entirely restricted to the battlefield and the racetrack. The contrast between the role of the humble ox, broken to

the servitude of the yoke, and the proud chariot pony, symbol of aristocratic power and pride, is nowhere better expressed than in the following lines in which man's greatest benefactor, Prometheus, proclaims that his was the hand that wrought these twin services to mankind:

I was the first that yoked unmanaged beasts,
To serve as slaves with collar and with pack,
And take upon themselves, to man's relief,
The heaviest labours of his hands; and I
Tamed to the rein and drove in wheeled cars,
The horse, the ornament of sumptuous pride.

> Aeschylus, *Prometheus Bound*, 462–66,
> trans. G. M. Cookson

Behind these proud words lie the economic facts which for many centuries determined the respective roles of ox and horse. Detailed discussion of the harnessing, feeding and other problems which had a bearing on the relative efficiency of bovines and equines as load-pullers will be found in Chapter 10. Here we shall draw attention to a few points of major importance. The ox was in many ways a better proposition in an economy where food costs in relation to

39 Relief from the Vigna delle tre Madonne showing a horse mill. Musei Vaticani, Rome.

work output were of serious concern. Horses require a high protein diet from high-grade pasture, which is not found in quantity either in Greece or Italy. The best bloodstock for cavalry or chariot-racing was raised on the high plateaux of Anatolia or Spain. Oxen, on the other hand, could keep on going at their proverbially slow pace on a diet that was inferior in quality, but the constituents of which were readily available to Greek and Roman farmers. Their second advantage over horses was admirably expressed by the thirteenth-century agricultural writer Walter de Henley, when he observed that the ox is 'Manne's mete when dede'. Oxen also have the advantage of a better power-weight ratio, having a greater tractive capacity than horses of the same size. 'The one advantage that the horse had over the ox was speed, and it was precisely in those situations where speed outweighed everything else that the horse was used—in warfare and in chariot-racing. The high mobility of the cavalry gave that arm its particular role in battle tactics, and on the race course a chariot, made as light as possible, and drawn by a matched team of two or four horses, represented the ultimate in speed to the Greeks from the eighth century BC onwards, and to the Romans after them.' (Landels 1978, 14.)

So far we have looked only at the horse and the ox, and only in their respective roles as load-pullers; what about the other animals mentioned at the head of the chapter? Adherence to a strictly evolutionary order would have given priority to our second topic—the use of animals as load-carriers. The ancestors of our modern equines, the wild horses and asses of prehistoric times, must have been broken in as mount and load-carriers for their owners many thousands of years before the first suitably curved branches and the first crudely shaped wooden frames were attached to them by primitive harness to form the earliest ploughs and sledges. In conveying produce or manufactured goods by land in the Mediterranean region the mule, hardier and more sure-footed than the horse, equipped with balanced panniers of soft

basketry had, and still has, a major role to play in the mountainous regions where wheeled transport is out of the question.

It is worth remembering that in 1881 the wheeled vehicle was still unknown in Morocco, that it only appeared in the Peloponnesus in the twentieth century ... Braudel (1973, vol. I, 283)

Jack donkeys of the best breeding strains for producing top quality mules commanded very high prices, and mule-breeding establishments, like the one owned by Marcus Varro the agronomist in the fine upland pastures of the Reate district, evidently brought handsome profits in replenishing the stocks of mules which, according to the same authority, handled the bulk of the grain and other produce. In carrying capacity three pack-mules were deemed equivalent to one wagon, as shown by the recently discovered inscription from Salagassos in Pisidia (see *JRS* 66 (1976), 106ff).

At the foot of the load-carrying list comes the abused, overloaded and ubiquitous donkey; with muscular power rated at only half that of a mule and one quarter that of an ox (see p. 129 and Forbes, *Studies* II, 85), the donkey was, and is still 'the symbol of everyday life in the Mediterranean' (Robert Ricard, quoted by Braudel, *op. cit.*, 226). Cato's inventory of stock and equipment for a medium-sized estate specializing in vine-growing, calls for three of them. Reference to the provision of a 'donkey for the mill' brings us to one of the few important industrial activities in which animal muscle power was applied on a large scale, namely the rotary mill for grinding corn, known commonly as the 'Pompeian' mill from the very large number of these machines that here found in *situ* on the premises of the numerous bakeries in that city. The evolution of this machine from the rotary hand-mill is fully discussed in Chapter 6. Here we are concerned with its place in the story of the development of power resources to meet new or increasing demands. The appearance on the scene of the early rotary hand-mills during the course of the third century BC coincides with the arrival of the professional miller

39

(pistor) 'and its evolution may well be related to the production of flour on a larger scale than needed for home use only (Forbes, *Studies* III, 149). To the same period belongs the introduction into the grain-milling process of the much larger rotary mill, the donkey-mill or Pompeian mill, which appeared in response to the increasing demands of the growing urbanized populations of the late Roman Republican period, as well as those of the new commercial farm enterprises based on slave labour. The mills found *in situ* at Pompeii are of two sizes, the smaller size for donkeys, and the larger for mules or worn-out horses. Forbes (*Studies* II, 81) estimates the energy output of a donkey-mill as varying between 0.4 and 0.5 HP, which represents about four times the power output of a man turning a capstan. The cheapness and the availability of donkeys account for the fact that they continued to operate for many centuries after the introduction of the water-mill, the most primitive form of which, the undershot wheel, is alleged to have given an output at least six times as great.[89]

In considering the role of animals as sources of energy in the all-important agricultural sector, we must emphasize the key position of the ox as the chief working animal on the farm. The energy needed for its most important task, that of ploughing the fallow between crops, will have varied enormously according to the type of soil encountered. To prepare a seed-bed in the light soils of southern Italy only light stirring of the surface was needed, and it could be done with a pair of small oxen, or even with donkeys, attached to a plough small and light enough to be carried on the ploughman's shoulders, but in many areas of the north, as Pliny tells us, no less than eight pairs of heavy steers panted under their yokes (*HN* 18.170). To maintain his teams in first-class condition for their work was the first care of the diligent manager, and some ingenuity had to be exercised to keep up their vital energy during the long summer drought, and at other times when the necessary energy-producing foods for this remarkably efficient working animal were

scarce. The animal-powered grain-harvester, a Gallo-Roman invention which does not seem to have spread much beyond its birthplace, which was somewhere along the common frontiers of France, Luxemburg and Belgium, is described elsewhere (Ch. 6). Two interesting features entitle it to a place in this chapter. The first of these is the method of harnessing, the second, the mode of application of the power. The machines (there were several varieties of them) were pushed through the cornfield by a single animal (mule or ox) harnessed in shafts, which were extended to the rear to make room for a steersman. It is interesting to note that the single animal in shafts is traditional to northern Europe, contrasting with the Mediterranean tradition of paired animals attached to a pole; shafts are also equally suitable for pushing a load as for pulling it; and the pushing arrangement was essential to the task, since the standing corn must not be trampled before it has been reaped. A similar arrangement of the motive power appears in the first modern reaper; Patrick Bell's machine of 1830 was powered by two horses working in shafts from behind. Landels, who makes no mention of this machine, nor indeed of any other applications of power in agriculture, such as the use of animal-powered grain-mills on the farm, refers in another chapter to another use of animal energy, which surely demands treatment under the heading of power and energy sources, namely the ox- or donkey-powered bucket-wheel or wheel of pots, used in Egypt probably from as early as the Ptolemaic period, and still used in the Nile valley, under the Arabic name of *sakiyeh*, for lifting water from the river to the *40, 41* irrigated plots. Mentioned for the first time in documents of the second century BC was the bucket-wheel normally powered by animals working a horizontal capstan, with a crown and pinion to convert the horizontal motion to the vertical wheel. The simple type used in Egypt and elsewhere in the Middle East consists of a chain of earthenware pots strapped to the rim, but Vitruvius mentions an improved version in which the shape of the buckets, and their angle

40 One of seven animal-operated *sakiyehs* between Puerto d'Andraix and Andraix town, Majorca.

41 *Sakiyeh* from Puerto d'Andraix showing the vertical wheel and bucket-chain.

of tilt were such as to prevent any discharge of the contents before they reached the top of the wheel (10.4.3).[90] In view of the frequent references in the papyri to the animal-powered wheel of pots, still popular in Egypt, it is surprising to learn that Vitruvius knows only the tread-wheel as source of power.

The last item falling for discussion in this section is a very curious animal-powered warship, with a transmission system similar to that of the bucket-wheel. This invention, which forms part of an anonymous military work of the fourth century AD known as *De Rebus Bellicis* uses the same form of geared transmission to convert the horizontal motion of pairs of oxen working on a capstan into the vertical rotation of pairs of paddle-wheels, but with the power line reversed. Landels, who has examined the main technical difficulties which would have to be overcome to make the invention work, concludes that while theoretically feasible, the whole idea was not very practical (1978, 16). In view of the space required for the unspecified number of pairs of oxen, the power-weight ratio would have surely been too low to make the vessel an effective competitor with warships driven by oars and sails.

Manpower and animal power compared

Before going on to look at the power resources provided by water and wind, we may find it useful to compare the muscular power of men with that of animals, taking note of those advantages and disadvantages which may have affected their respective deployment in the classical world.

The first and most obvious advantage of men over animals is the much greater range of tasks which they can perform. This is particularly evident in the area of land transport, as Landels points out, 'The human porter (in Latin *saccarius*—"sack man") is much more adaptable in every way than a vehicle or pack animal. He can make access for himself to the hold of a ship, can climb ladders, go along narrow paths or gangways and is, so to speak, self-loading and self-unloading.' (1978, 170). In addition to this manœuvrability, man is also remarkably versatile. Thus, where manpower resources are limited, and the operations needed change rapidly from day to day, as in cargo-handling in a seaport, or in building-construction, men can readily transfer from one activity to another without the complexities of reharnessing or of retraining that would be needed if animals were employed. In this connection it is worth noting that many of the highly sophisticated machines that have recently come into use in factory or farm to replace the ever more scarce manpower have been developed as a result of close study of the human muscular system.

On the other side of the balance sheet are the greater physical strength of working animals and their greater capacity for sustained physical effort. Thus one horse of the size used in the classical world could do the work of nine men, a mule that of about five. In porterage, these differences stand out sharply; a mule with side panniers could carry a load of 140 kg (*c.* 300 lb) for eight hours in a day, doing the work of three men (see p. 129).

Water power

The energy available in perennially flowing streams, once it was harnessed in the service of man, has provided him, over a span of many centuries, with an accessible, and in some areas an abundant source of cheap power for driving all manner of machines, from small grain-mills housed beside some modest mill-lade to the mighty cotton- and woollen-mills of nineteenth-century England.

The story of the development of this source of power from its humble beginnings can be partially pieced together from a number of references which, taken cumulatively, suggest the following likely order of appearance in the Mediterranean region of machines using water power to drive them: (1) 'undershot' water-wheels fitted with paddles, used to raise water irrigation by bucket-chain; (2) the same type of

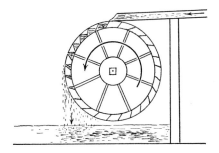

42 Undershot water-wheel, above; overshot water-wheel, below.

wheel fitted with gearing to convert the horizontal to a vertical drive for grinding grain (the water-mill); (3) the more powerful and efficient 'overshot' wheel, which requires a higher velocity than (2), provided by a 'mill-race' or aqueduct. Types (1) and (2) are described by Vitruvius (*De Arch.* 10.5), and the undershot wheel is often referred to as the 'Vitruvian' wheel. 7 There is some evidence that the overshot wheel was known at the time when Vitruvius was writing, but it was evidently the undershot type that was in demand, and a comparison of the two will offer some reasons for the preference. As Landels has shown (1978, 21 f.) the efficiency of an overshot wheel can be raised to as high as 70 per cent compared with only 22 per cent at best for the undershot type. This factor matters enormously where fuel costs are high, but running water costs nothing, and an undershot wheel requires no elaborate construction work, and no engineering works to increase the rate of flow. In fact, Vitruvius' mill was geared down instead of up, and will, therefore, have ground very slowly indeed; but it could still have been preferred to its more efficient competitor because it cost much less to install and did not need a high velocity supply.

Other applications of water power

In mediaeval Europe, the spread of water-mills for grinding grain was soon followed by the adaptation of the mill to other industrial purposes; these included saw-mills, hammer-mills for iron-working, and mills for use in the tanning of leather. The only evidence we have from the classical world of water-wheels being applied to any other operation apart from water-lifting and corn-milling is a brief, but very clear, if poetic, description of a water-driven saw-mill at work on the Ruwer, a tributary of the Moselle, in a well-known poem addressed by the fourth-century poet Ausonius to his native river:

56

> He [that is, the river Ruwer], as he turns his mill-stones in furious revolutions, and drags the shrieking saws through smooth white stone, hears from either bank an endless din... Ausonius, *Mosella* 361–64

It would surely be unreasonable to expect the poet to tell the reader whether the sawing motion was achieved by means of a cam and lever or a crank and connecting-rod, but there is no ground whatever for suspecting an interpolation, as one leading authority on mediaeval technology has suggested.[91] One might have expected other applications, but apart from Ausonius' saw-mill there is nothing in the records. There is also a gap in the history of the water-mill itself, extending from the time of Vitruvius to the multiple grain-mills at Barbegal near Arles in France, which are dated to the late third or early fourth century. The fashionable explanation for what appears to be a long time-lag between invention and widespread application is twofold: (a) lack of rivers with a substantial and perennial flow; (b) animal mills being much cheaper to construct, and there being no shortage of animals to drive them. It should, however, be pointed out that water-mills appear in the *Edictum de pretiis* (Price Edict) AD 301, and are likely therefore to have been fairly common by then; the absence of any evidence for the sort of rapid development that occurred in the eleventh and twelfth centuries as part of a real industrial revolution seems to provide a good example of failure to exploit an invention for lack of the necessary motivation. Roman engineers could also have developed other applications of the water-wheel to industrial processes, had they been encouraged to do so. Where the demand existed, technical resources were more than adequate to supply what was required, as in the military sphere, where such missile weapons as the torsion catapult and the stone-throwing ballista were developed to a high technical standard, and produced in large quantities. The later history of water-driven machinery also provides evidence of a tendency well known to students of the history of technology, that of a well-tried device to resist encroachment by one that is technically superior, and possessed of much greater potential, when it might well have been expected to fade out. The later history of the water-wheel provides an excellent example of the tendency; having failed throughout classical antiquity to develop its great potential as a prime mover for machinery, once it had broken loose from these restrictions, the water-wheel continued to make headway, competing successfully against the windmill (first used for milling grain about 1300 AD, and for pumping water a century later, and surviving the introduction of steam power for driving machinery). Its principal asset in the struggle lay in the fact that, where perennial streams were available, as in many parts of north-west Europe, it provided a continuous source of low-cost power; there was no reduction in output, and no vibration. Throughout Greek and Roman antiquity, however, the demand for increased mechanization was negligible, except in the areas of state service and public works; the demand for clothing, footwear and domestic utensils which, in modern times has led to mechanization and mass production, was, with some exceptions (above, n. 48) met by local craftsmen, working in small workshops to satisfy local needs.

Wind power

The earliest device designed to use the wind as a propellant was the sailing ship. From the idea *43*

43 Roman freighter under full sail, full and bye, and bellying the mainsail. Beirut Museum.

of setting a piece of canvas so as to enable a vessel to move over the water at a suitable angle to the wind to that of setting vanes at a suitable angle on a wheel, using the power of the wind to drive a piece of machinery does not seem a very big step, but the only classical reference to the use of wind as a propellant occurs in Hero's *Pneumatika* (1.43), where a tiny windmill is connected by means of a primitive rocker-arm arrangement to an air-pump for blowing an organ.[92] It remained an isolated phenomenon; no one conjured up the idea of building a really large version with a power output big enough to enable it to compete successfully with existing units based on human or animal muscle-power.

Steam power

In another chapter of his *Pneumatika* (2.11) Hero describes a steam-operated toy, consisting of a hollow ball which is made to revolve on a pivot by steam raised in a cauldron, on the lid of which the mechanism is mounted. A fire is lit under the cauldron, and steam passes through a pipe into the ball, from which it escapes through a pair of opposed bent tubes: the principle is the now familiar one of jet propulsion. The rotating ball is capable of very high speeds (a model made by Landels achieved about 1,500 rpm), but even a large scale-model would, it seems, have been quite impracticable; according to Landels' calculations, the fuel consumption needed to bring it to the level of work of one man would have been astronomical!

Compressed air power

The expansion of a gas when its temperature is raised can generate considerable power. Several of Hero's 'toys' such as the magic temple whose doors opened automatically to the visitor after he had made his preliminary burnt offering on the altar outside, were based on the operation of this principle, but for the same reasons as those given in the preceding section, there was no development in the direction of an efficient steam engine (Appendix 4).

PART II

6 Agriculture and Food

Agricultural technology

The methods by which farmers cultivate the soil are broadly determined by variations of terrain and climate, and by the availability of labour. The peasant-proprietors of early Greece and Italy, who worked their small-holdings with the help of their wives and children, aided usually by a slave or two (see Varro, *De Re Rustica* 1.17), owned no machinery, apart from an ox-drawn plough, and perhaps a roughly made brushwood harrow. Both countries are mountainous, and level plains are rare. In Italy, as the Elder Pliny reminds us, 'man has such a capacity for labour that he can perform the work of oxen—at all events mountain folk dispense with this animal and do their ploughing with hoes' (18.178). On the slave-run plantation, while the organization differed profoundly from that of the subsistence farmer, the tools remained the same. The organization of the servile labour force was closely geared in with the operations of the seasonal calendar, with the aim of keeping the 'vocal implements' fully employed 'round the clock'. Even at the seasons of haymaking and harvesting, when the tasks of ingathering were beyond the capacity of the permanent workforce, the use of labour- or time-saving devices appears rarely and in isolation, the employment of casual labour hired for the purpose being the norm. Within these limitations, however, the manual implements themselves show numerous varieties of the basic patterns. Some of these modifications reflect the complexity of the tasks to be performed

by a single operator, and double-ended and even multiple tools make their appearance; others appear in response to the need to speed up the work, while many emerge as a result of local preference.[93] In Roman practice, technical development is thus common, while innovation remains rare. Among tools of the double-ended class the most interesting combinations are those in which one end consists of a hatchet-blade, balanced on the other end by a pick, which might be either straight or crooked, and narrow or broad in profile. Both the narrow type (our 'pick-axe'), and the broad type (our 'mattock') are still widely used (e.g. by road-repairers and woodsmen). The pick-axe is well known as the excavator's and miner's pick. On the land its uses were manifold, and included chopping and rooting out tree-stumps after felling, as well as removing dead wood during ploughing operations in the vineyard. The mattock was also much used in orchards and vineyards, especially for digging round tree-boles and vinestocks, and stripping off surface rootlets. The outstanding example of technical development is the vine-dresser's knife *(falx* **44** *vinitoria),* which evolved from a simple 'bill-hook' into a multi-purpose implement, each separate part of which had its own specific function, as Columella explains (4. 25):

Now the shape of the vine-dresser's knife is so designed that the part next to the haft, which has a straight edge, is called the *culter* or 'knife' because of the similarity. The part that is curved is called the *sinus* or 'bend'; that which runs on from the curve is the *scalprum* or 'paring-edge'; the hook which comes next is called the *rostrum* or 'beak', and the figure of the half-moon above it is called the *securis* or 'hatchet'; and the spike-shaped part which projects straight forward from it is called the *mucro* or 'point'.

44 Vine-dresser's knife *(falx vinitoria),* the multi-purpose instrument of Columella.

Each of these parts performs its own special task, if only the vine-dresser is skilful in using them. For when he is to cut something with a thrust of the hand away from him, he uses the *culter;* when he is to draw it towards him, he uses the *sinus;* when he wishes to smooth something, he uses the *scalprum,* or, to hollow it out, the *rostrum;* when he is to cut something with a blow, he uses the *securis;* and when he wants to clear away something in a narrow place, he makes use of the *mucro.*

This highly specialized instrument belongs to a large class, the *falces* or 'hooks', spanning a wide range of activities in meadow and cornfield, vineyard and orchard, and reflecting the variety and complexity of manual operations on the farm.[94] The number of lopping and pruning instruments emphasizes the importance of trees and shrubs and their products in the agricultural economy. Thus a mixed farm with wine as the main crop required not only heavy timber for fencing, and light timber for hurdles, but large quantities of props, stakes and frames, as well as osiers for tying the vines to their supports. In addition, the leaves of trees and shrubs were cut to supplement the inadequate supplies of grass and legumes for animal fodder. The manufacture of agricultural implements and machines was evidently very much localized: Cato advises his readers where they can buy the best in each category:

> ... iron tools, hooks, spades, mattocks, axes, harness ... should be bought at Cales and Minturnae; spades at Venafrum; carts and threshing-sledges at Suessa and in Lucania ... yokes from Rome will be the best (*De Agri Cultura,* 135)

Ploughs

The most important piece of agricultural machinery, as well as the oldest, was the plough. Greek and Roman ploughs belong to the major class known as breaking ploughs or *45* ards; being symmetrical in their design, they differ from our mouldboard ploughs in that they stir the soil to a shallow depth by throwing it up on both sides of the dividing share, and are thus eminently suited to the prevailing soil conditions of the Mediterranean region. They

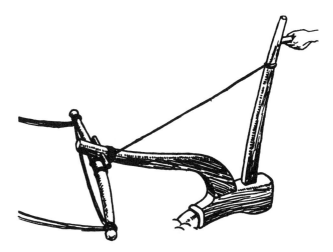

45 Sole-ard (Algerian) closely resembling Roman design, but harnessed for working with a single animal. No Roman original has survived.

were of wood, except for the attached share, and simple joinery (chiefly by means of mortise and tenon) sufficed for putting the parts together. Their basic design has remained unchanged and wooden ards may still be seen at work in the more remote parts of Greece and Italy (for detailed discussion of the various types of ard and their parts see White 1967, 126 ff.). Unlike the mouldboard plough these symmetrical ploughs cannot form ridges. Where these were needed for the purpose of covering the seed or for drainage detachable wings were used to throw up the soil on either side of the share and make a 'ridge-and-furrow'. Ridging boards appear to be a Roman invention. From the available evidence it would appear that the evolutionary progress of the front of the ploug ran from no protection (suitable enough for the very loose and friable soils of Mesopotamia) through the detachable and replaceable share, first of wood, later of iron (earliest surviving examples date from *c.* 1000 BC in the Middle East) to variations in ploughshare design reported in the mid-first century AD by Pliny (18. 171–72). The most important of these is a sleeved iron share furnished with a vertical *46* edge and horizontal cutting edges along either side, designed to cleave heavier soils and at the

59

46 Socketed ploughshare, fitted with sharp edges at either side for cutting off weeds: Pliny's fourth variety.

same time to save labour at a later stage by cutting off the roots of the weeds. Pliny concludes his account of the development of the ploughshare as follows:

Recently a contrivance has been invented in the Raetian area of Gaul, consisting of the addition to this type of plough of two small wheels ... the share is spade-shaped (18. 172)

That this important technical improvement should be reported from one of the Alpine provinces is logical enough; the addition of a wheeled forecarriage would be a natural response to the challenge of heavy clay soils. The Raetians may well have got their wheeled plough from the Celtic north, where it was destined to open up to cultivation vast areas of potentially fertile loams which the classical scratch-plough was unable to exploit. Pliny's account is thus of some importance: he shows awareness not merely of the evolution of the ploughshare, but of a connection between the technical advance represented by the double-bladed share and the invention of a plough with a wheeled forecarriage.

Labour-saving machines

Growing vines was, and still is, a capital-intensive undertaking. The wine farmer, having acquired suitable land, will have to provide capital for buying the plants, together with the necessary stakes and props. He will have to wait several years before his vines begin to produce. Vine-growing is also labour-intensive, needing about three times as many labourers per unit of cultivation as on a farm where only grain is grown. But as we can see by looking at one of the ancient seasonal calendars of operations such as we find in Columella's Book XI,

all the numerous tasks save one can be readily fitted into the regular work programme. The exception is the vintage, where speed is essential if losses are to be avoided, as the fruit goes rotten on the vine. This operation was let out to contractors, who supplied their own labour (see Cato 147 for a sample contract for selling the grapes on the vine). Where the main crop was grain recourse was had at harvest time to those gangs of itinerant casual labourers who have been, from early times, a familiar feature of the Mediterranean agricultural scene, and have only recently been displaced by the introduction of mechanical methods of harvesting. Outside the Mediterranean perimeter, however, there are signs of innovation and development. The Elder Pliny, who has an eye for technical advances (see above, on plough-types, p. 59) mentions two important items, both from the north-western sector of the Roman Empire, an 'improved' Gallic scythe for mowing hay *16* (18.261), and a harvesting machine for grain, of the type known as a 'header' or 'stripper', from the fact that the action of the machine is to strip *47* off the heads of grain in contrast to modern reaping machines, which cut off the plant at ground level. Pliny gives no details of the design, but the fifth-century writer Palladius provides a full description. The discovery in 1958 at Buzenol in Belgium of a sculptured relief containing a substantial portion of a reaper of the type described by Pliny (18.296) has aroused fresh interest in these machines, and led to more informed and intensive study of the technical and economic questions they pose, and to the discovery of fresh evidence. Since we can now identify two distinct varieties of these machines and possess at least four representations of them.[95] it is surprising that any scholar should either doubt 'whether Pliny's famous and controversial reaper was ever used' (Mossé 1969, 35), or deny that any further development in design took place (Hodges 1971, 199). Much misunderstanding has been caused by the fact that the literary evidence does not match that of the surviving monuments: Pliny's very brief descrip-

47 Reconstruction of harvesting machine type I *(vallus)* combining reliefs at Arlon, Buzenol and Trier. Width of frame 1 m. 20–1 m. 30, diameter of wheels 70–75 cm.

tion refers to the 'frame' type represented on the monuments, while Palladius' lengthy and precisely detailed account (*Op. Agri.* 7.2. 2–4) of the 'tumbril' type·does not fit the monuments. Pliny's machine (White 1967, fig. 118) was steered from behind, the shafts being long enough to accommodate both driver and animal (mule or donkey), the shafts being raised or lowered to suit the varying height of the crop. Palladius' machine, on the other hand (White 1967, fig. 119), was powered by an ox harnessed to a pair of much shorter shafts, the operator steering from the side.

The successful operation of a heading machine pushed from behind involves the solution of a number of technical problems. These include (1) a suitable shaping and arrangement of the teeth to avoid (a) clogging of the blades, (b) loss of grain from stalks too low to engage the teeth; (2) some device to prevent the wheels from becoming entangled by stalks wrapping themselves round the axle; (3) a harness and control system enabling the animal to maintain a steady, even pace; (4) continuous operation of the machine, avoiding frequent stops for clearing the blades and emptying the container. How successful were the designers and operators of these machines in solving these problems? First, Palladius' account (confirmed by the Buzenol slab) calls for blades that are upturned at the tip; this feature would promote a scooping action, reducing clogging and loss of grain; secondly, the Buzenol slab clearly shows that the edges of the frame were splayed out so as to protect the wheels from the wrap-

ping action mentioned in (2) above; the third point is strongly emphasized by Palladius. As to the fourth point, the Buzenol slab depicts the machine at rest, the operator being engaged in clearing the blades with the aid of a spatulate stick, which indicates an obvious defect in these machines, the effect of which could only be assessed by constructing a working replica. The earliest reference to a machine of this type occurs in Pliny the Elder (mid-first century AD), and the latest in Palladius (more than three hundred years later). It would indeed be surprising evidence of technological arrest if no changes in design had taken place over a period as long as this; in fact, we have two distinct spe-

48 Buzenol harvesting machine showing a workman cleaning the interstices of the blades.

61

cies of machines, the larger tumbril being described in detail by Palladius but otherwise unrecorded, while the lighter frame is known to have been made in more than one variety. In 1963 a mutilated slab in the museum at Trier was identified as part of a different type of 'frame', equipped with a 'grass box' resembling that of a garden lawnmower in place of the underslung container.[96] None of the three extant representations of Pliny's machine can be accurately dated so that at the present stage their relationship can only be surmised: the Trier machine may be an improved version of the Buzenol type, the container being redesigned to increase its size and thus reduce the number of stoppages during the harvesting operation, while the much larger Palladian machine may be regarded as a more radical change, employing a much more powerful animal.

Both at the beginning and at the conclusion of his account Palladius stresses the economic advantage of the mechanical harvester over manual methods:

In the more level plains of the Gallic provinces they employ the following short cut or labour-saving device for harvesting. With the aid of a single ox the machine outstrips the efforts of labourers and cuts down the time of the entire harvesting operation *(Op. Agri.* 7. *init.)*

In this way, after a few journeys up and down the field the entire harvesting process is completed in a few hours *(ibid., ad fin.)*

If only the author had rounded off this excellent account by telling his readers the size of the field and the number of journeys required to complete the job!

Techniques of threshing and winnowing

The first process after harvesting the grain is that of separating the ears from the straw, the next that of removing the chaff from the grains. Threshing methods in classical times included the age-old techniques of treading out the corn with animals, and beating it out with wooden flails. The simplest of several improvements on these methods[97] was the animal-drawn 'drag' (White 1967, fig. 115)—a heavy board with flints embedded on its underside (Latin *tribulum*), which still survives in many parts of the Mediterranean region and the Middle East. Another improvement was the 'Punic cart' (White 1967, fig. 117b), which broke away the straw, not by a rubbing action, as with the drag or sledge, but by a set of toothed rollers passing over it. Varro, the first writer to mention the device, notes that the driver sits on it (to provide the necessary weight), and that it was in use in the Province of Nearer Spain (which suggests that it derived originally from the advanced farmers of Carthage). To get rid of the chaff both Greeks and Romans used two different instruments; either a winnowing shovel (Gr. *thrinax, ptyon,* Lat. *vallus, ventilabrum*) to toss the threshed material into the air in a gentle, steady breeze, which carries away the chaff, leaving the grain to fall on the threshing-floor, or a winnowing basket (Gr. *liknov,* Lat. *vannus*), which was shallow and open at one end; the winnower shuffled the contents to and fro.[98] The former method, still widely used in the East, is obviously unreliable: 'If however there is no wind for several days, let the corn be separated from the chaff with winnowing baskets' (Colum. 2.20.5). In Britain, until the comparatively recent introduction of forced-draught winnowing, the basket reigned supreme—but for the opposite reason: the winds are too gusty.

Storage of grain

Stored grain, if it is not to deteriorate, must be kept dry (with a moisture limit ranging between 10 per cent and 15 per cent), cool (if possible below 60°F or 16°C); it must also be protected against attack by vermin. Granaries on the upper floors, and with through ventilation, were favoured in Greece and Italy, while both natural caves and underground silos are reported from other parts of the Mediterranean region. The Romans laid stress on careful surfacing of walls and grouting of joints. Here and on the threshing-floor the surface was bonded with *amurca,* an important by-product of oil-making; in addition to its bonding capacity, it also gave off a nauseous odour, which may have acted as an insect-repellent (White 1970, 189; 196 f.). Where large quantities of grain were kept in store as at Ostia and Rome, or at the legionary headquarters on the frontiers, there was the additional problem of enabling the walls to withstand the powerful lateral thrust of piled, unbagged grain. Buttressing and other structural arrangements are discussed in detail by Rickman 1971, 79, 131, etc.

Food technology

The diet of classical Greeks and Romans consisted in the main of food-grains processed into porridge and bread, supplemented by vegetables, fish, poultry, spices, and to a lesser extent meat.[99] Honey was virtually the only sweetener. In this chapter we shall be mainly concerned with the technical problems related to the primary process of converting these food-grains into meal and flour, and the secondary one of converting the latter into porridge and bread. In view of the overwhelming importance of wine as a beverage drunk by all classes, and of olive oil as an essential feature of the classical diet, attention will also be given to the techniques employed in the making of wine and oil.

The two principal food-grains were wheat (in several varieties of both husked and naked types) and barley. In ancient Greece, barley, which for a long time was dominant both as animal fodder and as human food, was made, not into bread, but into the 'kneaded thing' (Gr. *maza*), a sort of cake. Barley is much less suitable for making bread than wheat, since the attachments that enclose the grain (the glumes) cannot be removed by threshing; barley was subjected to a preliminary roasting before hulling; this process will have largely destroyed the gluten content of the grain, making it unsuitable for leavened bread (Moritz 1958, xxi). It is, therefore, not surprising to find barley-bread described as 'fodder fit only for slaves' (Hipponax of Ephesus in Athenaeus, *Doctors at Dinner* 7, 304b). The Roman equivalent of *maza*

was *puls,* a pap prepared not from barley but from the husked wheat called *far* or emmer. This cereal suffered from similar disadvantages in bread-making, but long after it had been generally superseded by the more easily processed naked wheats, *puls* made of wheaten flour continued to be eaten with vegetables, and it continued to be poor men's food in the heyday of the Roman Empire. (See Tables 1 and 2, pp. 221 ff.)

How the food-grains were processed

The processing of the various food-grains, whether the end-product was porridge or bread, comprised a number of tasks, all of them tedious. Before the introduction of animal power into the milling process, the various operations were all carried out by hand, normally by women. In the case of naked wheat, the grain must be cleaned by sieving, ground in a handmill, and either boiled into porridge or kneaded and baked into bread. If the grain was barley or emmer, three processes preceded the grinding, viz. (1) roasting or parching, (2) hulling with pestle and mortar, (3) pounding and sifting to produce groats (see Pliny 18. 105 ff., who explains (112) that the husked grain 'is

49 Wooden pestle and mortar (African) for pounding maize. The mortar consisting of a hollowed out tree-trunk is still common in many parts of Africa (cf. *ill. 45*.). Mortar diameter 25 cm; depth 70 cm; pestle 1 m. 20 long.

50 Terracotta from Thebes, late 6th century BC, depicting a man working a saddle quern *(mola trusatilis)*.

pounded in a wooden mortar to avoid the pulverizing that results from the use of a stone mortar, the motive power for the pestle being provided, as is well known, by the labour of chain-gang convicts. After the grain has been stripped of its coats, the bared kernel is broken up with the same instruments'). The wooden mortar consisted of a hollowed out tree-trunk set on a base and bound to prevent splitting, the pestle (Roman *pilum*) consisted of a long pole, tapered at both ends, with a handgrip incised in the middle; identical in shape to the well-known soldier's throwing-spear, to which it gave its name.[100] Ill. 49 shows a heavy pestle and

mortar of this type used in many parts of Africa for pounding maize.

While husked grains were pounded, naked grains were rubbed out by grinding between rough stone surfaces. The abundance of surviving millstones and parts of mills, as well as models and representations of different types of milling equipment, make it possible to study in some detail the development of the various devices from the primitive saddle-quern to the highly efficient water-powered rotary mill. More particularly we can trace the evolution of the grinding process itself, from the use of un-grooved surfaces through a variety of grain-rubbers and their grooving arrangements designed both to secure a more even flow of the grain and to improve the grist. The introduction of rotary motion into the process led to a succession of improvements in milling, and to the harnessing of much greater sources of power. The first of these was the rotary hand-mill *(mola varsatilis)*; the details are not clear and we have no firm dating for its first appearance; milling between a pair of round, flat stones now gave way to milling between an upper concave stone and a lower convex one; the process is virtually continuous, the operator's one hand feeding the grain through an aperture in the upper stone, while the other rotates it by means of a vertical wooden handle

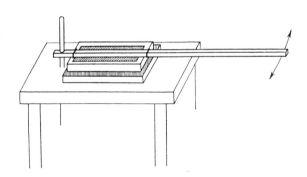

51 Diagram to illustrate operation of Olynthian grain-mill: the mill has become a machine in which the slit controls the feed and the grooving the particle-size.

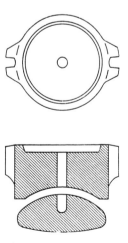

52 Diagram showing the parts of a hand-mill (Cato's 'Spanish mill', *De Agri Cultura* 10. 4.).

socketed into the stone (exactly as described by the anonymous author of the poem *Moretum (Country Salad)*). The sloping surfaces caused the grain to pass continuously between them without the necessity of feeding back the unground grain, as with the pushing-mill and its successor the hopper-rubber.

This was the first stage in a series of improvements which made it possible to apply animal-power to turn the upper stone, and later, by the insertion of a simple gearing system, to employ the waterwheel, which had been first used for lifting water, to drive the upper millstone from its centre.

54 The donkey mill *(mola asinariae)*, a familiar sight to every visitor to Pompeii, consisted of a bell-shaped lower stone (the *meta*), which remained stationary, while the hollow, hour-glass shaped upper stone (the *catillus*), was made to revolve it. The motive power for the various sizes of mill (two sizes are featured among the large number found at Pompeii), was provided by donkeys, mules and worn-out horses. The grain, dispensed from a hopper, was ground as it descended between the upper and lower stones, while the flour collected round the base. The lamp-stand shown on the bas-relief (ill. 39) indicates that there was continuous production round the clock. Recent excavations at Morgantum in Sicily yielded a number of small, man-operated vertical rotary mills, the design of which so closely resembles that of the animal-powered 'Pompeian' mill that it may provide an important link between the latter and the earlier horizontal rotary querns. The earliest application of waterpower to milling appeared not long after the first appearance of the simplest type of waterwheel, the undershot type (see table on p. 194). Within less than a century of the arrival of the
56 former in Italy, we find a description of the water-mill in the pages of Vitruvius (*De Arch.* 10.5.2):

Water-mills are turned on the same principle and have the same construction [as water-wheels], except that to one end of the axle a toothed drum[i] is attached; this is set vertically on its edge, and turns

53 Upper stone of small mill found at Saalburg, Germany, showing dovetail and handle socket.

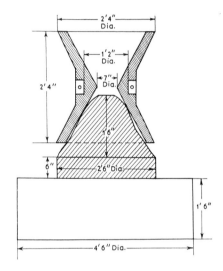

54 Section through the stones of a 'normal' Pompeian mill. The measurements are taken from a mill in the Casa di Sallustio, Pompeii.

55 Mills from the bakery in the Vico Storto, Pompeii.

65

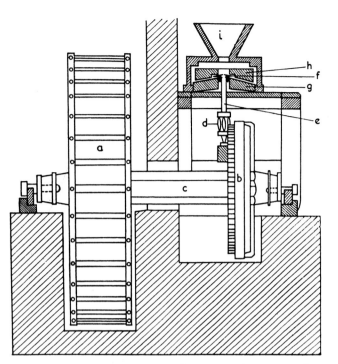

56 Vitruvian water-mill as commonly represented. *a,* water-wheel; *b,* vertical drum; *c,* axle of *a* and *b; d* horizontal 'drum'; *e,* axle of *d; f,* iron dovetail, *g,* and *h,* millstones; *i,* hopper.

with the wheel. Close to this drum is another larger one,[ii] toothed in the same way and set horizontally, which engages with the first drum. In this way the teeth of the drum fitted to the axle, by driving the teeth of the horizontal drum, set the grindstones revolving. A hopper suspended over the machine supplies the grain for the mill, and by the same revolution the meal is produced.

He means a 'crown wheel' with teeth at right angles to the disc; [ii] This means that 'the millstone was geared down, and turned more slowly than the water-wheel' (Landels 1978, 24).

A gear ratio of less than 1 : 1, though out of line with later European mills, which ran to as high as 2½ : 1, is consistent with the flow of an undershot wheel, which lacked the jet effect produced by the familiar 'mill-race' of the later overshot type. A possible reason for the slow 'take-on' of the water-mill[101] may well have been the low output, compared with that of the overshot mill; in the absence of precise gear

ratios it is misleading to offer statistical comparisons, such as that a water-mill could grind about forty times as much grain per day as a donkey-mill, which ignores essential variations in the gear ratio of the former and the size of the millstones of the latter![102] The application of water-power to the tedious and time-consuming process of milling marks an important stage in technical development, but it did not start an immediate revolution and put its slower rivals out of business. The *Edictum de pretiis* of AD 301 lists the maximum permitted prices of four types of mill, the horse-mill (1500 *denarii*), the donkey-mill (1250), the water-mill (2000) *56* and the hand-mill (250), implying that all four *52* devices were in common use at that time. The situation is not at all surprising. The rate of development and spread of a new process is affected by a variety of factors; these include (a) prime cost compared with that of existing devices; (b) availability of the required source of power (undershot water-wheels need a con- *7, 42* tinuous supply of flowing water, which in the Mediterranean area was subject to considerable seasonal fluctuations, whereas animals were readily available); (c) availability of the product to the consumer (animal-mills and water-mills were fixed, whereas hand-mills were portable). In fact, all three types of mill survived the fall of the Roman Empire in the West. Moritz (1958, 138–39) notes that much effort was required to suppress the use of hand-mills in the Middle Ages, that animal-mills survived as a standby until modern times, and that the cutting of Rome's water supply by the Gothic invaders in the sixth century AD, led to a food crisis which was solved by a further invention, that of floating mills mounted on anchored barges which could rise and fall with the changing level of the Tiber.[103]

Numerous estimates have been made of the relative efficiency of ancient mills. In spite of the large quantity of surviving querns, mill- *50* stones, and complete donkey-mills, the only satisfactory experiments so far are those carried out on a Romano-British quern (details in Moritz 1958, 151 ff.). The reconstructed donkey-mills in the Naples Museum, based on a now discarded theory, are valueless (Moritz, pl. 4 and pp. 64 ff.). Forbes (*Studies* III, 94) gives an output of $3\frac{1}{4}$ bushels per day for a donkey-mill, but gives no dimensions! Jasny (1944) has shown that the high price of wheat, together with the very low power-output per bushel required for grinding the coarse meal (no more than $\frac{1}{2}$ h.p. per bushel compared with $1\frac{1}{2}$–2 h.p. in a modern mill) meant that grinding costs will have amounted to not more than 10 per cent of the total. This would partly explain the survival of other methods. Donkeys and worn-out horses continued to plod around the bakeries for many centuries after the introduction of the water-mill (see Tables 1 and 2, pp. 221–22).

As for the quality of the product, we are fortunate in having a large number of animal-powered mills still *in situ* on various Roman sites, and we also have a series of calculations made by Pliny of the various products obtained from common breadwheat and durum wheat respectively. The opportunity of testing Pliny's figures against those obtained from a reconstructed mill has only recently been taken up, and the only findings available at present are those made more than twenty years ago by Moritz; but those were based on experiments with a hand-mill and have only limited value.[104] Nor is the literary evidence helpful: we have only a single surviving description of the grinding process, and that also with a hand-mill.[105] One thing is absolutely clear: the product of the process, whether by hand-mill, animal-mill or water-mill, was a coarse meal, not a flour. The refuse had to be removed by filtering the meal through a sieve; and here we do have evidence of technical improvement:

There were no bakers at Rome until the war with King Perseus, more than five hundred and eighty years after the foundation of the City Romans used to bake their own bread, and this was usually women's work, as it still is in most countries ... sifters made of horse-hair were invented in the Gallic provinces, linen sifters for ordinary and top-grade flour in Spain, and sifters of papyrus and rush in Egypt (Pliny, *HN* 18.107–8)

The finest of these sieves (the horse-hair and the linen) will certainly have produced a much finer flour than the best Greek product: Greek sieves 'would probably have achieved little more than the elimination of large pieces of bran, together with unground or partially-ground grains' (Moritz 1958, 163).

Technology of wine- and oil-making[106]

The extraction of its juice from the grape, and of its oil from the olive-berry involve a number of operations, for which quite distinct instruments were used. For one stage, however, that of pressing the partly treated pulp, the same range of pressing devices was used for both. It is convenient as well as logical, therefore, to deal with both in the same chapter.

The making of wine involved three processes, treading in the vat, pressing out of the juice in the press, and fermentation in storage *60* containers. The purpose of treading, which was quite distinct from the next stage of pressing, was to remove some of the juice by a gentle squeezing process, producing a mush, and facilitating the next stage, that of mechanical pressing by lever- or screw-press, without loss of the juice. The juice expressed by treading was boiled down and used to improve the flavour of poor wine. Next, the residue from the treading was strained, placed in flexible baskets, and put under the press. The juice was then conveyed by gravity flow into large storage containers, made either of earthenware or wood, for the process of fermentation.

The lever-press, the screw-press and the wedge-press, are in the rare category of inventions of the classical period. The earliest Greek *57* version of the lever-press consisted of a wooden beam anchored into the ground and weighted at the other end, and pulled down by a rope. The earliest Roman version described in detail by the Elder Cato (it is, in fact, an oil-pressing outfit comprising four lever-presses) was a massive affair, the 25-ft levers operating inside twin verticals no less than two ft thick (Cato calls them 'trees'), mortised into the walls *58* of the pressroom. The pressure was applied

57 Primitive Greek lever-press with stone weight.

58 (below) Catonian press-beam and capstan, from the model collection in the Museo Nazionale, Naples.

manually through rawhide ropes attached to capstans turned by hand-spikes. In the process of squeezing to express fluid, it is the final inches of pressure that count, and the simple Catonian press with pulleys, ropes and manually operated capstan is not the most efficient way of doing the job. The next stage was to find a better method of raising and lowering the huge heavy lever. In the third book of his *Mechanike*, Hero of Alexandria describes four different types of press, a lever- and capstan-press, a lever- and screw-press, and two varieties of direct screw-presses. The first of these works on a totally different principle from that of the Catonian press; instead of using their muscle power to bring down the lever with ropes, the men use the capstan to lift a heavy weight which is then suspended over the press-beam, and forces the latter down to press the fruit. This device, to judge from Drachmann's tentative reconstruction (we have only an Arabic translation of the text, and two bas-reliefs which appear to show portions of the press)[107] does not represent an improvement on the Catonian; moreover, as the author points out, it is dangerous to operate:

if a handspake should break (during the hoisting process) the stone would come down and injure the workmen; also, if a handspake should (under the heavy strain) slip out of its slot, the same misfortune might befall them.

Drachmann (1932, 64 f.) cites some evidence to support the suggestion that the machine was

fitted with a primitive pawl, designed to prevent the capstan running backwards if a handspake collapsed or slipped out, but this is no more than a plausible conjecture. The first definite advance was embodied in a Greek invention, dated a century later than Cato, and described by Pliny (18.317). The new element was a spar, fitted with a screw-thread, fixed to

the floor and passing through the free end of
60 the lever. On the screw above the lever was a
nut which was made fast to the lever. The spar
must have been secured to the floor so that it
could revolve, but not move upwards. The
screw was slotted to take handspakes. This
screw-and-lever-press had one obvious defect:
since the end of the lever forms an arc as it
comes down, the screw will tend to be forced
out of the vertical. An improved version, also
mentioned by Pliny, had its screw attached to a
stone weight hanging clear of the floor, thus
getting rid of the friction and jamming associated with the prototype. It also simplified the
lifting part of the process, the screw being
turned in the opposite direction, and the weight
lowered to the ground.

59   Hero's lever- and screw-press, which he describes as strong, safe and easy to operate, may
well have possessed all the above qualities, but
we cannot test his opinion, since the text at this
point is not easy to interpret, and in some places
unintelligible (there is a full discussion by
Drachmann 1932, 70 ff., with figs 23 and 24).
Resembling Pliny's 'Greek' press, it differs in
having a screw fastened to the lever, with a
long nut attached to the other end of it, which
is fitted with handspakes for turning. Attached
to the other end of the nut is a heavy weight; as
the handspakes are turned, the nut 'swallows'
the screw above it, allowing the weight to
lower the beam. An obvious disadvantage lies
in the limited travel imposed by the design:
since Pliny's lever is internally threaded, the
screw can be moved the full travel. In Hero's
press, however, the long nut severely reduces
the travel. Drachmann, who examined the evidence in great detail, concludes that Hero's
press would not have been as efficient as Cato's
58 lever and capstan. Among the defects noted by
him are (1) the reduced travel, (2) the long
screw hole with a solid base which could only
be cut by cutting the block of wood in two
lengths, and fashioning the threads in two
halves by hand—'a most laborious way'
(Drachmann 1932, 70), (3) reduced pressure,
since only the weight of the stone was available.

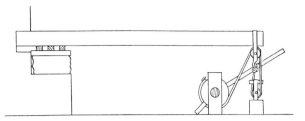

59 Hero's lever-and-screw press with hanging stone
weight.

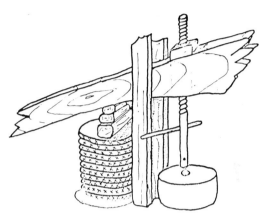

60 Lever-and-screw press in use in Morocco, showing
weight to be attached to screw, capstan and coiled
weight enclosing berries.

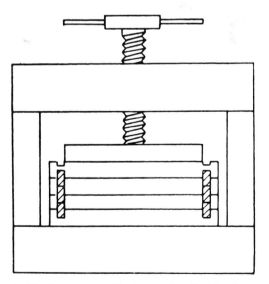

61 Single-screw portable press as described by Hero:
berries enclosed in interlocking boards.

The next advance was to do away with the lever altogether, using the screw to apply direct pressure on the fruit. Direct screw-presses are mentioned both by Hero and Pliny, but each of these accounts contains difficulties of interpretation, that of Hero being complicated by a combination of difficulties, which include the terminology of the parts of the presses, corruption in the manuscript, and the fact that the third book of the *Mechanike* survives only in an Arabic version. Hero describes two presses, a portable single-screw press in a frame, resembling a letter-press, and a second type which had twin screws and a moveable beam. Drachmann (1932), a work which is still indispensable, sought to establish a connection between the appearance of Hero's screw-presses and that of the screw-cutting machine for making internal screw-threads, for the making of which he gives detailed directions (*Mechanike* 3.21), and held that they could not have been developed before its invention around AD 50. This attractive hypothesis must, however, be rejected; Hero's small portable frame-press will have needed a screw-cutting machine for making the female screw-threads; but large wooden nuts of the type needed for Pliny's screw-press could be threaded without any such device. Indeed, the main difficulty was not how to construct such presses, but how to take up the counter-thrust when the screws were at full squeezing pressure. Hero's small presses will have easily absorbed the reverse thrust within their rectangular frames, but the severe pressures developing in a large wine- or oil-press of this type will have required much stronger frames; and there is evidence of this in Pliny's account of recent developments in press design, which brings to an end his description of the various operations of the farmer's year (18.317):

In the old days they used to pull down the press-beams with ropes and leather thongs, using handspakes. Within the last century the type of press known as the 'Greek' has been invented, with an upright beam fitted with grooves that run spirally.[i] To this 'tree' some attach a block of stone,[ii] others a box full of stones which rises at the same time as the tree; the latter arrangement being the most highly approved. Within the last twenty-two years a scheme has been invented, involving the use of short press-beams and a smaller press-room, with a shorter upright beam running straight down into the centre, placing drums on top of the grape-skins with their full weight bearing sown, and piling a heap of stones on top of the presses.

The more recent of the presses mentioned here is clearly a direct-screw press, like Hero's single-screw frame press, the problem of reverse thrust being solved, in a somewhat clumsy fashion, by the piled-up stones *(congries)* surmounting the frame (Drachmann 1932, 149). Pliny's 'superstructure' press represented economies of size and space; Drachmann (1932, 58) regards it as an inexpensive solution of the problem of reverse thrust mentioned which has been mentioned above.

The wedge-and-beam press

A form of the Catonian lever-press appears on a wall-painting in the House of the Vettii at Pompeii; on another wall in the same house is a representation of a press of a quite different design, in which the extraction of the juice is effected, not by the action of a lever and capstan, but by that of wedges and hammers. Its appearance in such a setting seems to be *prima facie* evidence that it was not uncommon, but no identifiable remains are attested, nor is there any reference to such a press in literature.

Placed on a stone foundation are two narrow, vertical timber frames which face each other. In the stone is a shallow depression, with a channel and a spout, through which the oil passes into a basin. The depression is filled with pulp from the mill, and on it is placed a plank, above which are alternate tiers of wedges and planks. The wedges are driven in by two *amorini,* one operating from each side, wielding

---

[i] The 'Greek' press is a lever-press, the drum and capstan being replaced with a screw-threaded vertical, which is drawn down or up by a capstan at its foot;
[ii] Reading, with Hörle (*RE* s.v. *torcular,* col. 1740), *stela* for codd. *stella.*

long-handled mallets. A wedge-press would be very efficient, provided the frames were strong enough to take the strain, but apart from a similar press on a mural from Herculaneum, which has no platform and only three rows of wedges against the four rows of the Pompeian specimen, we have no surviving reference to the device.

Technical problems connected with oil-making

The processes involved are more complicated than those required in wine-making. Whereas the grapes can be severely squeezed without removal of skins or pips, the olive kernel must first be removed (crushing would adversely affect the flavour); in addition, the berry contains, as well as the oil, a watery fluid *(amurca)* which must also be removed to avoid contamination; Columella (12.52.6–7) mentions four devices for removing the kernel: the oil-mill *(mola olearia)*, the revolving mill *(trapetum)*, the 'clog-and-vat' *(solea et canalis)*, and the the bruiser' *(tudiculum)*. The first three are set out in order of preference with no technical descriptions, while the fourth, presumably the only method unfamiliar to Columella's readers, is given a very inadequate description, followed by comments on the difficulties involved in its use. The revolving mill (2) is described by Cato, with full instructions for making it, with an estimate of cost, while the oil-mill (1), for which there is no other literary reference, is known from several relief sculptures, and from similar machines still used many centuries later. Both devices work on the principle of a pair of vertical rollers turning on a central spindle in a mortar, the surfaces being set at a distance sufficient to cause the kernels to be released without crushing. The millstones of the oil-mill (1) are cylindrical, and the surface of the mortar is flat. The millstones were adjustable. The revolving mill (type 2) consisted of a pair of stones with convex surfaces set vertically, the curvature being made to match the concave curvature of the mortar in which they revolved. A big drawback was that adjustment of the stones upward

64
63

62 Wedge-and-beam press, showing wedges of timber being driven in to separate the intervening boards, the bottom one of which squeezes the fruit. From a wall painting at Pompeii.

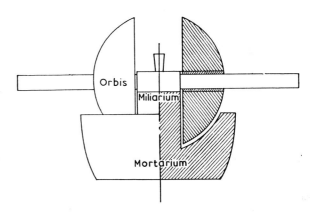

63 Reconstruction of Greek oil-mill *(trapetum)* from Olynthus. In the middle of the circular mortar stands a solid column in the top of which is set an upright *(columella)* fastened with lead. Over the pivot a wooden beam is fitted which carries two plano-convex millstones.

71

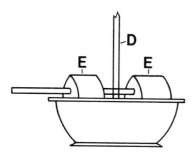

64 Columella's *mola olearia*. From the relief in the Palazzo Rondanini, Rome. The column *(D)* has two bearings, enabling the round millstones *(E)* to be adjusted for height and so prevent crushing of the kernel.

provements. Columella gives good reason for preferring the oil-mill:

oil-mills make the operation easy to organise; the stones can be lowered or even raised to suit the size of the berries, and so prevent crushing of the kernel, which spoils the flavour of the oil (*RR* 12.52.6).

This is the only literary reference we have to this type of mill; but we do know that it later replaced the inferior and more expensive Catonian type. Two monuments, the Rondanini relief in Rome and a sarcophagus from Arles, depict the Columellan mill; in his detailed study of the evidence Drachmann (1932, 42 f.) draws attention to a defect in the design of the Rondanini specimen, which he regards as a halfway house between the old Catonian mill, and that of Columella: wear on the vertical pivot by the revolving axle could cause the stones to wobble; a slight wobble would be enough to crush the kernels along with the berries; the Arles sarcophagus shows a development in this part of the mill in the large, vertical beam, with a bearing at either end which replaces the simple pivot-pin.

or downward was impossible save by structural alteration (see Drachmann 1932, 4); also the curved stones could only do their milling effectively at the height for which they were designed. Any change in the height of the axle would throw the matching curves out, and upset the whole process. The design was inflexible, and, therefore, incapable of technical im-

65 The Rondanini relief, Palazzo Rondanini, Rome.

7 Building

Building technology is concerned with two things: first, the use of available materials, in the form of wood, stone, brick, concrete and metals, in the construction of buildings designed for specific purposes, which is what is known as 'materials technology'; secondly, the technical problems posed by the use of these materials in carrying out the designs of the architect, which is what we call 'structural engineering'. These problems involve questions of stress and strain, of thrust and counter-thrust, of tension and compression. If these questions are not answered correctly—if the architect fails to do his sums properly, or if the builder in his turn fails to carry out the instructions of the architect, the building may fall down. The results can be more serious than those discussed in the section on tunnelling inaccuracies in the chapter on hydraulic engineering (Ch. 12, pp. 158ff.).

In handling this topic it will be important for us to discover, if we can, what sort of instructions were given by the architect, and how the foremen, the stonemasons and the bricklayers carried them out. The results, as we shall see, are often surprising, and by no means in line with what we should regard as normal building practice. Our first task is to examine our sources of information. These comprise literary evidence, building inscriptions and the remains of the buildings themselves. Here we must notice that the greater part of the literary evidence comes from the Roman side, while in the case of inscriptions the situation is reversed; we have very little on Roman methods, but from Greece we have a number of surviving documents, including some specifications for buildings, and some instructions for the transport of materials to the site. Finally, we have the buildings themselves, which constitute the largest body of evidence. They include identifiable ground-plans of buildings that have not survived, some structures that survive more or less complete, and a number of unfinished buildings and reject items which throw much light on the order and method of working on the site. An obvious snag here is the fact that roof structures are difficult to establish, since so many buildings have lost their roofs. Before we can get down to the job of finding out how the builders coped with their construction problems we must first determine what sort of structural problems they were setting themselves. The easiest approach to this question is to take two fundamental problems that faced both architects and engineers: first, how to bridge a gap (e.g. put a road across a river); secondly, how to roof a space in order to protect the contents of a building, whether human or inanimate, from the elements. In the earliest phases of building technology these problems are separate—arches are employed for spanning rivers, and later for carrying the visible portions of aqueducts across valleys in multiple arcading, while spaces are roofed over by means of structures that are basically tents. With the discovery of a convenient method of stringing arches together in line to form a three-dimensional vault (the barrel- or tunnel-vault), we arrive at a jumping-off point in the direction of bold and imaginative methods of construction—barrel-vaults piled one on top of another to take the enormous weight of the stone seating of the Colosseum, or intersecting barrel-vaults providing the Baths of Diocletian with a vast central space unencumbered by supporting columns. Ordinary dwellings of the classical period were of simple construction,

81

88

73

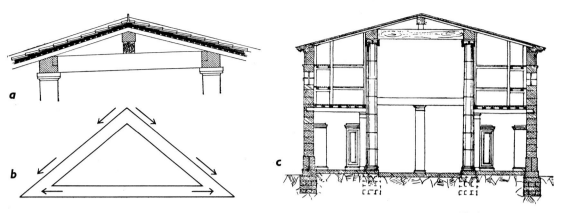

66 *a,* Roof construction of Philo's arsenal at Piraeus; *b,* tie beam truss; *c,* transverse section of the arsenal.

and problems of thrust and stress were negligible. The walls had virtually no footings, and the easiest way to burgle a Greek cottage was to dig under the outside wall. The standard term for a burglar in Greek is 'wall-excavator' *(toichoruchos),* and making a hole in a roof was a matter of prising off a few tiles, as in the New Testament incident of the man with the palsy (Mark, 2.4). As for public buildings, the structural basis even of an elaborate temple such as the Parthenon is very simple,[108] consisting of a rectangular inner shrine—the house for the statue of the deity who was worshipped in it, built up of walled courses of dressed masonry on three sides, and provided with an entrance porch or doorway made up of two vertical members (the jambs) with a horizontal member (the lintel) above. This technique is usually referred to as the 'post-and-lintel' system. The inner shrine was surrounded by a continuous colonnade, and surmounted by a pitched roof, covered with tiles. This method of building imposes some constraints: there is first of all the limited bending strength of the masonry lintel under loading; in the Parthenon many of the architrave blocks (the horizontals above the colonnades) are cracked; but they are unlikely to fall down unless shaken by earthquake or by a direct hit from a powerful bomb.[109] Indeed, a considerable part of the structure survived an internal explosion when the temple was used as an ammunition store in the seventeenth century

during a war between the Venetians and Turks. There are severe limits to the permissible width of horizontal blocks supported on columns. The Parthenon architraves measure a fraction over 4 m or about 14 ft. This will impose no restrictions so far as the length of the building is concerned; the architect can simply multiply the number of columns on the sides without any loss of stability. But suppose he wishes to increase the overall dimensions; making the building wider involves increasing the span of the roof, and this imposes another constraint. Here the limiting factor is the skill available for the carpentry required to make the roof.[110]

The earliest Greek buildings were made of mud-brick walls set on stone plinths and covered by simple thatched roofs. In course of time thatched roofing gave way first to wooden shingles and later still to earthenware tiles, while masonry began to be substituted for timber in columns and beams. To support the increasing weight of the roof, a type of timber framing was developed, consisting of a collection of pieces of timber resting one upon another, the loads on which were carried vertically downwards, and were not redistributed so as to pass into the walls. In technical terms, Greek roofs were, up to a late period, 'untrussed'[111]. This meant that increased loading of the roof could only be met by providing beams of vast section both for the long ridge-pole and for the transverse beams. This emerges very

74

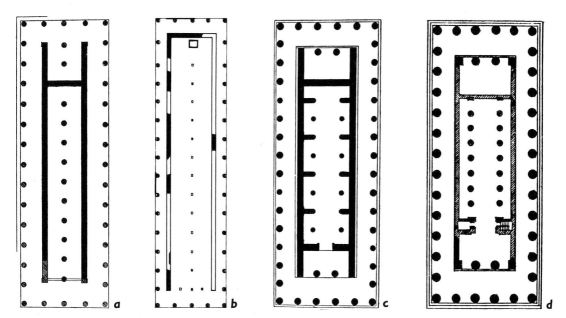

67 Ground plans of Greek temples: *a,* Thermum, Temple of Apollo; *b,* Samos, first Temple of Hera; *c,* Olympia, Heraion; *d,* Paestum, second Temple of Hera.

clearly in an important surviving document, the specification for the naval arsenal at Piraeus, the port of Athens, designed towards the end of the fourth century BC by the architect Philon.[112] The roof specifications are as follows:

And he will join tightly together lintels of timber, and lay them on the columns; width $2\frac{1}{2}$ feet, height $2\frac{1}{4}$ feet on the higher of the two faces. Number: 18 in each row.

And he will lay cross-beams on the columns above the passageway, width and height to correspond with that of the lintels.

And he will lay ridge-beams, width 28 inches, height 22 inches, not counting the chamfer, on a support set on the cross-beams, 3 feet long and 1 foot 5 inches wide.

And he will fasten the ridge-beams to the cross-beams by means of struts; and he will lay rafters 10 inches thick and 15 inches wide, at 20-inch intervals. On these he will put battens, 8 inches wide and 2 inches thick, at 4-inch intervals.

On these he will lay wooden sheathing, 1 inch thick and 6 inches wide, nailed on with iron nails. He will then put on a covering layer, and tile with Corinthian tiles fitting closely into each other.

The above is an extract from what the document describes as 'Description of the stone arsenal for the tackle, by Euthydomos, son of Demetrios, from Melite, and Philon, son of Execestides, from Eleusis'. The document speaks for itself. Lacking any notion of transferring the weight of the roof to the side-walls, the builders had no option but to increase the size of the supporting timbers; even the lightest of the arsenal timbers, the cross-battens, are massive, and the total weight of the timber was staggering! It would be easy to draw from the evidence of Philon's arsenal the conclusion that the constraints we have been discussing produced a stagnation of building technique. This would be only partially true: as we shall see, considerable ingenuity was shown by Greek builders in finding workable solutions to their structural problems. The chance survival in two religious centres, one in the northern part of mainland Greece, and the other in the island of Samos, of the ground plans and portions of the structures of successive temples erected on the same site, have made it possible to study the technical difficulties of roofing, and the solutions attempted by successive generations of builders. The earliest buildings at Thermum, chief religious centre of Aetolia, were houses

66

75

with a very distinctive ground plan shaped like a hairpin, with long central rooms, a deep porch at the front and an elliptical apse at the rear; they probably had pitched roofs of thatch, projecting over the walls to form eaves. A later example on the same site, probably a temple adapted from a house, had a primitive elliptical colonnade around it consisting of 36 wooden columns set in stone sockets. Still later came the temple of Apollo, one of the oldest Doric temples in Greece, with a rectangular ground plan that looks forward to the complete classical type, but showing affinities with its predecessors in having a triangular gable only in front, and a hipped roof at the rear.[113] Similar technical developments can be traced in the first temple of Hera at Samos (prob. *c.* 750 BC), which was simply a narrow hall, over 100 ft long, with a row of wooden posts down the centre supporting a primitive roof of thatch or clay, a very unsatisfactory roof-support system for a building designed as a shelter for the cult-image of the patron deity, which had to be displaced to an awkward side position![114]

67a

67b

These primitive roof-systems created no technical difficulties, but the arrival of roof-tiles of terracotta in the late seventh century, and the development of a matching gable at the rear created new problems in roof support.[115] The ground plan of the temple of Hera at Olympia (early sixth century BC) illustrates the steps that were now being taken.[116] In the interior of the building were two rows of eight columns designed to ease the load on the main transverse beams supporting the horizontal ceiling and the sloping roof above it. The roof was tiled, and the gables at either end were decked out with enormous circular ornaments (the *acroteria*), measuring more than 2 m in diameter, making a considerable load for an untrussed roof to carry. We shall next examine the second temple of Hera at Paestum in South Italy, dated to about 460 BC, and one of the best preserved of all ancient temples. The interior, which has fortunately survived almost intact, has a double range of superimposed columns, the sole purpose of which was to support the ceiling and

67c

67d

roof, which, as we have noticed earlier, has a direct vertical thrust. We must remember, however, that this clutter of internal columns did not adversely affect the functioning of the building: a Greek temple was simply a shelter for the cult image of the god, so that the architect was not required to provide any space inside for worshippers. Indeed, there were temples from which the public were totally excluded. Externally the roof of an early classical temple, complete with its array of terracotta tiles, decorated tile-ends (*antefixes* and *acroteria* must have been aesthetically satisfying in the extreme,[117] but the entire system of roof support is still based on very primitive techniques.

Up to now our attention has been concentrated on technical problems related to the building of temples. Historians of Greek architecture have tended to focus most of their attention on religious buildings, and particularly on the masterpieces of fifth-century Athens, to the neglect of secular structures, treating the end of the classical period as the beginning of decline. In fact the developments that took place in many parts of the Greek world during the fourth and third centuries presented architects and planners with a range of quite new structural problems, one of which will now be examined in some detail. The informal political atmosphere of classical Athens meant that many communal activities, including the meetings of the sovereign Assembly, took place in the open air; with an average of 180 days of sunshine out of 365, a roofed parliament building was unnecessary. For mass meetings, involving large numbers of people, the theatre was available, except for the few days in the year when it was occupied with the dramatic competitions. The much smaller numbers of the executive Council could easily be provided with a roofed structure. But the planners responsible for the new cities which were founded in increasing numbers from the fourth century onwards were often required to provide the smaller cities with roofed buildings capable of housing the whole Assembly. The traditional way of solving this type of problem was to extend the post-

68a and-lintel system and multiply the number of
supporting columns as necessary. The limitations of the traditional system are painfully
obvious; it is not a happy arrangement if speak-
ers on the platform are obscured from the
audience by a forest of supporting columns. The
Assembly Hall of the federal capital Megalopolis
in Arcadia (a new foundation dating from 371
BC), shows a neat way of solving the problem of
68b masking. The Thersilion,[118] as the Assembly
Hall was called, measured 50 × 63 m or
172 × 218 ft, giving a total floor area of over
3,000 sq m, or 35,00 sq ft, and a seating capacity
of about 6,000. If all 10,000 representatives
turned up, together with their 50 councillors,
there will have been severe congestion. The 67
columns supporting the roof were ingeniously
arranged in five concentric series, each column
being fixed at one of the intersections of a series
of lines radiating from the speaker's rostrum,
(which was placed somewhat to the south of the
centre), so as to provide a maximum unim-
peded view from any part of the Hall. The floor
was also raked towards the platform, an essential feature in an auditorium of these dimensions. The Thersilion was destroyed when the
city was sacked in 222 BC, and apparently not
rebuilt. The excavations of 1890 revealed no
evidence about the roof structure of this
remarkable building. Another interesting
example of what could be achieved within the
limitations of the traditional system was a large
68c hotel *(katagogion)* erected during the fourth
century at the great international healing centre
of Epidauros.[119] Its ground plan was rectangular, consisting of four courtyards surrounded
by colonnades (ten columns to each side, making a total of 156 columns). The buildings were
two-storeyed, with twenty rooms on each
storey, providing a total of 160 rooms. Terraces
formed by the roofs of the colonnades gave
communication between the rooms of the upper

68 *a*, Eleusis, Hall of the Mysteries, Iktinos' original
design (*c*. 430 BC); *b*, Megapolopolis, the Thersilion;
c, Epidauros, the Hotel *(katagogion)*.

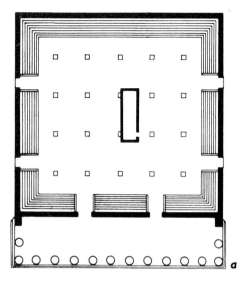
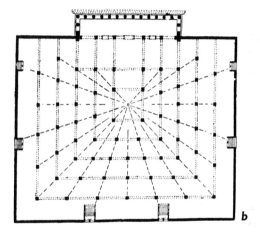
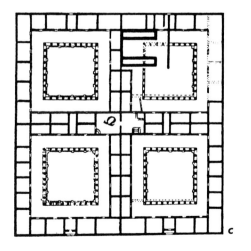

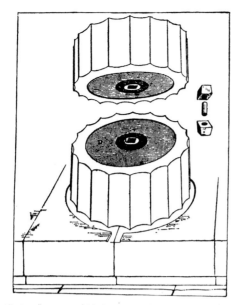

69 Construction of columns with centering pin. Both surfaces of each drum were counter-sunk to take square wooden plugs 10–15 cm square and 8–10 cm deep, into which the wooden pins were inserted to give accurate centering.

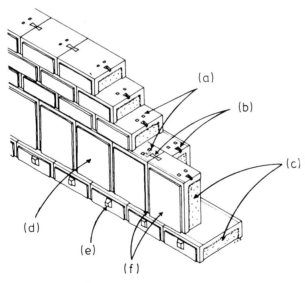

70 Early Greek construction techniques: *a,* U-shaped holes; *b,* dove-tail metal clamps; *c,* rebating *(anathyrosis)* to make unmortared edges fit snug; *d,* slabs *(orthostates)* set on edge; *e,* tenons for ease of handling; *f,* preliminary dressing.

level. The ground plan closely resembles a ludo board.

Greek building methods

Greek builders laid their masonry dry, without mortar between the joints. Cramps and dowels were therefore essential for tight jointing. For the accurate setting of column-drums, each drum was provided with a square counter-sinking on both sides; into this was fitted a hardwood plug, which was bored to receive a circular wooden pin, thus ensuring, by means of a simple device, the accurate centering of the drums. For bonding horizontal courses of masonry, metal clamps, inherited from earlier civilizations, were employed. These were commonly made of iron, or of iron cased in bronze, and held in place by lead solder. Slots of the required shape, commonly in the form of a letter T, were sunk in the ends of adjoining blocks to receive the clamps.[120] Occasionally iron 'bearers' (often incorrectly called 'reinforcing beams') were used. Mnesicles, the architect of the great entrance system of the Acropolis at Athens (the Propylaea), used 6-foot beams to transmit loads of 64 tons from the marble ceiling beams to the columns on either side.[121] The architect, it would seem, had insufficient confidence in the load-bearing capacity of his architrave, but rather more in his iron beams, which had a low safety factor, as compared with that employed by modern architects.[122] But this confidence seems to have been fully justified!

Problems of transport and handling

From the latter part of the sixth century blocks were being cut in the quarry to match the size required; the method (adopted presumably from Near Eastern practice) was to 'block out' by channelling, using wetted wedges to detach the blocks from their beds; the technique remained unaltered until the Roman period, when the demand for very hard stone and for thin stone veneers brought on some changes. It has commonly been assumed that heavy blocks

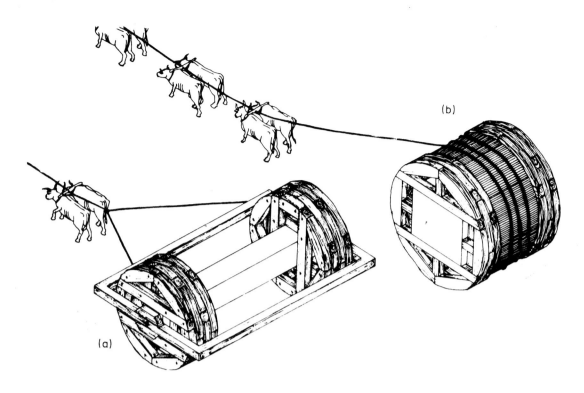

71 *a,* Cradle for moving heavy beams. Metagenes, son of Chersiphron, architect of the temple of Artemis at Ephesus (560 BC) had architraves serve as axles: his father had started the idea by pivoting the column-drums and putting them in circular frames. The wheels were *c.* 4 m. in diameter. *b,* Colossal stone transport. Paconius (early 1st century BC) tried to dispense with the diameter pivoting, but his great reel was too narrow in relation to the diameter (4 m against 5 m) so that it could not be steered.

were normally transported on rollers, as depicted on Assyrian and Egyptian reliefs; but there is strong evidence for the use of heavy ox-drawn wagons, the transport of extremely heavy blocks being achieved by multiplying the spans of oxen required; thus we learn from the well-known building inscription from Eleusis that as many as thirty-seven yokes were used to move heavy blocks of marble from Pentelikon to the site.[123]

Vitruvius (10.2.11–14) reports on an interesting series of innovations in this area of technology; he begins with Chersiphron's simple *71a* but ingenious 'cradle' for moving heavy column-drums, which consisted of a wooden frame into which the drum was fitted with gudgeon-pins, the whole contraption resembling a garden-roller. He concludes with a description *71b* of Paconius' 'giant' 'cotton-reel' for moving architrave blocks. The notion was a plausible one, but there was a snag, which proved to be fatal: the 'cotton-reel' being too short for its diameter (a mere 4 m against 5), the roller could not be steered straight, but slewed off course as the rope uncoiled, and had to be continually pulled back into line. As often in the history of invention, disaster struck the inventor: 'Thus, by drawing to and fro, Paconius got into such financial difficulties that he went bankrupt' (Vitruv. 10.2.14).

Cranes of the 'shear-legs' type (so called *72* from their resemblance to a pair of shears—, operated by compound pulleys and winches, came into common use in the late sixth century.[124] These useful devices, whether worked by men on a capstan, or more efficiently on a *3, 74* tread-wheel, displaced the earlier method of lifting by ramp, which was prodigal in its use of manpower, and therefore, as Coulton points out (1977, 144), much less suited to the Greek city-states than to the eastern autocracies from which it was derived. There are also indications

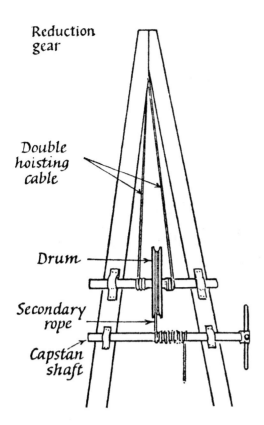

Reduction gear

Double hoisting cable

Drum

Secondary rope

Capstan shaft

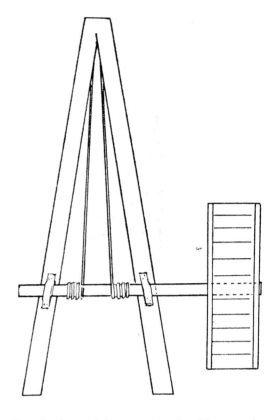

72 Crane fitted with reduction gear. An improved version of Vitruvius' 3-pulleyed crane *(trispaston)* with a drum and secondary rope.

74 Here the drum of the crane is replaced by a treadwheel operated by men from inside.

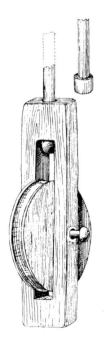

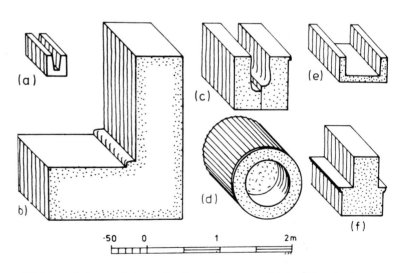

73 Pulley and sheave from Zugmantel, West Germany.

75 Weight-reducing techniques for heavy beams. *a*, Prinias, Temple A (630 BC), U-shape hollowing; *b*, Syracuse, Temple of Apollo (early 6th century BC), architrave cut down to L shape; *c*, Delphi, Athenian Treasury (570 or 490 BC), U shaped architrave; *d*, Delphi, Naxian column (570 BC) partly hollowed; *e*, Bassae, Temple of Apollo (430–400 BC), ceiling beam widely hollowed; *f*, Stratos, Temple of Zeus (*c.* 320 BC), cross-beam cut down to inverted T.

80

that building methods were being modified in order to avoid the use of blocks too heavy to be lifted by crane (Coulton, *loc. cit.*).

74 By the Roman period the demand for heavier building units had already led to the introduction of the 'winding-drum' crane, which was capable of increasing the mechanical advantage by a factor of five, and of the more efficient tread-wheel; we have no dates for either of these important improvements, but both may well have been products of that rapid technical advance which began in the fourth century and reached its climax in the third (see *Sources of Information,* s.v. Ktesibios and Archimedes). Another way of dealing with the problem of weight was to bring it within crane capacity by hollowing; from as early as the late seventh century we find lintels cut down to 75 U-shape, L-shaped architrave blocks, and a hollowed-out column-drum (Coulton 1977, 145 f.).

Dressing and finishing

The technical implications of building in rectangular forms without mortar are considerable. Inserting mortar between the blocks irons out any minor inaccuracies in the alignment of adjoining faces. To get over this difficulty Greek builders reduced the area of contact between faces to a narrow band about 7 cm (3 in) wide at the edges, the remainder of the facing surfaces being recessed and thus out of contact.[125] All joints, horizontal and vertical, were left with a slight bevelling to prevent slipping when the blocks were brought together. The blocks could be set directly in their final positions, and little was needed in the way of blocking up, apart from some manoeuvring with crowbars, a process which was made easier by the provision of holes and handling bosses. The projecting bosses found on blocks discarded on site, and often classed among the lifting devices, have the wrong profile and are not arranged to suit the centre of gravity (Coulton 1977, 4 ff.); they were presumably used for manoeuvring. In fact, all the various lifting devices share a common basis—a preoccupa-

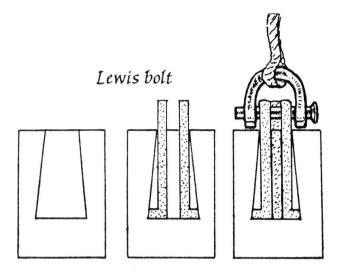

76 Aids to lifting blocks showing Lewis bolt. An alternative to projecting tenons (see *ill. 70*) but probably not widely used since sockets took time to cut and because of the risk of splitting when wedge is used.

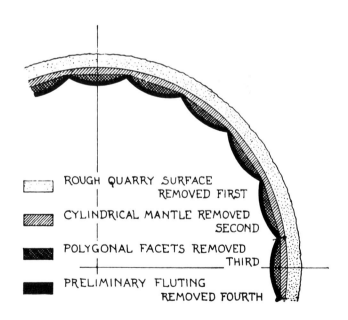

77 Section of drum to show the various processes and layers. The razor-edged projections *(arrises)* of the Doric Order are actually .08–.09 cm.

78 Tenons for lifting column into position on the Athens acropolis. The one on the right has tenons, the left-hand one shows centering of drums.

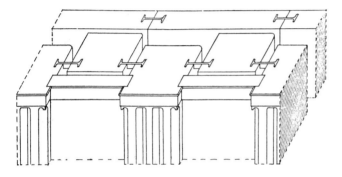

79 Triglyphal frieze of the Parthenon: axonometric view, showing the thin slabs of the metopes mortised into rectangular grooves cut into the sides of the triglyph blocks. Note also the 'dumb-bell' clamps.

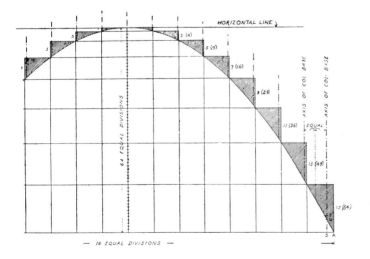

tion with the problem of removal of the device after the stone has been fixed in place (see Dinsmoor 1950, 174).

The same care was exercised in centering and finishing the column-drums; the expense accounts of the temple called the Erechtheum at Athens, together with the evidence of the unfinished temple at Segesta in Sicily, make it easy to analyse the various processes. The horizontal surfaces were worked in four concentric zones, the outermost and innermost at the level of the joint, the second being roughened to give a better bond, and the third recessed to reduce the area of contact. From the vertical surface no less than four layers were successively stripped off to produce the final fluted surface. The results of these precise and painstaking methods are remarkable: after the lapse of centuries some Greek walls have become fused. Students of the Parthenon drew attention long ago to the fusion of some of the joints of the temple platform, and to certain columns which, though pushed out of the vertical position from above, have retained their rigidity, implying that the drums which made up the column had grown together.[126] *77*

Viewed as a whole Greek building techniques are largely a matter of practical geometry executed on the blocks and drums in the course of erection with the aid of very simple instruments, cord, plumb-line, level and square, and based on the fewest possible specifications. We know from surviving building inscriptions that the architects did not issue the builders with detailed specifications.[127] For example, Doric temples have a regular arrangement at the level of the frieze, consisting of alternating triglyphs *79* and metopes. We know that the builder was not given precise dimensions, but worked from the size of the metopes.

Using these as a module, he did the necessary multiplication sum, and ended with a piece

80 Diagram for constructing stylobate curvature. The result was a parabolic curve. The platform may be compared to a rectangle cut from the surface of a melon: in the Parthenon the rise is 6 cm on the front, 13 cm on the flanks.

of three-dimensional geometry. Nor were the masons presented with blocks of identical dimensions; it is by no means easy to do this without sophisticated sawing equipment, particularly when the material is marble, and a lot of trimming was evidently done at the erecting stage. The actual measurements of the five architrave blocks that form the east front of the Parthenon (regarded as a masterpiece of optical and other refinements) show appreciable variations in width—the first four are respectively 4.256, 4.232, 4.332 and 4.302 m wide, and the last is only 4.151. I think we must infer that the masons, receiving blocks from the quarry that did not precisely correspond to the required measurements, used their common sense and did minimal trimming on all except the last, making it much the shortest.

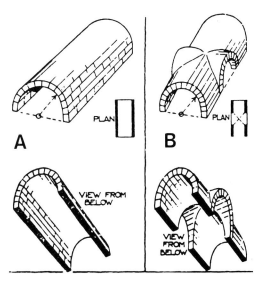

81 Roman barrel-vaulting showing (*a*) single, (*b*) intersecting vaults.

Roman building technology

82 It has been said that the Romans had no architecture; what their builders produced, according to this view, were imitation Greek façades, backed by brick and rubble. The emperor Augustus, we are told, found Rome a city of brick, and left it a city of marble (Suet. *Augustus* 28), but most of the marble was a mere skin, a falsework, concealing the practical, but inelegant, work of the engineer. There is a kernel of truth in the assertion. A more important difference, and one which had a bearing on the development of Roman building technology, may be seen by comparing a Greek with a Roman theatre building. The auditorium of a Greek theatre seems almost to grow out of the hillside which provides support for the seating at minimum cost; the Roman theatre on the contrary is a solid, unified architectural entity: the terraced hillside of the Greek theatre with its detached scene-building was transformed into a single, compact, independent structure containing auditorium and stage within a semi-cylindrical unit. The same difference may be observed in the architectural treatment of the heart of the city, the Greek *agora* and the Roman *forum*. Even in the carefully planned Hellenistic city,

with its chequer-board street plan, and modular arrangements of house blocks and public buildings, the *agora* is still an open space, colonnaded but not walled off, while the imperial Roman *fora* are areas of closely defined and concentrated function, deliberately screened from other activities. The construction of these complex enclosures set new problems for their designers. But the Romans possessed one great natural asset in the supplies of volcanic dust found near Naples (the *pozzolana* of today), which enabled them to make waterproof cement of amazing strength. For the rest, they exploited with skill and increasing boldness three important structural forms, the tie-beam truss (possibly an invention of the late Hellenistic phase of Greek architecture) (see ill 66b), the arch and the vault (both inherited by Greece from the Orient). The inventors of the timber truss had used it only timidly and for special *61b* purposes, while their use of arches and vaults was confined to openings or passages.[128] In Roman hands (see Appendix 9) the arch appears as a means of liberation from the constraints of the post-and-lintel system, while the barrel *81* vault, which was no more than a continuous series of arches, became the means of enclosing the required space in a continuous curve; and the piercing of these vaulted walls led on to the fantastic splendour of the Baths of Caracalla (AD 215), to reach its culmination in the domed magnificence of Hagia Sophia. Roman builders were extremely fortunate in having at their disposal a wide variety of materials, including volcanic tufas, limestones, marbles, clays, brick and mortar (see Vitruvius, 2.3–7). The use of baked bricks, which dates back a long way in Mesopotamia, comes quite late into Graeco-Roman practice, and does not seem to have been widely used in Italy before the end of the Republic. Vitruvius' chapter on brickwork (2.3) is entirely concerned with sun-dried brick. Intense volcanic activity in prehistoric times in central Italy provided local builders with a fine range of building stone, possessing varying degrees of durability and workability.[129] They included the very easily worked volcanic tufas (formed by compression of volcanic dust), and the harder limestones, such as those from Gabii and Albano, both still in use under the names of *sperone* and *peperino* respectively, as well as the famous travertine (from Tivoli), of which both the Colosseum and St Peter's are built. Its only

defect was that it was not fire-resistant; after the Great Fire of AD 64, which destroyed more than one-third of Rome, Nero ordered rebuilding in *peperino* (Tac., *Ann.* 15.43). Mortar, as we have seen, played virtually no role in Greek architecture. When the Greek builders of later times combined hydrated lime with rubble as a building material, they often used too little lime, and had to use very solid stone facings (Robertson 1945, 233). The chief disadvantage of ordinary lime mortars as jointing for stonework lay in the fact that the strength of the joint did not match that of the stone (with the more primitive mudbrick it was possible to make a wall of homogeneous strength, which is the main reason why dome construction came so early in the East). It was some time during the third century BC that the Romans made the momentous discovery of pozzolana. The combination of what is, in fact, a very finely ground silica with lime and water produces, by a chemical process not yet fully understood, a cement which combines immense strength with both waterproof and fire-resistant properties.[130] The Romans now possessed the means of making their joints as strong as the material joined (Appendix 9).

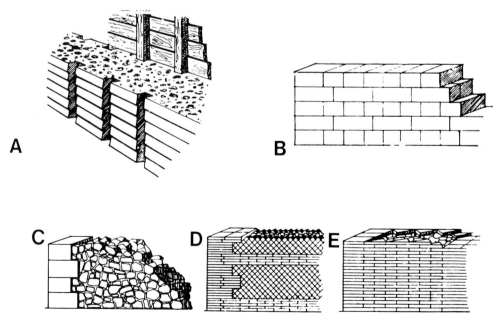

82 Types of Roman wall construction. *a,* concrete; *b, opus quadratum; c, opus incertum; d, opus reticulatum; e, opus testaceum.*

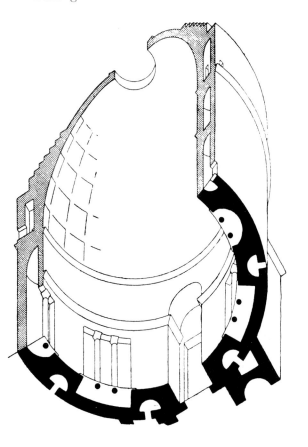

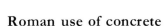
83 Pantheon, Rome. Sketch showing construction. This Roman temple (to All the Gods) is the antithesis of a Greek one: with a completely closed interior, a circular plan, a domed roof, and specifically Roman in structure, materials, and system of proportion.

84 Pantheon, Rome. Interior restored. The vast rotunda is in brick. Its diameter, 43 m, is exactly matched by the height of the dome. Six of the eight splendid niches are flanked by a pair of columns which appear to support the dome. Painting by Giovanni Paolo Panini, *c.* 1740. National Gallery, Washington, DC. Samuel H. Kress Collection, 1939.

Roman use of concrete

Recent research into Roman building technology[131] has made it possible to give a coherent account of its development, from the first hesitant use of a new and revolutionary material to the mastery of technique displayed in the many great surviving examples of arches, vaults and domes. Space considerations have limited our discussion to the first two. Of the extant writers, the first, Vitruvius, wrote at a time when the new cement was still being employed sparingly and tentatively, while Faventinus (early fourth century), and Palladius (perhaps half a century later) belong to a period when the great age of imperial concrete structures was already at an end.[132] At first it was used as little more than a superior kind of mortar. The earliest dated concrete structure in Rome was the platform of the temple of Concord (121 BC). The concrete used in this building has been shown to be of poor quality, confirming Vitruvius' attitude of extreme care and even mistrust of what was still in his time a fairly recent innovation. Vitruvius approves of two main uses of concrete, for the construction of foundation courses, where it superseded the old system of square masonry with rubble infill, and for the building of city tenement blocks *(insulae)*. In another passage, at the beginning of a chapter

85

82d on walling (2.8.1.) he compares the new 'network' *(opus reticulatum)* system of concrete construction, that is, pozzolana cement with an aggregate made up of lumps of tufa or brickbats 'which everyone uses now', with the old
82c method of 'random work' *(opus incertum)* 'which is not pretty to look at, but stronger'. The 'network', he says, 'is likely to cause cracks because it has the beds and joints scattered in all directions' (if that is the correct meaning of the phrase he uses). The concrete for walls was cast between timber shuttering, with the vertical supports for the latter on the inside, as we know from the marks that can still be seen on surviving buildings. As in masonry construction, builders took advantage of the various kinds of stone available when making concrete, the aggregates ranging in weight from heavy solidified lavas used in foundation walls to lightweight tufas and pumice for the upper reaches of vaults and domes, where reduction in weight was crucial. Tests carried out on the dome of
83 the Pantheon in Rome show successive reductions in the weight of the aggregates up to the
89 top of the dome, the famous 'window' being set in an area of almost pure cement. Broken brick and tile, which have great bonding strength, were also frequently used. Research on Nero's Golden House shows that a great deal of broken material from the devastating fire of AD 64 was incorporated into the concrete.[133] The method varied from random spacing of the stones in the prepared wooden 'forms' to the very careful methods used in the early second century, when the stones were often laid by hand in regular rows, and carefully spaced in the mortar in order to achieve an even distribution of the load—clear evidence of increasing mastery of the required techniques. In addition to 'random' work and 'network' facing to the concrete core, a third method, known as 'brick-
82e work' *(opus testaceum)* became very popular during the Empire. In this system the facing was made up of triangular bricks with their apices pressed into the concrete core. Externally the surface closely resembled that of a modern brick wall. What is really important, as

Ward Perkins has emphasized, is not the variety of treatment given to the outer facings, 'but what was going on within the concrete core' (1977, 247).

Roman arches and vaults

We now pass on to areas of more remarkable technical achievement, the construction of arches and vaults. Neither the arch nor the simple barrel-vault was a Roman invention: the Roman contribution, which was of profound importance to the future development of architecture, was twofold, comprising both structural innovation, and steady if unspectacular advance in the use of new materials. The application of concrete to these structures, while it disposed of numerous problems of thrust and loading, involved the builders in new problems relating to the handling and application of the material while still in the liquid state. It was easy enough to make the wooden shuttering needed for wall-construction in light-faced concrete, but it was a different matter when arches and vaults were being built by this method. The story that has emerged from recent studies is one of cautious advance by trial and error, but we have no record of how many vaults collapsed during the process of mastering the new techniques. We look first at the development of the arch. The true arch with its wedge-shaped components (known as *voussoirs*) is a late–comer in western architecture, making its first appearance in sixth-century Egypt; there were no large arches in Greece before the third century BC, nor in Italy before the second. The Milvian Bridge, which carries the main north road across the Tiber two miles north of Rome, contains the oldest surviving bridge arches of Roman workmanship; its date is 109 BC The 60ft-arches are entirely of masonry, but three kinds of stone were used, each chosen to suit the requirements of that particular portion of the structure; tufa was used for the core, travertine for the vulnerable voussoirs, and peperino for the facings. Two cost- and time-saving devices are worth noting: first, the arches in many extant bridges and aqueduct arcades are set back from the line of the sustaining piers, enabling the builder to erect the necessary wooden centering without first putting up scaffolding to carry it; secondly, economy of timber for the centering was achieved, as may be seen in the most famous of all surviving Roman aqueduct arcades, the Pont du Gard near Nîmes in Provence. Here the parallel sets of arches in three tiers, which carried the water across the River Gard at a height of 160 ft, were simply placed side by side without being keyed together: this meant that one timber framework wide enough to support one of them would serve for an entire tier. The slight loss of strength incurred does not seem to have adversely affected the stability of the structure! The

81 standard Roman form of vaulting, the barrel- or tunnel-vault, is no more than an arch extended in depth. But whereas the centering for masonry arches was of lightweight construction, the mass casting of poured concrete needed in vaulting demanded both first-class carpentry and a variety of reinforcing

85 devices.[134] These elaborate structures were the basic elements in the creation of new and exclusively Roman patterns of architectural experiment. Vitruvius has provided a detailed formulation of the whole process, beginning with the making of the wooden framework (7.3.1–2):

Set up horizontal furring strips at intervals of not more than 2 feet, using cypress for preference, since fir is quickly spoilt by decay or age. Arrange these strips so as to form the shape of a tunnel, and make them fast to the joists of the floor above, or to the roof, if it is there, by nailing them ... to ties fixed at intervals. Next take cords made of Spanish broom, and tie Greek reeds, previously pounded flat, to them, conforming to the required curvature. Next on the upper surface of the vaulting spread a layer of mortar made of lime and sand, to catch any dripping from joists or roof ... Then when the vaulting is fixed and interwoven with the bundles of reeds, coat the under surface with roughcast rendering, follow this with a layer of sand mortar, and finish off with marble plaster.

This passage illustrates both the variety of rein-

85 Nîmes, so-called 'Baths of Diana'. Conforming in overall plan to the normal basilica pattern, this late 2nd-century hall shows remarkable originality in the roofing system and its interior decoration. The barrel vault roof is formed of large blocks set parallel to the major axis and supported by ribs which rise above the column.

forcing material used, and the cautious attitude of those who used it. In spite of their growing mastery of concrete, Roman architects never seem to have abandoned the 'back-up' technique of built-in ribs.

Development of Roman vaulting[135]

The earliest concrete vaults and domes were presumably supported on full wooden centering until they had dried out; the centering was then removed, and the structure either stayed

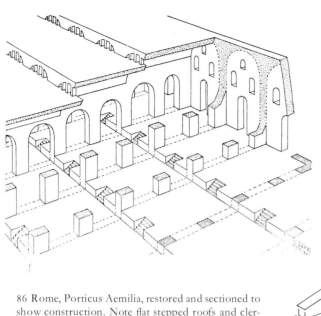

86 Rome, Porticus Aemilia, restored and sectioned to show construction. Note flat stepped roofs and clerestory lighting.

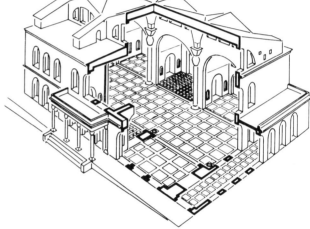

87 Basilica of Maxentius in section to show construction. The remarkable combination of three groined vaults roofing the nave and two barrel vaults covering the side sections provide the largest hall built in antiquity. Its area of 80 by 51 m. is divided only by four columns.

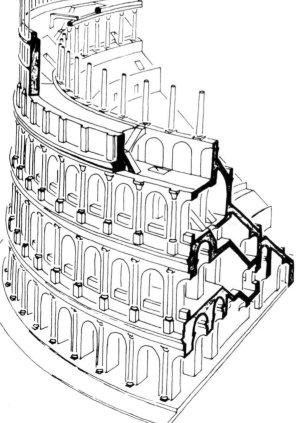

88 Colosseum, Rome. Auditorium in section to show construction. Successive tiers of seats are supported by tiers of passage-way arches as shown by black lines.

up or fell down! We have no information on the failure rate, but we must assume that, as with all processes of trial and error, it remained high, until stability was improved by strengthening either the foundations or the mortar, or by reducing the weight of the aggregates, as indicated earlier in this chapter (above, p. 86). The substitution of rows of tiles over the scaffolding in place of a complete timber centering system appears as early as the reign of Nero. The concrete was laid on top of the tiles, which remained in position when the scaffolding was removed. The total amount of space that could be covered by a continuous barrel-vault was unlimited, but its width was, of course, restricted to the span of the arch. By the first century AD this restriction had been lifted with the invention of the intersecting barrel-vault, but this *81b* advance, which paved the way to spectacular achievements in spatial planning, was preceded by a very interesting experiment in the use of multiple barrel-vaults, the building known as *86* the Porticus Aemilia,[136] which was built in 193 BC, and restored twenty years later. The Porticus Aemilia was a long, narrow warehouse (487m × 50m or 1600 ft × 200 ft) made up of six rows of contiguous barrel-vaults, whose long sides were pierced by rows of arches so as to form a continuous open space. Each side consists of three shallow terraces, which enabled the architect to provide the building with clerestory lighting, somewhat resembling that of a modern single-storey factory building, which points the way to the great developments that lay ahead. By the first century AD builders had discovered that two vaults could be made to intersect at right angles without any danger to the stability of either. Such 'groined' vaults could be used to roof a large rectangular space, the roof supported only by piers at each of the four points of intersection. Once this scheme had been put into practice, it was a simple matter for the architect to minister to the expanding tastes of emperors obsessed with notions of size and splendour by multiplying the number of intersecting elements, creating such vast and imposing structures as the Baths of Caracalla or

89 Pantheon, Rome. Sketch showing the brick arches in the walls and the coffering of the dome. The interior arches are functional, directing the thrust on to the pillars which project beyond the niches.

the Basilica of Maxentius.[137] Vaulting could *87* also become curved or circular; applied to the design of theatres a system of superimposed continuous barrel-vaulting could be used to provide both support for the auditorium and an ample supply of exits at suitable levels. The external walls of the semicircular or elliptical auditoria of theatres and amphitheatres could now be furnished with tiers of continuous arcading, in which the classical orders could be displayed in new and magnificent settings, as in that masterpiece of imperial magnificence, the Flavian amphitheatre better known as the Colosseum. The wide range of available building materials mentioned earlier was employed with great skill,[138] and on the whole with economy. Two exceptions, the brick ribbing and the reinforcing of concrete vaults with built-in relieving arches, as in the walls and dome of the Pantheon, can be justified by the resulting stability of the structures to which these methods were applied; the Pantheon, for example, is the oldest major roofed building in the world still standing more or less intact after a life of more than 1800 years! As for the brick ribbing, different theories have been advanced to account for this peculiar type of reinforcement, if that is what it is. The effect of it is to divide the vault into a series of small compartments. Was it perhaps introduced to enable the whole vault to be quickly erected in skeleton form, so that a number of gangs of workmen could then move in and fill the several compartments simultaneously with concrete, and thus speed up the whole process? The method will have reduced the amount of timber needed for the centering; is it possible that a shortage of labour had begun to make itself felt by the time the system became widespread? This is one of many aspects of the history of building technology where further research is needed.

We may fittingly conclude this chapter by examining briefly some of the organizational problems of the construction industry. Masons share with smiths and metalworkers a tradition of closely guarded craft skills and processes handed down from master to apprentice; and much attention has recently been given to these craftsmen and their work (e.g. by Mossé 1969 and Burford 1972); while the art of the carpenter has received rather less attention. In an interesting review of the supporting services needed by Roman vaulted architecture W. L. McDonald (1965, 143 ff.) has pointed out that a mastery of construction in wood was an indispensable basis. There is plenty of evidence for the re-use of shuttering, framing and formboard work, making it probable that some degree of standardization was achieved. The history of shipbuilding and military technology provide good examples of the solution of difficult problems in wood construction, but large-scale vaulting set new problems in carpentry; and it is surely a measure of their increasing mastery of this craft that the catalogue of technical improvements attributed to the Romans includes five items from the carpenter's tool-kit, namely the auger, the brace, the countersinker, the frame-saw and the gimlet.[139] But the overriding impression conveyed by the study of the various advances in building technique mentioned in this chapter is that these advances were governed, not by inventions, but by the 'chance discoveries and rule-of-thumb wisdom of successive generations of contractors'.[140]

89

8 Civil Engineering and Surveying

This branch of engineering is concerned with the construction of major utility projects such as dams, reservoirs, roads, bridges and harbour installations. The now familiar division of the subject into its 'traditional' three branches, civil, mechanical and electrical, to which have been added chemical and electronic, belongs only to the last two centuries. In antiquity, and right up to the time of Leonardo and beyond, engineers were all-rounders, as may clearly be seen by looking at the table of contents at the beginning of Vitruvius' treatise *De Architectura;* here, in addition to the expected chapters on the building of temples, public baths, theatres and private houses, instructions are given for the making of all manner of mechanical contrivances from cranes to catapults.[141] The word *architekton* has a much wider connotation than its English offspring: thus the term is applied by Herodotus to Eupalinos of Megara, who around 550 BC designed and 164 supervised the construction of a water-tunnel to supply the city of Samos.[142]

When engineering problems crop up in popular discussion of how people lived and worked in classical times, the questions most commonly asked are either in the category of 'How on earth did they manage to get those huge columns up there?' or in that of 'Why did they go to the trouble and expense of building those splendid arcaded aqueducts to carry the water across that river when they could easily have piped it across if only they'd known about water finding its own level?' The answer to the first question might be that with multiple pulleys and rather a lot of rope one man can raise a load of up to two tons. He would also need a lot of time! The usual answer to the second question has been to explain that they were, in fact, well aware of the principle involved, but that the pressures in the bottom of a U-pipe at the depth required on many sites were too great for the available materials. However, the probabilities are that the choice of method was governed by economic considerations.[143]

In this branch of technology, Greek achievements were unimpressive when compared with the progress made by their hydraulic engineers, and completely overshadowed by the striking successes of the Romans. The chapters devoted to Greek engineering in a recent popular handbook[144] include full treatment of building technology, military engineering and naval architecture, with detailed accounts of the advances

90 Diversionary dams in the Wadi Megenin, Tripolitania: *a* and *b,* Roman; *c,* modern. As the photograph shows, all three were eventually rendered useless by the course-changing habit of the wadi.

91

made in the two last-named branches during the Hellenistic period, but very little on civil engineering, apart from the harbour works at Alexandria, which were designed by Sostratos of Knidos, who was also responsible for designing the first and most famous of all lighthouses, the Pharos (see below, p. 105).

Roman engineers, on the other hand, made important—and very durable—contributions in the field of communications, outstandingly in the construction of roads, bridges and barrage dams, less spectacularly in that of harbour installations, where the Greeks had already made significant progress.

The road system of the Roman Empire was not only of major importance in itself, as a network of vital communications, but its execution involved the recruitment of increasing numbers of highway engineers, together with others qualified to construct the necessary bridges, viaducts, cuttings, embankments and tunnels.

In studying the history of road transport, due consideration must be given both to the economic requirements of supply and demand, and to the nature of the traffic carried: rough trackways may satisfy the needs of undeveloped rural communities at little expense, but all-weather routes for heavy vehicles involve heavy outlays in materials and manpower both in construction and maintenance. The predecessors of the Greeks, both in Crete and on the mainland, needed good land communications, and their roads, to judge from extant remains of bridges, culverts and other features, were built to a high standard. But with the collapse of Mycenaean power, the roads fell into decay, a fate which was later to overtake the much more elaborate road system of imperial Rome.

Greek roads

These roads, even those of imperial Athens, were of a very low standard. Being unsurfaced, and altogether lacking any system of storm-water drainage, they were unsuitable for wheeled traffic at most seasons of the year, and

91 Rutted road in Greece, showing a siding branching from the main road to enable vehicles to pass.

impassable in winter, when they degenerated into watercourses. The realities of the road situation are splendidly illustrated for us in one of Demosthenes' *Private Speeches,*[145] in which the owner of a farm abutting on to a road is engaged in a civil action for damages against a neighbour, whose farm adjoins his, but stands on higher ground. The complainant alleges that the defendant has diverted to his own use the water running off the public road, thus depriving him of his fair share of that useful, and in most parts of Greece, far from plentiful, commodity! The exceptionally rugged terrain made road transport for commercial purposes even less important by comparison with transport by sea than elsewhere in the region. Most roads were mere tracks, suitable only for pedestrians and pack animals. It was far cheaper to send goods from port to port over what Homer aptly calls the 'watery paths' (*Od.* 3.71). Two exceptions must be noted: first, the so-called 'rutted' roads,[146] a technical development in the form of grooves cut in the natural rock wherever a dirt road crossed an exposed rock surface—implying a standard track width for wheeled vehicles; secondly, a primitive form of 'macadamized' road made of quarry chippings rammed into the subsoil.

91

106

It is commonly stated, mainly on the strength of the numerous disparaging comments made by the traveller Pausanias, that Greek roads continued to be bad right through into the Roman period. But there is a passage in the historian Livy, in which a Roman envoy, despatched to the Macedonian court at Pella just before the invasion of Asia in 190 BC, comments favourably on the state of the roads: 'they had bridges at river-crossings, and were metalled *(munitas)* where the passes were difficult'.[147] Paved roads had existed in the Near East from as far back as the second millennium; and Assyrian achievements in road-construction are well attested for the reigns of Sargon (722–705 BC) and Sennacherib (703–681 BC). It is not unlikely that the Roman invaders learnt something about the construction and surfacing of roads!

Roman roads

The first point to observe in studying Roman roads is the fact that they were planned to facilitate the movement of troops, and that—at any rate until the later Empire—those troops consisted mainly of infantry. Road construction was governed by the need to provide a firm *92* footing for legionaries marching in all sorts of weather conditions, not a surface soft enough for draught-animals, which were not normally shod.[148] There is clear evidence of improvements in the surfacing of roads from gravel to paving blocks of flint or basalt, secured by kerbstones. Military needs also determined the alignment of roads. If the prime consideration had been the convenience of vehicular traffic, gradients would surely have been avoided by the provision of cuttings or, where the gradient was at all severe, by taking the road along the contour. The demand for an all-weather road involved the construction of embankments across marshy tracts, and the filling in or bridging of depressions. Roman road engineers were fully aware of the difficulties imposed by the varying nature of the terrain through which the line of communication was to pass; and the

92 The Via Appia just outside Rome showing the paving blocks and kerb.

93 Section of the Via Flacca at Pisco Montano, near Terracina. The depth of the slice removed from the cliff (126 Roman feet = 121 English feet) was carved by the engineers on the cliff face.

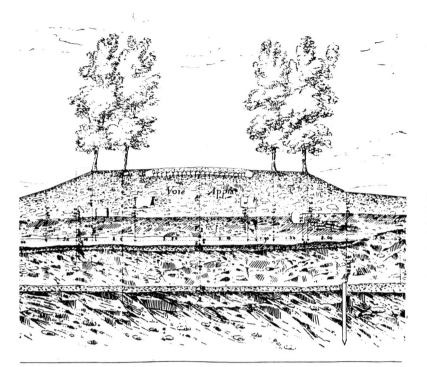

94 Section through the Via Appia, made in 1813, showing foundation bed of earth and clay, laid down on the Pomptine plain by Appius Claudius in 312 BC; above this the road-bed, made up of a mixture of flints and pebbles pounded together, not cemented, and flanked by two rows of kerbstones. The early carriageway is shown buried deep beneath the successive resurfacings topped by a layer of the familiar paving stones, also flanked by kerbstones.

notion that a typical Roman road was wide, straight and flat is not supported by the findings of research.

Methods of road construction

It has been commonly argued that the Roman idea of a road was technically unsound, being that of a wall buried horizontally in the ground (Forbes 1938, 138). At the opposite extreme we are invited to believe that 'the Roman roads were often no better than the Greek or Persian roads that had preceded them' (Hodges 1970, 188). Though neither of these views is tenable, it has unfortunately not been easy to get at the facts; until recently few roads have been carefully sectioned to determine how they were built, 'although they are carelessly cut through when channels, foundations, etc. ... are being dug (Sterpos 1970, 25).[149] Secondly, surviving sections are difficult to date, and finally we have only one surviving Roman description of the making of a road, and that in a poem! Vitruvius has nothing to say on the subject; but he has written at some length on the techniques employed in the construction of terraces and floors (*De Arch.* VII.1). Four successive layers

are prescribed: first, a layer of stones puddled in clay; next, one of concrete made up of small stones; then a finer concrete of gravel, topped by a paving of slabs or stones. Here, surely, is the source of the erroneous 'buried wall' theory mentioned above. A deep section made across the Via Appia in 1813 revealed no trace of 94 cement, nor of any of Vitruvius' layers. More recent work on the Appia (Fustier, *REA* 62 (1960), 95; 63 (1961), 27) shows layers of agglomerate of varying sizes packed with clay, and topped by basalt, but not a sign of cement! One point emerges: there was no 'standard' Roman road; more than half a century ago Stuart Jones (1912, 46 f.) found enough evidence from Roman Britain to support his opinion that varieties of construction were almost infinite: his list includes a stretch running below what is now Edgware Road, London, consisting of a single layer of rammed gravel, flanked by dwarf walls of concrete, topped by a paving of large flints set in lime grouting. Another section on the road from the Kentish ports to London, near Rochester, consisting of close-set wooden piles, above these a 1 m layer made up of a mixture of Kentish ragstone and broken tile (note the sensible use of local materials!); above that a 12 cm layer of rammed earth; above that again a mixture of gravel and earth, topped by polygonal paving. Other studies have shown even more clearly than Stuart Jones that Roman road engineers varied both their materials and their methods, even on the same road, according to the degree of support in the subsoil, and the availability of suitable materials. Robert Chevallier (1976, 87) illustrates the point from a recently published survey of a road in the Haute-Savoie region of France,

the road is made of successive layers of gravel, often attaining one metre in thickness, the result in many cases of a number of resurfacings. People have spoken in the past of consolidated gravel, and we have often noticed ourselves that the gravel roads in *Boutae* (Roman Annecy) had a binding of compacted clay and thin mortar so as to make up what we would call today a rough concrete.

Where a road was required to cross marshland, the need to set the road on a raised causeway was recognized early; where the terrain was waterlogged, as in the Hautes-Fagnes region of Belgium (Mertens, *ap.* Chevallier, 89–90) other methods were used. The Via Mansuerisca was begun with a complete timber framework pinned to the ground with vertical stakes, the cross-beams being slotted to take lines of joists, which carried the sides of the highway. The joists 'bore a transverse "corduroy" of tree-trunks and on this in turn there lay limestone flags cemented with clay, covered again by the road-metalling of gravel and pebbles' (Chevallier 1976, 90). In fairly close proximity to Rome widths are known to have varied from 3.50 m. (11½ ft) on the Via Tiburtina to 7.50 m. (25 ft) on the Via Salaria. Valuable evidence on the history of particular Roman roads has recently come to hand as a result of systematic surveys such as that carried out in the Roman Campagna by Shelton Judson and Ann Kahane (*PBSR* n.s. 27 (1972), 107 ff.). Here it has been possible to trace the very ancient network of roads and trackways radiating from centres such as Gabii and Tusculum, centres which in imperial times had declined in economic importance to make way for other centres linked by different lines of communication. More important for our purpose are the results of deep excavation of successive layers of road construction where the line of communication has remained unchanged; a recently excavated section of the military highway that linked the legionary fortress of *Deva* (mod. Chester) with *Segontium* (Caernarvon) in North Wales has revealed a kind of palimpsest, with three roads superimposed one upon the other. The top layer consisted of a narrow roadway, 2.7 m (9 ft) wide, made of cobblestones roughly laid in a bed of clay; below that was a slightly wider

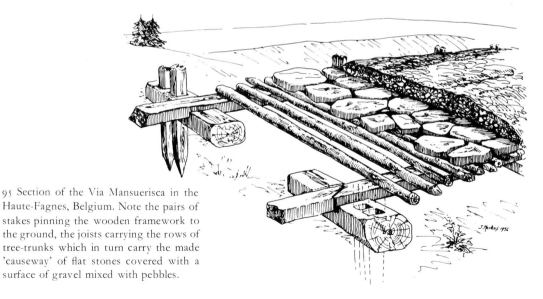

95 Section of the Via Mansuerisca in the Haute-Fagnes, Belgium. Note the pairs of stakes pinning the wooden framework to the ground, the joists carrying the rows of tree-trunks which in turn carry the made 'causeway' of flat stones covered with a surface of gravel mixed with pebbles.

96 Roman road on Blackstone Edge, Yorkshire, showing the well-preserved pavement of stone setts 4.9 m wide supported by kerbs and the central stone trough. The heavy wear on this was probably caused by friction from the brake-poles of wagons descending the 1 in 4 gradient.

roadway laid with the same materials, and below that again an 18-ft roadway well cambered, and provided with gutters, beautifully paved with cobbles set in a mixture of clay and laterite, and as smooth as an average modern tarmac road. This road has well-marked ruts, about 1 m apart, as in several of the streets of Pompeii. Among the more remarkable stretches of well-paved Roman highways still in good condition is a section of the important 96 road that ran over the Pennines at Blackstone Edge, between Halifax and Rochdale, linking the two northern garrisons at York and Chester. Here the surface layer of well-fitted stones is divided in the centre by a shallow stone trough up the steep bank, succeeded on the level moor above by a continuous line of slabs, but with no troughs. We have no precise information about braking on ancient vehicles, but loaded animal-drawn wagons will have required some means, however primitive, of keeping them under control on steep descents, so that the suggestion offered by Margary (1973, pl. XII(a)), that the central trough was 'perhaps intended for the brake-pole of carts on the very steep hill', seems plausible.

We now turn to the very scanty literary evidence. This consists of an account, written by Statius, to commemorate the construction of the Via Domitiana, which ran from Mondra- *101* gone in Campania to the nearby port of Pozzuoli, and was completed in AD 95 (*Silvae* 4.3). The writer, who was evidently interested in technical devices,[150] has managed, in a mere sixteen verses, to give a comprehensive and compelling account of how the task was carried out,

First comes the task of preparing the ditches,
Marking the borders, and deeply as needed,
Delving into the earth's interior;
Then with other stuff filling the furrows,
Making a base for the crown of the roadway,
Lest the soil sink, or deceptive foundations
Furnish the flagstones with treacherous bedding;
Then to secure the roadway with cobbles
Close-packed, and also ubiquitous wedges.
How many hands are working together!
Some fell the forest while some denude mountains,
Some smooth boulders and baulks with iron;
Others with sand that is heated, and earthy
Tufa, assemble the stones of the structure.
Some with labour drain pools ever thirsty;
Some lead the rivulets far to the distance.

This eloquent description does not, of course, cover the entire process; the Via Domitiana was an unusual road, and the poet was not speaking generally. But it does give the reader the opportunity, all too rare in the history of classical technology, of seeing the road-building gangs at work. The present state of research does not provide enough material for anything like a history of road-making in Roman times, but certain stages in its evolution can be discerned; the position may be briefly summarized as follows: for a long period prior to the construction of the Appia (begun in 312 BC) roads consisted of light foundation layers topped by a gravel surface. Foundations will have been made stronger to carry the flint paving which was introduced over a short section of the *92* Appia in 296 BC. There is some evidence of improvements both in the surfacing and in the consolidation of the surface by means of kerbstones and wedges. Most of this comes from inscriptions on milestones, which also furnish the dates of major repairs and improvements.[151]

Recent work on Roman roads may well lead to fresh thinking on land transport problems and on the purpose served by the road network. It has long been known that the 'made' section of a Roman road, whether finished in gravel or stone, was flanked by levelled verges, which together exceeded in width that of the 'carriageway'. It has also been pointed out by engineers that the paved carriageways do not appear suitable for the animal-drawn vehicles that passed over them. It is now being suggested that the paved portions were not used by carriages; first, because the vehicles themselves, lacking any form of springing, would not have stood up to the inevitable battering, and secondly, any vehicle which encountered even a modest military detachment en route, would be forced off the road (see above, p. 93, on the width of Roman highways). The problem would disappear if we assumed that wheeled traffic did not, except in cuttings or on viaducts or bridges, use the metalled surface, but ran along the verges or 'hard shoulders'. Such an assumption (and at present it is no more than that) would also account for the fact that transport animals were not normally shod. That wheeled traffic ran over the stone setts of city streets as at Pompeii and other Roman cities is, of course, undisputed: the evidence is there in the deeply worn indentations. Then as now, city streets had raised sidewalks, not verges on which vehicles could move. But there is plenty of evidence on inter-city highways of 'ruts' not naturally formed by traffic, but deliberately made, especially in mountainous sections, where the purpose appears to have been to prevent a wheeled vehicle skidding off into space, as for example on the Via Flacca at Pisco Montana near Terracina. The problem raised here is taken a stage further in Chapter 10, where it is discussed in relation to the design of wheeled vehicles.

75

Bridges and viaducts

The weight-supporting arch was already an important element in Roman architectural practice

97 Section of a Roman mountain road, Via delle Gallie, near Donnaz, in the Val d'Aosta, Italy. The road has been carved out of solid rock and the roadway provided with artificial ruts.

by the mid-third century BC, and it was not long before the technique was employed to carry highways across rivers, or depressions too deep for embankments. The English reader, familiar with those long stretches of undeviating highway which characterize the lowland zone of Roman Britain, may easily forget that the pioneers learnt their craft and developed their ingenuity in solving road-construction problems in the very different terrain of peninsular Italy, where a great deal of the work consisted in making embankments, cuttings, and hillside traverses, with the necessary excavation, infills, and supporting walls. Although the bridges of a great highway like the Flaminia, many of 98 which are still in use, rank with the highways themselves as impressive witnesses to the skill of their designers, surviving texts tell us nothing at all about the bridges or the methods used in building them. Fortunately it is possible, from the large number of surviving examples, to establish the basic methods, and to

98 Bridge over the Tiber at Narni, on the Via Flaminia. Only one arch of the original four (all unequal in span) survives.

study, at least in part, the developmental aspect of bridge design. In spanning a river, the aim of the bridge-builder is to provide a safe roadway with the minimum of obstruction to the flow of the stream; he may also be required to provide free passage for shipping beneath the bridge. These factors will affect the design in varying degrees. Roman arches were almost invariably semicircular in shape; this imposed an awkward constraint on the design of single spans, forcing the roadway to rise somewhat towards the centre, as in the bridge at Rimini (Sterpos 1970, 42). A high proportion of surviving bridges have only a single span, a fact which has suggested that engineers wanted to avoid making supporting piers in the river-bed. But Vitruvius' brief account of harbour works (*De Arch.* V.12) contains instructions for the making of coffer-dams, using pozzolana cement (see Ch. 9), and there are several bridges with multiple spans, including bold examples like the bridge *98* over the Tiber at Narni, where the largest arch has a span of 32 m, and the overall height is

about 30 m. Only one arch now remains, so that we can only imagine its original splendour, a sight which inspired the historian Procopius to make the following restrained comment:

Caesar Augustus constructed this bridge in time past, a sight worthy of close examination, because of all the vaulted structures known to us it is the highest (*Gothic Wars,* 1.17)

Surviving bridges with multiple spans suggest another explanation for the high incidence of single spans, namely the fact that the size of the supporting piers severely restricted the flow between them. Fresh light has now been thrown on the techniques employed in bridging by underwater surveys carried out recently at Minturnae,[60] where the Via Appia crossed the Garigliano (anc. *Liris*). The bridge appears to have been in continuous use for almost eight centuries (from *c.* 295 BC to *c.* AD 491). The striking fact is that this major bridge, built originally of timber, remained essentially a wooden structure throughout its existence. Unlike most Italian rivers, which are highly seasonal in flow, the Garigliano has no low-water period when it would be easy to construct masonry piers in the bed; the depth of the river at this point (9 + m) would make caisson construction difficult. To make things worse, the bedrock slopes downwards at an angle of 45° to the line of the river. On all the above counts the Minturnae crossing called for a wooden bridge supported on piles rather than a masonry one supported on stone piers. Where river-bed structures could not be avoided, Roman engineers built the necessary piers at irregular intervals, in order to take advantage of the terrain. The search for the easiest place for footings offers a reasonable explanation for the uneven spacing of the arches, the largest span being placed at the deepest point, which is not necessarily the mid-point of the river. With this type of construction a further restriction develops under flood conditions, as the water rises above the springs of the arches, threatening the stability of the structure. Many surviving bridges have been equipped with stormwater

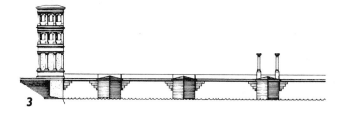

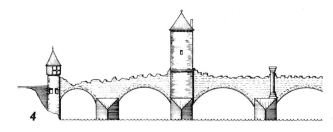

apertures in the masonry between the arches. As for the actual arches, we can detect a number of structural improvements, designed either to strengthen the arch itself or to 'key' it more closely into the rest of the wall. These improved techniques include the use of 'headers' and 'stretchers' instead of plain voussoirs, as in the viaduct at Ariccia, dated to the second century BC (Sterpos 1970, 65), multiple rows of voussoirs, as at Suessa Aurunca (Sterpos, 68–69), and voussoirs that extend into the adjoining masonry, as at Porto Terres, Sardinia (Sterpos, 53, lower). Care was also taken to prevent either bulging or squeezing of the structure by equalizing the distance separating the head of the arch from the roadway above it, as well as the horizontal distance between arch and abutment. Systematic work on the two adjoining bridges over the Moselle at Trier shows, in a dramatic way, the enormous technical improvement in setting and consolidating the piers, and this bold piece of engineering forms a fitting conclusion to our survey:

While the earlier bridge had limestone piers resting on ironshod oak piles rammed into the river-bed, parallel to the first bridge a more adventurous technique was adopted. Double-walled wooden caissons, the interior space packed with clay to make them water-tight, were so constructed that within them the mud and gravel could be dredged away and the massive stone foundations of the new piers laid directly on bedrock. The resulting solidarity scarcely needs stressing. Of all Roman bridges over Rhine and Mosel this one alone survived into the Middle Ages.[155]

Cuttings, embankments and tunnels

The varied skills of Roman engineers can still be seen in many surviving road cuttings and embankments. Here is Amedeo Maiuri's description of a very striking example, the deep, curved cutting on the road linking ancient Capua with Pozzuoli (anc. *Puteoli*), known as the 'split mountain' (Montagna Spaccata):

99 Reconstruction of the Roman bridge at Trier. *1,* Ancient pile supports, 1st century AD; *2,* Roman bridge, first state; *3,* Roman bridge, second state; *4,* Roman bridge, state after the vaulting in 1400 till its restoration in 1968.

A Roman cutting, using the technique of a military trench with straight sides, covered with that close, fine reticulate of the Augustan age which ... serves not just to conceal, but to compress and contain in the closely-woven mesh the firm concrete wall of stone and cement which does the job of supporting the embankment ... the Romans did not omit to do what a good modern constructor would do, to turn some of the arches against the thrust of the hill-side ... the walls have resisted the thrust of the tufa cliff and the impact and the jostling of the vehicles for almost 2,000 years! (A. Maiuri, cited by Sterpos 1970, 77)

Road tunnels were used very infrequently, and only when the lines of a major highway would

100 Forlo tunnel on the Via Flaminia, dug in 76–77 AD as the inscription at the NE entrance attests. The ancient tunnel (still used by the successor to the Flaminia) is 38.3 m long, 5.5 m wide and 5.95 m high.

101 Section of tunnel on the Via Domitiana. Notice the Gothic-style ribs.

otherwise have had to be diverted by a long detour, as in the Forlo tunnel on the Via Flaminia near Pesaro. This tunnel, which was cut through solid limestone for a distance of 40 m (130 ft) is still used by the modern highway. More famous was the road-tunnel between Pozzuoli and Naples, the 'Crypta Napolitana' (Via Domitiana); a letter from the philosopher Seneca (*Letter to Lucilius* LVII) describes, in highly rhetorical phrases, some of its bad features, one of which, the rising dust, will have been particularly unpleasant for 'rush-hour' travellers: *100*

101

Nothing is darker than these torches, which do not permit us to see among the shadows, but only to see the shadows. Besides, even if there were light, the dust would black it out.

Aqueducts

The term is popularly applied to those elegant and often spectacular arcades which still survive in many parts of the former Roman Empire as impressive testimony to the skill of Roman construction engineers. In fact, the term refers strictly to the channel by which water is led (lat. *conducere,* hence the English term 'conduit'), from its source by gravity flow to city, town or farm. The technology is fully discussed in Chapter 12 on hydraulic engineering (pp. 157ff.), but at this point we may appropriately mention the three types of structure that might be called for, according to the variations in terrain through which the aqueduct was required to pass. The open channel system[154] meant that the selected gradient for the flow had to be maintained, no matter what rises or falls occurred, between source and point of delivery, in the level of the terrain. An intervening mountain could either be bypassed by channelling along the contour, or pierced by a water-tunnel. The first of these methods involved much structural work, since the channel would need to be enclosed in a low wall of masonry, strongly embedded in the hillside (the Latin term for it, *substructio,* is very appropriate), so that it would not easily be carried *102* away by landslides after abnormal rains. Tunnelling, which, though also laborious, was both more economical in materials and presented fewer difficulties to those responsible for setting and maintaining a suitable gradient for the flow, seems to have been the more favoured method. The advantages of tunnelling were strikingly illustrated some years ago in the province of Epirus in north-west Greece. During the construction of the new highway from Jannina to Preveza the construction engineers carelessly sliced off the side of a sloping valley floor. Heavy rains then carried away hundreds of tons of topsoil, exposing the line of inspection shafts belonging to the aqueduct supplying water to the city of Nikopolis, some twenty miles away, and making possible a detailed examination of the structure of the shafts and the methods used in constructing them.

Dual-purpose shafts?

Opinion is divided among students of ancient surveying on the question whether tunnelling was commonly based on optical sighting over the mountain, followed by shaft-sinking, or on a primitive form of triangulation round its base, as described by the engineer Hero in his book *172* on the *dioptra* or plane-table. In either case the method employed by both Greek and Roman engineers was the same as that used in mining, viz. to excavate the tunnel in both directions from the base of the shaft, a channel being subsequently sunk in the floor, correctly aligned to *174* the desired gradient by means of the *chorobates* or *dioptra* (see Ch. 12). Vitruvius, who is our sole surviving source of information on the

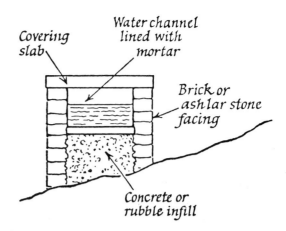

102 Cross section of a *substructio* in a Roman aqueduct.

construction of water-tunnels, recommends that the shafts should be sunk at intervals of 116 ft (35.5 m.), a spacing which suggests that they must have been intended to serve more than the initial survey purpose. In a well-argued discussion of this problem Landels (1978, 39) points out some of the advantages to be gained from the point of view of maintenance, after completion of the tunnel. These include ease of access for inspection in case of subsidence, leakage or flooding, and a means of reducing air-pressure 'if the inflow of water increased very sharply (for instance, after a freak rainstorm) and filled the whole tunnel'. He also points to a more efficient deployment of the workforce during the construction phase if several shafts were being sunk simultaneously, compared with conventional tunnelling, which severely limits the number who can work on the face at the same time *(loc. cit.)*.

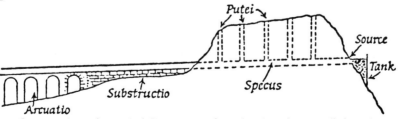

103 Cross section of a typical Roman aqueduct showing the tunnelled section *(specus)* with its inspection shafts, the intermediate section *(substructio)* supported in brickwork and the familiar arcaded section which takes over above a height of around 2 m.

The visible sections of an aqueduct take one of two forms. Where the ground slopes below the level required to maintain the gradient, the conduit is carried on a solid masonry support *102 (substructio)*. At heights above 2 m this is replaced by the more economical arcading, consisting of external courses in brick or squared masonry, with an infill of rubble or concrete. The safe height limit for single-tier arcades, above which the columns would be liable to buckle, seems to have been about 21 m or 70 ft (see Appendix 11).

Dams

The process of damming a stream so as to restrict its flow and raise its level, forming a reservoir, which can then be tapped to serve a variety of purposes, from powering grain-mills to providing towns with domestic water, is very ancient; indeed beavers, if not the inventors of the process, were certainly practising this form of hydraulic engineering long before man first appeared on the scene. For the most important developments in this department we must look to those areas where flood-control on the one hand, and the conservation of water for irrigation on the other, were crucial to the livelihood of the peoples of arid and semi-arid regions, and to the Roman engineers who planned and carried out the necessary works.[155] The numerous surviving dam-structures testify both to the ingenuity of the engineers in finding solutions to a wide range of problems, and to their capacity for making technical improvements. Of outstanding interest are the large storage reservoirs at opposite ends of the Mediterranean, those of Mérida in western Spain *105* (Proserpina and Cornalvo) and of Homs in *104* Syria (the famous Lake of Homs, with a continuous history of nearly 1700 years!). Less spectacular but of almost equal interest are the wadi-dams of Tripolitania. We shall begin our *90* survey with the Lake of Homs.[156] The first thing to be noticed is the sheer size of the project. Built by the emperor Diocletian's engineers in AD 284, the dam was nearly 2 km (1¼ m) long, creating a reservoir 40 km² (15 sq m) in area, making it the largest man-made lake up to that date. It was a dual-purpose project, designed to provide domestic water for the city of Homs and irrigation water for the fields around it. The core was made up of rubble formed of locally quarried basalt (the engineers had chosen a convenient basalt sill as the base for the dam wall), both sides being faced with cut blocks of basalt of pyramidal shape, the tapered faces being incorporated into the core, exactly as in the brick/concrete system known as *opus reticulatum,* and the flat surfaces exposed and grouted with a hard mortar. In his detailed account of this remarkable dam Norman Smith (1970, 42) points out that there have been so many repairs and modifications over the cen-

104 Cornalvo dam near Mérida (Emerita Augusta) Spain. View of the water-face. The water-tower originally had apertures for controlling the outflow. Length 200 m (656 ft), height (at centre) 20 m (66 ft). Probably later than the Proserpina dam.

turies that we cannot be certain of its original size and form; nor is there any prospect of clearing matters up, since a new dam, completed in 1934, rests firmly on top of the old. But 'the builders of the new reservoir can be confident of one thing at least: during nearly 1700 years of use, the Roman reservoir accumulated virtually no silt' (Smith, *loc.cit.*). Turning now to Mérida (the first-century veteran colony of *Emerita Augusta*), we find that our two dams are part of the most remarkable assemblage of well-preserved Roman remains outside Rome itself, which include a superb Roman theatre, an amphitheatre, a circus and two road bridges, one nearly half a mile long. Both are described in detail by Norman Smith (1970, 43–47). The most interesting structural feature of the Cor-

104 nalvo dam (the later of the two) is the arrangement of intersecting internal walls which divide the interior into a 'series of deep masonry boxes' filled with stones or clay—an unusual and effective way of ensuring a stable barrier. Second in importance at Cornalvo is the large square tower, with sides of 4.4 m ($14\frac{1}{2}$ ft), which reaches almost to the height of the dam-crest. The apertures made at different levels were evidently provided so as to ensure a supply of water when silt had blocked the outlet at the base. Out of action during the Middle Ages, the dam was again brought into use in the eighteenth century, not as in Roman times for domestic water, but to drive grain-mills and for irrigation; it is still used today for the last-named purpose.

105 The other Mérida dam, the Proserpina, which appears to be earlier in date (early second century AD, consisted of a huge unfaced earth embankment on the downstream side, the inner face being equipped with no less than nine masonry buttresses, to prevent it being toppled by the great weight of the embankment when the reservoir was low. The outflow from the Proserpina reservoir was controlled by a pair of tunnels fitted with sluices; but there is clear evidence of silting-up; in the seventeenth century the outlet tunnels were raised, so as to clear the accumulation, which was by then several

105 Proserpina dam, near Mérida. View across the water-face, showing the powerful buttresses (projecting 3 ft at the top and 12 ft at the bottom. Length 430 m (140 ft); height at centre 12.2 m (40 ft). Probably early 2nd century AD.

metres deep. To build a dam like Proserpina requires little engineering expertise, and plenty of manpower. In this respect it resembles the numerous dams constructed in India and other over-populated areas of the developing world. All have the same major drawback — heavy siltation. The dam at Cornalvo seems to indicate technical advances over Proserpina both in structure and in measures to combat the effect of silting.

All the Roman dams so far identified in North Africa were, with one exception, multipurpose structures, providing flood-control, water-storage and soil-conservation. Thus the wadi-dams of Tripolitania studied by Vita- *90* Finzi,[157] were designed to hold back the heavy seasonal wadi-flows and irrigate the adjoining fields. In the absence of silt-traps, the inevitable siltation led to an important change of use, the area behind the dam becoming a fertile cropland, and the wadi run-off now providing the essential irrigation water for the crops. Where water-storage for domestic purposes was the aim, the problem of silting was met by constructing dams higher upstream. This provided both the silt-trap to protect the water-supply, and the water for irrigating the fields after the seasonal flooding.

Choice and use of materials in dam-construction

Norman Smith (1970, 38) has laid stress on the resourcefulness of the Roman dam-builders in selecting the combination of materials appropriate to the purpose. A notable example, the Kasserine dam in south-west Tunisia, was built in the second century AD to provide a water-supply for the town of Cillium. The dam consisted of a rubble-and-earth core with a facing of masonry blocks. Since it was designed to pass the overflow over the crest, the top was faced with stone to prevent scour, and the dam wall curved to give more room for the overflow (for details see Smith 1970, *loc.cit.*). The inner faces of these masonry dams were protected against seepage by a powerful sealant consisting of a combination of hydrated lime and crushed brick/pottery which the Romans called *opus sig-ninum*. The outer faces were normally stepped, and many have buttresses for added strength (Smith 1970, pl. 14).

The remains of the wadi-dams, on the other hand, show a very different treatment; since water was not to be allowed to pass over the crest, masonry spillways were built into the end of the dam at a lower level, the flow being controlled by "stop-logs" whose fixing-slots are often clearly visible' (Smith 1971, 37).

Tradition and innovation in dam-construction

In constructing diversion dams for irrigation, and wadi-dams for soil retention, Roman engineers were using established techniques, but carrying them out on a larger scale. However, in two other areas, with the use of dams for flood-control and for supplying water to urban communities, they were breaking fresh ground (see Moore 1950, 98, 102). Projects such as the Lake of Homs and the Cornalvo *104* dam described above give evidence of the ability to organize engineering construction on a big scale. In addition the proven durability of many of the surviving dams seems to be due to two important developments: first, better methods, especially in the use of water-resistant mortars as sealants; secondly, the use of spillways to arrest erosion.[158]

Harbour works[159]

The earliest man-made harbour works of the classical period were discovered at Delos, in the shape of a 300 ft long mole formed from large blocks of local granite and dated to the eighth century BC. The lowering and placing of these blocks, weighing about 10–12 tonnes apiece, at the seaward end of the mole, where the channel was 10 m deep, will have required the use of cranes, which do not seem to be attested, at least in the building industry, before the sixth century. Eretria in Euboea had a very long jetty, 700 m long and 20 m deep, while an important early development at Corinth was the creation of a new, artificial harbour at Lechaion, on the Gulf, probably the work of the energetic ruler, Periander (*c.* 627–586), and designed to meet the growing needs of the city's expanding trade with Sicily and the West. The much older port of Kenchreae, on the Saronic gulf, was a natural harbour protected by two convenient promontories. By the fifth century, the basic facilities needed to provide safe anchorage in bad weather had been provided. Piraeus, which grew rapidly from the time of Themistocles as a naval harbour with docks and ship-sheds (for the architecture of the latter, see *74* Ch. 7, p. 75f) became, in the fourth century, the chief commercial port of Greece, with quays and warehouses. Piraeus had only one basin suitable for commercial shipping, but Alexandria *13*

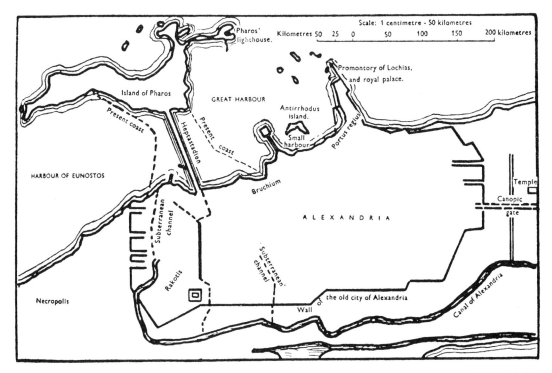

106 Harbour plan, Alexandria, showing the two harbours and the causeway *(Hepstastadion)* which joined the mainland to the island of Pharos.

106 had both an east and a west harbour, as had Syracuse, Cyzicus and Tyre. The most remarkable feature of the facilities at Alexandria[160] is the advantage taken by the early engineers of certain natural features, notably the long east-west reef, which made a solid base for the great breakwater designed to protect incoming shipping from the prevailing north-west winds. The early harbour works, dating from about 2000 BC, were subsequently submerged by underwater erosion, but portions of the enclosed harbour, covering about 300 acres, can still be seen in calm weather beneath the surface. The main component, a massive mole, 600 ft (208 m) wide and about one mile in length, linked the mainland to the island of Pharos, forming two basins; it was built to a depth of 36 ft (103 m), requiring vast quantities of stone to be dug out, transported to the site and laid in place. The site had two drawbacks, however: first, dangerous reefs making the entrance hazardous—hence the later lighthouse on the island;

and secondly, insufficient depth of water in the basins. Alexander chose a new site, on a neck of land between Pharos and Lake Mareotis, the latter forming a natural wet dock, while the island became the site of a splendid multi-towered structure with a light burning on the topmost storey—the world's first lighthouse. Built on the eastern extremity of the island which gave it its name, the Pharos, designed by Sostratos of Cnidos, is said to have taken nineteen years to build. The diminishing tiers rose from a 100 ft square base to a height variously estimated at 300 or 450 ft. The light, said to have been visible for nearly thirty miles out to sea, was designed to give safe entrance, not, as with its modern successors, as a warning to keep clear of danger.

The lighthouse was examined by the Arab geographer Edrisi in the thirteenth century, who noted that the blocks were bonded with molten lead—the method used on the second Eddystone lighthouse five centuries later. Of the

105

107 Roman lighthouse at Brigantium (La Coruña) NW Spain. Much restored over the centuries, this lighthouse preserves the essential structure which closely resembled that of the Pharos at Alexandria, having the lowest section quadrangular, the intermediate octagonal and the highest cylindrical.

many successors of the Pharos built by Roman engineers the best preserved is that of Coruña *107* in the north-west corner of Spain, while foundations and portions of the lowest tier of lighthouses may still be seen on either side of the English Channel (at Dover and Boulogne), and at the great North African port of Lepcis Magna *108* in Tripolitania.[161]

Roman harbours

The most important single technical development in harbour construction in the Roman period was the introduction of pozzolana, which was found in large quantities around the harbour of Pozzuoli beside Naples (about the *114* effects of it on building technology, see Ch. 7, p. 83). The need for a second port nearer to *83* Rome than Pozzuoli had long been felt when the emperor Claudius took the first step in the conversion of the open roadstead of Ostia at the mouth of the Tiber into a safe enclosed basin, offering protection from dangerous offshore winds. Much time and trouble was saved, as Pliny explains, by a device both economical and technically sound:

Claudius of Blessed memory, having retained for several years the ship specially ordered by Gaius for importing the obelisk (from Egypt), and which was the most remarkable vessel ever seen at sea, had towers constructed in it of pozzolana, brought it from there to Ostia, and sank it to help form the harbour 36.202).

The remains of the harbour, now three km from the coast, have for centuries formed part of a private estate, and have only recently been made available for research by the development of the new Rome airport at Fiumicino.[162] Unfinished at the death of Claudius, the new harbour works were completed by Nero, who celebrated the event by the issue of a fine *sestertius*, *110* which shows the quays and warehouses, as well as a number of vessels riding at anchor in the basin.

108 Lepcis Magna, showing remains of the lighthouse, 21.2 m square at its base and at least three-tiered, which flanked the 80 m– wide harbour entrance.

Many other harbours of the Roman period remain largely or wholly unexplored, but several seasons of work on the twin basins of ancient Carthage have produced very interesting results. Recent operations by Laurence Stager of the University of Chicago, who has been excavating the Roman commercial harbour works under the auspices of the internationally sponsored 'Save Carthage' programme, *111* have exposed the massive Roman quays, and behind them, the remains of warehouses, a flagged piazza flanked by colonnades.[163] The unexpected appearance in the rubble infill below the surface of the piazza of fragments of urns and stelae, along with charred bones, shows that the engineers had no compunction about defiling the contents of the sacred Tophet, the sanctuary lying immediately to the west of the basins, where the notorious sacrifices of children had once taken place. Full publication will doubtless shed considerable light on the development of this great seaport.

112 Recent work at the Etruscan port of Cosa,[164] *113* the hinterland of which has been the scene of a massive combination of field surveys and excavation of an important agricultural complex, provides an instructive example of a variety of engineering skills applied to the solution of a nexus of problems on a difficult site. The aim of the various works that are now being revealed was to convert a primitive off-shore anchorage into a protected harbour, and at the same time to develop a first-class sluice or artificial water-channel between a lagoon and the sea; the development also included the establishment of a

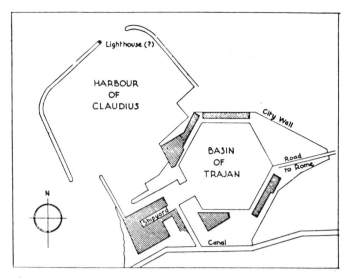

109 Diagram of the harbour of Portus at Ostia, built by Claudius (AD 46). The hexagonal inner basin with its quays and shipyards was added later by Trajan (AD 98–117).

110 Bronze *sestertius* of Nero, AD 64–65; obverse, bust of Nero; reverse (shown here) the Claudian harbour of Portus, with ships at anchor or discharging at the quays; around the edge, warehouses. The coin is inscribed POR (TVS) OST (IENSIS).

111 Carthage, commercial harbour. In the background the naval harbour, drained for excavation; the central breach in the harbour wall marks the channel linking the two harbours.

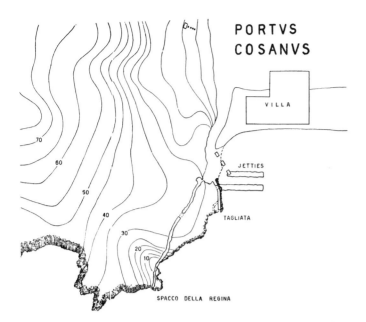

PORTVS COSANVS

VILLA

JETTIES

TAGLIATA

SPACCO DELLA REGINA

112 Harbour plan, Cosa, showing the two jetties and the two channels. Spacco della Regina, a natural cleft (260 m long) cleared to prevent silting of the harbour: the shorter Tagliata, an artificial channel which replaced the Spacco (see *ill. 113*).

113 fishery. A glance at the map will show the main natural features of the harbour area, which include a very long natural sluice (260 m long and 30 m deep at its deepest point) and a lagoon which lacked any control over the tidal movements which filled and emptied it. The earliest scheme, dated to around 170 BC, included scouring the sluice and levelling its bed, and cutting an outlet channel linking the lagoon with the harbour. These improvements will have served to control the level of the lagoon (especially at very high tides) and prevent excessive wash, and at the same time to stabilize the fishing grounds (which depend on the maintenance of appropriate water-levels and temperatures). The excavated rock was economically used to construct the first stage of the enclosed harbour. A century later, the collapse of the sluice-way walls made a revised scheme necessary: the old channel was replaced by an artificial one, and a new dike was built, which enabled the salt content and the temperature of the fishery basins to be controlled.[165]

Another innovation made possible by the waterproof cement from Pozzuoli was the use of arcaded moles designed like aqueducts. One of these appears as a prominent feature in a well-preserved wall-painting from Herculaneum.[166] Casson (1971, 368) notes that the *114* innovation was of limited importance, being restricted to the Campanian coast and the early imperial period; the point was made by Vitruvius: in a short chapter (5.12) on the construction of harbours and shipyards he gives instructions for the making of coffer-dams, using a strong cement of pozzolana and lime, followed by an alternative method for use where tides or currents are too strong to allow the coffer-dam to hold together, 'but in places where this powder is not found, the following method must be employed'. Why the restriction? Were there transport difficulties, or did the powder lose its strength rapidly, so that it had to be applied within a limited time after being dug?

Harbours in decline

The harbours of the Mediterranean region fall into a number of categories: (a) open roadstead—satisfactory if required to meet the needs of small-scale coastwide shipping. The sea is virtually tideless, and small vessels can be off-loaded and loaded in the surf; (b) natural harbours consisting of a bay, with headlands; these were often extended to enclose a larger expanse of protected water, e.g. the twin harbours, of Kenchreae and Lechaion at Corinth; (c) man-made harbours, with moles projecting at either side of the planned harbour, e.g. the Claudian harbour at Ostia; (d) two natural harbours *109* formed by a projecting headland of hard rock, giving safe anchorage from storms coming from more than one direction; 'hammer-head' promontories of this type are common, e.g. Cyzicus; (e) inner basins formed from natural lagoons or low-lying ground—common on the east coast of Italy, e.g. Cosa. A common requir- *112* ement here is a canal beside the river estuary, giving access to the basin, e.g. the Claudian harbour at Ostia; (f) deep-water harbours formed

113 Cosa harbour. Northern end of the later of the two sluiced channels, known as 'Tagliata' or the 'Cut'. It was 4–5 m wide and 70 m long, part trench and part tunnel, and wholly man-made (see plan, *ill. 112*).

114 Wall-painting from Herculaneum, now in the Museo Nazionale, Naples, depicting the harbour of Puteoli (Pozzuoli), surrounded by warehouses. In the foreground, an arcaded mole projects into the sea. Inside the harbour, ships at anchor or berthed at the quays.

by mole and diversion of the river to run alongside a second mole equipped with quays, e.g. *117* Lepcis Magna.

Two major natural activities create problems, which can impair the efficient operation of a harbour, and even render it useless in course of time, unless steps are taken to deal with them. These are, first, the depositing of silt at river-mouths; river-ports get silted up, and become unusable. The classic example is Ostia—the Roman city is now two miles from *109* the sea, and the basins are now high and dry (see below). Secondly, the depositing of sand, caused principally by the construction of moles and jetties projecting into the sea; dredging is essential to maintain sufficient depth both in the approach channel from the open sea and at the quays. Here the prime case is Lepcis, which, it appears, had a very short life before the sand-deposits made it useless; here the effect was aggravated by the fact that the Wadi Lebdah, *90* which had been made to change course, resumed its old channel, and left the splendid quays high and dry, as they are today. In the *117*

115 Ostia, Piazzale delle Corporazione. Square-rigged Roman freighter about to pass the Claudian lighthouse and enter the harbour of Ostia: with its 200 m-wide entrance, Portus had an area of nearly two-thirds of a square mile; for the tiered lighthouse see *ill. 107*.

115 case of Ostia, we have some valuable evidence on the development of silting. In the years 337–41 AD Lucius Cornelius Madalianus, prefect of the corn supply *(praefectus annonae)*, authorized an overhaul of the Claudian basin, which included work on the moles, on the lighthouses, and dredging work. R. Meiggs (1973, 2nd ed. 166) notes that the mole in question can be seen from the air projecting 300 m from
109 Trajan's inner harbour into the Claudian harbour at right angles to the communication canal. This mole was designed to protect Trajan's harbour from sand; it seems that the left-hand mole had collapsed and was letting in sand. By the tenth century Trajan's basin had become a mere pool, totally cut off from the sea and joined to the Tiber only by a ditch. As for Lepcis, the emporium for trade with the Fezzan, and an exporter of olive oil and grain, we learn that by the early sixth century, when Belisarius was engaged in the reconquest of the African provinces, the harbour was already silted up (A. Merighi, *La Tripolitania Antica* II (1940), 82; Procop. *Gothic Wars*, 14.6–11).

Canals[167]

In recent times canal-building has become an important activity, responding to a variety of demands in the economy: (1) as an extension to the already flourishing use of navigable rivers by linking major inland waterways; examples include the Canal de l'Est which links the French and German canal systems via the Saône and Moselle rivers (see Table 6); (2) as a bypass from the coast to inland centres, where the local rivers are unfavourable to through navigation; example, the Manchester Ship Canal, bypassing the upper Mersey (see Table 6, Nos 9, 12, 13); (3) to link important sea- or waterways separated by narrow isthmuses; examples include the Corinth and Suez Canals (see Table 6, No. 17); (4) a fourth type, the drainage canal for removing surplus water from agricultural land, controlling rivers liable to flooding, etc., has had a virtually continuous history from early times.

What was the situation in the ancient world? The subject has received scant attention,[168] in spite of the fact that many canals were constructed, all four types being represented in our sources, and reported as having been attempted, successfully completed or at least projected; the information is summarized in Table 6, which contains seventeen entries, the majority of which were complete, and operational. By their very nature, and that of the terrain in which they tend to be sited, canals, like harbours, need constant maintenance—protection of embankments against erosion, dredging to prevent siltation, and so on. Once neglected they fall rapidly into decay, and eventually merge into the landscape, leaving few if any traces. It is not surprising, therefore, that the evidence for canals is almost entirely literary; the lack of archaeological evidence may also explain, though it does not condone, the almost total neglect of this important means of transportation, which can be traced far back into pre-classical history, and which played a significant role in the commerce of the western provinces of the Roman Empire.

What were the technical problems involved in canal-building, and what success did their builders achieve in controlling the flow of water, and keeping the channels navigable? In both drainage and navigation canals difficulties may be encountered if there is an appreciable

116 Vast harbour basin of Caesarea, built by Herod the Great. The remains of ancient moles can be seen in the dark areas above and below. Exposed moles are more recent (lower and upper right).

117 Lepcis Magna, quayside. The harbour area of Lepcis extended over more than 100,000 m² (25 acres) and there were 1,200 m of quays filled with built-in mooring blocks. Behind were covered porticoes and spacious warehouses.

difference in water levels at the two ends. The movement of the water may need to be controlled by weirs, barrages or sluice-gates. In navigable waterways, again, the need to maintain a depth sufficient for the ships using the canal may require the installation of pairs of water-barriers or 'pound-locks', so called because the water is impounded, and the craft enclosed within the barriers raised or lowered. The lack of visible remains is a severe handicap to investigation, but the only surviving account, that of a projected canal of the first type, contains enough information to enable us to see what the technical problems were, and how they could be solved.

The two major technical problems are: first, how to carry boats over rising ground, which demands some kind of lock-system to impound water; and second, how to maintain a sufficient depth of water. Among the canal projects mentioned in the record are two that will certainly have required some lifting devices. The first is a passage in the Roman historian Tacitus (*Ann.* 13.53.3):

Lucius Vetus was making preparations to connect the rivers Rhine and Saône by means of an inter-connecting canal. His intention was to have merchandise passing through the Mediterranean, thence up the Rhone and Saône, running down from there into the Moselle, and so into the Rhine, and on from there into the Ocean. This would do away with all the difficulties of land transport, and there would be direct communication by water between the Mediterranean and the northern lands.

The most likely route for this 'Grand Junction' canal (Tacitus is silent on the point) would be that followed by the Canal de l'Est, constructed for the same purpose rather more than a century ago. Its total length is about forty miles, with a steepish ascent from the Moselle to the watershed (130 ft in two miles), and a gentle descent into the Saône valley (450 ft in thirty miles). Some form of pound-lock would seem to be needed for the short lift from the Moselle,

but there is no mention of anything of the kind: the objections to Vetus' plan were political, not technical.[169] The second reference consists of a series of exchanges by correspondence between the Younger Pliny, Special Commissioner for the province of Bithynia, and the Emperor Trajan, concerning a very much less ambitious project, but one that Pliny regards as economically very desirable. In his opening letter (10.41) Pliny explains the problem very succinctly:

There is a sizeable lake not far from Nicomedia (the capital), across which marble, farm produce, wood and timber for building are easily and cheaply brought by boat as far as the main road; after which everything has to be taken on to sea by cart, with great difficulty and increased expense. To connect the lake with the sea would require a great deal of labour, but there is no lack of it. There are plenty of people in the countryside and many more in the town, and it seems certain that they will gladly help with a scheme which will benefit them all.

Pliny goes on to request the emperor to send an architect or an engineer, whose first task will be to find out whether the lake is above sea-level (the locals reckon it at about 60 ft), but Trajan's reply, and Pliny's further comments (Letters 42 and 61) are concerned with maintaining the level of water in the lake when the proposed canal is built; neither correspondent refers to any technical problems connected with the 60 ft lift! Flood-gates and sluices were in common use, but, as Norman Smith points out (*TNS* 47 (1977/8), 83), they would scarcely have been of any use on Pliny's waterway. On the other hand, dry docks, essential for ship-repairing, are mentioned more than once in the literary sources,[170] and pound-locks are not far removed from these. Since neither of the two schemes came to fruition, the whole question might be dismissed as academic, but the subject of canals as 'an integral part of a transportation system that was vital to the running of an empire' (Smith, *art. cit.*) deserves more attention than it has so far received.

9 Mining and Metallurgy

This is a very wide-ranging subject, both as regards the variety of minerals extracted and processed, and in relation to the wide distribution of the deposits from one end of the Mediterranean region to the other. Detailed study will, therefore, inevitably be restricted, in the case of mining, to the exploitation of the *119* famous silver deposits at Laurion, near Athens, and to the gold-mining areas of north western Roman Spain; in the second part of the chapter attention will be concentrated on the production of copper, gold and iron. The reader who requires a comprehensive coverage will find competent surveys of both topics in volumes VII–IX of Forbes, *Studies*. J.F. Healy's monograph (1978) is rendered more valuable by reason of the accompanying distribution maps, and the excellent illustrations.

The most important minerals extracted were gold, silver, copper, tin, lead and iron. For the major sources of supply see Table 13. The workability of mineral deposits depends on two main factors: the geological setting in which the particular deposit occurs, and the way in which the metal is combined in the ore. Thus whereas copper and iron occur as chemically combined metallic lumps, gold and silver are found as native metals in quartz veins. Weathering of the enclosing rocks releases the precious metal, which is then carried away by streams, whence it finds its way into the river gravels, with a characteristic distribution known as 'placer' gold, the heavier nuggets being found near the source or 'mother lode', and the finer particles in the accumulations of sand farther downstream. These placer deposits can easily be made to yield up their gold by 'panning', a technique made familiar by the comparatively recent exploitation of regions such as the Yukon in north-west Canada. But where the gold lies embedded in the quartz, as in the rich reef deposit of the Transvaal in South Africa, heavy equipment is needed to extract the metal from the quartz conglomerate.

As for the three main sources of information, the weak and sporadic literary evidence is fully compensated for by the abundant survivals of ancient mining activity on the ground, in the shape of disused galleries, adits, shafts, spoil-heaps, and even the remains of mining equipment such as picks, sacks for carrying the *120* ore, and tread-wheels for draining the underground workings. Added to these is the rich array of artefacts of all kinds, including coins and jewels made from the precious metals, and everyday tools and implements in bronze and iron. Here too, as elsewhere in ancient technology, the absence of spectacular 'breakthroughs' should not blind us to those advances and innovations which can be pieced together from analysis of the visible remains, whether of mining installations, as at Laurion or Rio Tinto, or of furnaces and other visible tokens of metallurgical activity, supported here and there by literary evidence.

It is clear, for example, that ancient prospectors, however rudimentary their knowledge of geology,[171] were able to discover and exploit, within the limits of the available technology, the mineral deposits they sought; thus the gold deposits of the Sinai have been so effectively worked that today no gold remains; of copper, however, there are visible traces, and that metal was still being worked at the beginning of this century. It is worth noting in this connection that we have in the Turin papyrus a geological map, dating from the early thirteenth century BC, showing schist, granite, and, crossing both,

113

118 Corinthian painted *pinax*, Pentaskovfi, Corinth, showing a miner, *c.* 550 BC, extracting ore at the face and three others collecting and removing the ore in buckets.

strips of 'head' or 'combe' deposits, with marginal notes on the mines and quarries of the district.[172]

In prospecting, both Greeks and Romans relied almost exclusively on surface indications, chiefly colour; some literary references and site analyses suggest distinct progress in this department by the Romans, as intimated by the following passage from Pliny (*HN.* 33.67):

prospectors for gold begin by lifting the surface earth *(segullum)*, which is the name they give to the earth that reveals the presence of gold. This consists of a pocket of sand, which is washed, and from the sediment that remains an estimate is made of the lode.

In a later passage Pliny notes how copper was found beneath layers of silver and alum, showing a more sophisticated approach to the search than that provided by mere surface indicators.[173]

Greek mining for silver

Greek mining methods are well illustrated from the rich Lipsada mines near Stageira in southern Macedonia, which contained workable deposits of gold, silver and lead, as well as some copper. The main seam, some 18 m or 60 ft thick, extended for about 16 km (10 miles), and was intersected at intervals by numerous cross-seams. Early exploration showed that the richest deposits lay at the intersections, and shafts were later sunk at these points.[174] Technical developments designed to overcome the difficulties of mining at increasing depths can best be studied at Laurion in Attica,[175] one of the most extensive of all mining complexes. The first metalliferous ores to be recovered on this historic site (collected, not mined!) were the easily identifiable surface deposits of red oxide of iron, know from its blood-red colour as *haematite*. 'Silver', declared Pliny, 'is only found in deep shafts, and gives no signs to raise the expectation of its existence, giving off no glittering gleams like those that come from gold' (*HN.* 33.95). But at Laurion the topmost of the three silver-bearing zones lay in places so close to the surface that the earliest workings took the form of trenches and shallow caves, to be succeeded, as early as the sixth century BC, by open-cast workings, with short, almost horizontal, adits (1. 1). Exploitation of the second zone required shafts of increasing depth. The shafts, which were either rectangular or square in section—sizes range from 2.00 m × 1.90 downwards, with an average dimension of 1.90 m × 1.30 m—had a peculiar feature in common, 'every ten metres of depth the cross-section was turned through an angle of 8–10° so that the rectangle at the base of the shaft might be at right angles to that at the top' (Healy 1978, 80). Various explanations have been offered, none of them convincing. The rich vein discovered in 483 BC, the return on which was used by Themistocles to build the fleet that destroyed the Persian armada three years later, was part of the third zone, where the workings attained depths of 35 m. A century later, after a period of stagnation brought about by the exhaustion of Athens following the disastrous Peloponnesian War of 431–404 BC, activity was renewed, and shafts were sunk to a depth of around 100 m. Further development in the late

119

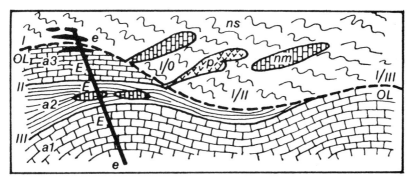

119 Laurion, cross-section of silver-bearing strata, showing the silver-bearing veins occuring between the alternating layers of limestone *(a1, a3, nm)* and volcanic schist *(a2, ns)* at *I, II, III* and *I/0, I/II* and *I/III.*

fourth century took the form of lateral galleries at the 50 m level, but there is no evidence of any attempts at linking the shafts, and the workings remained at this primitive stage; the galleries were too low to allow the miner to stand upright, and so narrow that he had only two working positions, either on his back, or on his side; no wonder progress was slow—at an estimated 10–12 cm in a ten-hour shift or around 12 m a month (Ardaillon, 25). Roman miners, it seems, were more ambitious, driving galleries with communication between them, and in some mines exploiting several levels with connecting shafts. It is possible that in these developments they were simply transferring and adapting to mining the methods they had earlier employed with success in southern Etruria and elsewhere in Italy to overcome the problem of waterlogging in valley bottoms by means of a network of underground drainage ways.[176]

Getting out the ore

The exploitation of the top layer at Laurion presented no difficulties. Access to the branches or 'stopes' of the second zone was provided by ladders, or more simply by notched tree-trunks fitted to the sides of the shafts, and thence on foot, and in a crouching position. Picks, chisels and wedges of wrought iron were used to break the ore away, and leather or esparto grass bags (1. 2), fitted with handles to carry it to the shaft, whence it was raised to the surface by porters climbing ladders fastened to the side of

the shaft, or hand-to-hand fashion. No improvements on this are reported from either the Greek or the Roman period; indeed, even more primitive methods are attested, Diodorus reporting (13.1) that boys were used to gather up the dislodged rock piece by piece, clambering into the working areas through tunnels formed by the removal of the ore, and Pliny reporting (*HN.* 33.71) that the miners themselves carried out the ore on their shoulders; no tramways or pit ponies to ease the strain on human muscles! Surviving equipment found in the form of drums (Rio Tinto in Spain) and hubs (Ruda in Romania) show that windlasses

120 Well-preserved miner's ore-bucket of esparto grass, with wooden cross-bar handle and reinforced sides. From Aljustral, Portugal.

115

had been introduced in some of the larger Roman complexes.[177] As to the methods of working, detailed studies at Laurion have made it possible to assess the technical problems that emerged there as the upper strata were cleared, and mining proceeded at lower levels. Many years ago, in a comprehensive study, the French archaeologist Ardaillon estimated that to dig a shaft 100 m deep with an aperture of 2 m² would take one man more than three years, working an eight-hour day for 365 days in the year; if all the rock consisted of the more easily worked schist, two men could do the job in twenty months. He did not, however, explain how two men could operate within such a confined space! He also established the fact that the earliest known shafts are younger than the galleries to which they give access. It looks as if the concessionaires of Laurion could not afford to provide the miners with more air or greater working space; their job was to pass along the cramped galleries and pursue what the French writer describes expressively as 'the fleeing wealth' (Ardaillon, 37; cf. pl. 3). But working at these depths set up problems of ventilation and extraction which could not be ignored; a miner at rest uses *c.* 780 litres of air per hour; at work these requirements are trebled. Twin shafts make their appearance, providing simultaneously both improved ventilation and easier removal of the ore. In the absence of a forced draught, however, the device of a second shaft would be only partially effective. No ancient account of shaft-sinking has survived, but Pliny, in a survey of natural water supplies, refers to fire-setting as a means of ventilation, without mentioning the obvious hazards involved; he also makes a passing allusion to the risk of asphyxia during well-sinking operations:

When wells have been sunk deep the well-diggers are killed if they encounter sulphurous or alum-laden fumes. A test for this hazard is to let down a lighted lamp and see if it goes out. If it does, vents are made in the side-walls of the well, to right and left, to take off the poisonous gas. Apart from these noxious substances, mere depth makes the air oppressive; it can be improved by fanning with linen cloths. (*H.N.* 31.49)

Roman gold mining

This aspect of the exploitation of important natural resources has, like other Roman achievements in technology, been commonly either taken for granted, or simply ignored (the topic is not even mentioned in Hodges 1970). Yet there is valuable information in Pliny's *Historia Naturalis,* and the recent publication of the results of systematic exploration of a number of sites in north-west Spain and Britain by Barri Jones and others[178] has opened a new chapter in the study of what the most recent contributor to our subject has rightly called 'this vast and important area of ... technological achievement' (Healy 1978, 13), providing both an opportunity for close assessment of the level attained, and a basis for further research. From the many sites investigated in what was the richest gold-mining area in the Peninsula, we have selected that of Las Medulas which lies *121* some 60 km to the west of the military colony of Asturica Augusta, the modern Astorga. The alluvial deposits here are very extensive, reaching an altitude of around 1000 m at their highest point. The site can be closely dated by artefacts and coins, and the surviving aqueducts, gullies, tanks and other features have made it possible to draw a very well-defined picture of the methods employed in recovering the metal. The site, which was exploited at various levels from the escarpment at the head of the valley, was served by no less than six aqueducts (the three largest being between 2 and 3 m broad), which brought the water from a source some 20 km distant, delivering on the site some 34 million litres (nearly 7 million gallons) per day. The water was used in at least two distinct ways: first, it was played directly on to the deposit in a continuous stream, in the same way in which ground sluices were used in nineteenth-century mining; by the second method, the water was directed into a number of large storage tanks, from which it was discharged in

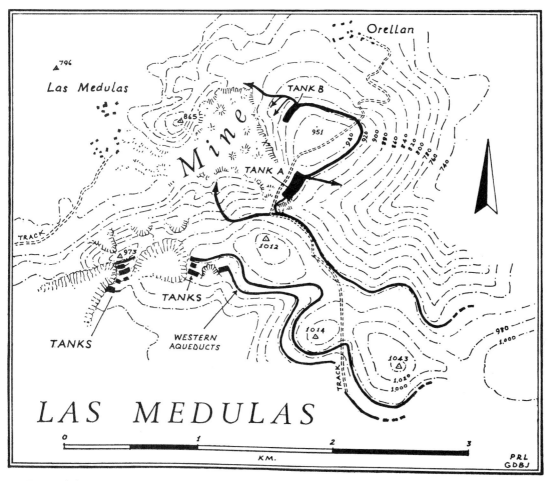

121 Las Medulas, Spain, mine complex. General plan showing the location of the mine, the aqueducts leading down to it, and the hushing tanks.

a series of waves, to remove the overburden. This latter process, known technically as 'hushing', is still used, and for exactly the same purpose, in modern tin mining, with the greatly increased velocity provided by jets of water discharged from 'monitors' (as for example in the tin mines at Jos in Nigeria). Removal by hushing is attested by Pliny in a lengthy account of gold mining in Spain (*HN* 33.66–78 (at 74)), which describes other technical processes related to gold mining; portions of the chapter are cited at the end of this section. Hushing not only faciliated the recovery of gold from deposits close to the surface; it also aided the prospector. As we have seen (above, p. 113) geological knowledge was very defec-

tive in antiquity. Had the sequence of the stratification been known, the mining of deeper and richer deposits would have begun much earlier; in fact, much of the prospecting was a hit-or-miss affair. But recent work at the Dolaucothi gold mine in North Wales[179] has thrown new light on the problem, suggesting that development there took place according to the following sequence, (1) small-scale attacks on the gold-bearing veins in the prehistoric era, stimulating (2) extensive Roman exploitation, using more sophisticated techniques.[180] In this phase, exploratory hushing exposed other neighbouring deposits, around which the main development took place. Further research in other known, but as yet unexplored, areas should

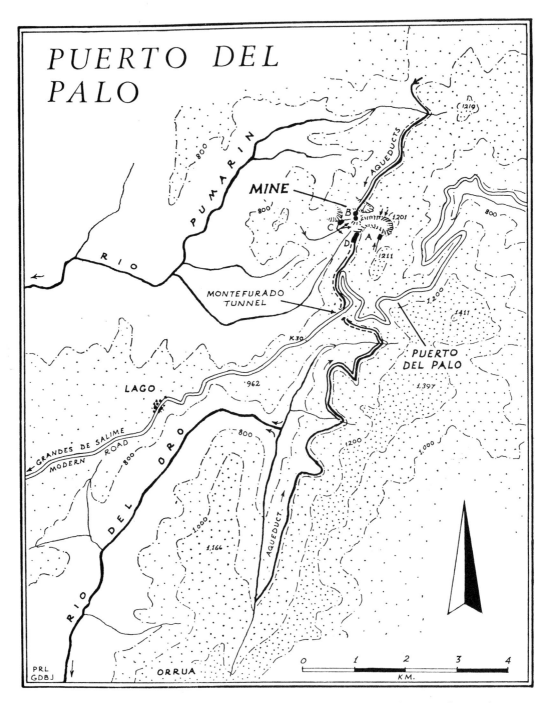

122 Puerto del Palo, Spain, mine complex. General plan showing southern aqueducts and the water-tunnel of Montefurado.

make it possible to plot the geographical distribution of the workings more accurately, distinguishing the zones of Roman exploitation from those of pre-Roman date, which were abandoned with the introduction of a much more efficient technology.

Hydraulics and mining

122 The predominant part played by hydraulic engineers in the development of Roman gold mining is strikingly illustrated from another site recently surveyed[181] which differs from that of Las Medulas in geological structure. Puerto del Palo lies to the north, about half way between Las Medulas and the Atlantic coast. In contrast to the alluvial deposits of the latter mine, it presents an excellent example of an open-cast vein mine, which is equally remarkable for the scale and sophistication of the hydraulic methods employed. Here two aqueduct systems have been traced, one of which made use of the water from an adjacent valley to that of the main site by means of a tunnel. The extreme height of the hard-rock deposits meant that local streams gave an insufficient head of water, so that the engineers proceeded to tap the more plentiful resources provided by the much larger catchment area of the adjoining valley. In the light of the evidence from these impressive workings it is no longer possible to brush aside Pliny's account of the extraction processes as highly coloured rhetoric. We conclude our survey with some extracts from his account, using the admirable if somewhat free rendering of Lewis and Jones, whose running commentary shows convincingly how closely an ancient text and an excavation can correspond.

Pliny on gold mining

(extracts from *HN* 33.70–76)

The third[i] gold-extraction method rivals the projects of the legendary race of Giants. By the light of lamps long galleries are excavated into the mountain. The lamps measure the shifts, and the men may not see the light for months on end. Miners of this class are termed *arrugiae*. In them sudden collapse can crush

the miners, so that diving for pearls or dyes was a safer job. As a protection the overburden is supported by rock arches[ii] at frequent intervals. Whatever the underground mining methods employed, hard quartzite masses will be encountered. These can be split with the help of firesetting involving the use of acid.[iii] More often firesetting in adits makes them too hot and smoky; instead, the rock is split by crushing machines incorporating 150 lb weights.[iv] The miners then carry the ore out on their shoulders, each man forming part of a human chain working in the dark ... Yet rock of this kind is considered relatively easy in comparison with *gangadia,* a form of conglomerate that is almost insuperable.[v] The method used in this case is to attack it with iron wedges and the crushing machine mentioned above ... When the work is completed, the miners cut away the roof props, beginning with the last. A crack, seen only by the watchman perched on the slope above, is the prelude to the collapse. With a shout or wave he orders the miners away and leaps to safety. The overburden falls apart as it collapses with an incredible crash and blast of air.[vi] Equally laborious and more expensive is the associated problem of running aqueducts mile after mile along mountain ridges to wash away mining debris ...[vii] The problems are innumerable; the incline must be steep to produce a surge rather than a trickle of water; consequently high-level sources are required. Gorges and crevasses are bridged by viaducts; elsewhere protruding rocks are cut away to allow the placing of flumes ... On the ridge above the minehead reservoirs are built measuring 200 feet each way and 10 feet deep. Five sluices about a yard across occur in the walls. When the reservoir is full, the sluices are knocked open so that the violent downrush is sufficient to sweep away rock debris.[viii] Another task awaits on level ground. Water conduits are cut in steps and floored with gorse, a plant resembling rosemary, that collects gold particles. The conduits are boarded with planks and carried over steep pitches. Thus the tailings flow down to the sea and the mountain is washed away.[ix]

NOTES

[i] The reference here is to deep mining; the first is to placer mining, and the second to open-cast. [ii] The 'pillar-and-stall' method still used in the deep-seamed coal mines at Witbank in South Africa. [iii] The word is *acetum,* which normally means vinegar. The method

is said to have been used by Hannibal in his passage of the Alps, and is commonly dismissed as ineffective. But any acid, if poured into surface cracks induced by firesetting, will produce splitting. [iv] The phrase used *(fractaria machina)* is not found elsewhere. It could imply a battering ram with an iron head; such a machine could dislodge pieces of the conglomerate rock mentioned below (72 *init.*) as needing iron wedges in addition. I think Pliny has mistakenly reversed the treatments. More appropriate treatment for the hard quartzite; conglomerate rock breaks up more readily [but not all types of conglomerate; some can be very hard, the degree of hardness depending on the degree of cementation—K.D.W.] [vi] 'From the mention of the watchman it appears that this is not a description of mistakenly reserved the treatments. More appropriate the creation of adits that are collapsed in the early stages of an open-cast.' (Lewis and Jones, *JRS* 60 (1970), 183). [vii] Using aqueducts to provide water for washing away debris is surely the process known as hushing (above, p. 117). [viii] 'Reference to the aqueducts serving Las Medulas (above, p. 116) shows that this description may not be as rhetorical as it first appears.' (Lewis and Jones, *loc.cit.*). [ix] Lewis and Jones note that Pliny has conflated two processes: (1) stepped washing-tables cut in the ground, as found at Dolaucothi, and (2) wooden washing-tables with boarded sides, as used in nineteenth-century Californian mining, and known there as 'Long Toms', 'cradles', or 'rockers'.

Metallurgy

References to washing equipment in the concluding section of Pliny's account leads naturally to a discussion of the intermediate processes between extraction and smelting. Strictly *124* speaking, the sifting and washing of ores are preparatory phases of metallurgy; but, as Forbes rightly points out, these operations 'were mostly conducted at the pit-head, and not in metallurgical centres' (*Studies*, VIII, 218). This is still the practice today, the 'mineral-dressing' process being carried out as near as possible to the pit.

Pounding

Although the preliminary pounding by hand was still practised in Roman *Dacia* (Romania), and as late as mediaeval times in Britain, it was

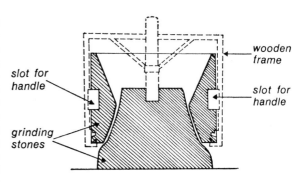

123 Ore-grinding mill in section. Similar in design to the 'Pompeian' grain mill and possibly derived from it, it has the same design of lower stone *(meta)* but two separate stones set in a wooden frame: each of the two wooden handspakes was turned by two other workers.

generally superseded by pounding in mortars with iron-clad pestles, and reducing to powder in hand-querns or hand-operated mills.[182] The *123* extent of the treatment at this stage depended on the amounts and varieties of impurities to be removed before smelting: 'The selected grains, when crushed, often contained other compounds, for example, in the case of argentiferous galena, varying amounts of quartz, blende, pyrite, fluorite, iron compounds and other impurities' (Healy 1978, 143).

Washing

The simplest method of removing the unwanted material from silver ore as described by Strabo (3.2.10) entailed as many as five successions of washing and crushing, the lighter waste materials being sieved out under water.[183] Not surprisingly, these primitive methods were found to be totally inadequate to cope with the through-put of argentiferous ore to be processed in the heyday of the Laurion complex; here several washing plants have been excavated, and recent experiments have shown that the techniques employed were remarkably efficient; water stored in cisterns was run off into long tanks, from which it was released in jets on to the pounded and granulated ore which had been previously shovelled into serrated troughs to be washed by hand. A system of channels and settling tanks enabled the ore to be graded automatically for density and purity; the water could be recycled, a very important feature where large quantities of water were needed in a virtually rainless area, as Healy points out in his detailed and well-illustrated survey (1978, 144–48).

Smelting

The conversion of the various metalliferous ores as recovered from the earth into refined samples ready for industrial or commercial use involved a great number of different, and in the case of iron, very complicated, processes. To illustrate the range and complexity of these processes we may notice that the related topics of mining and metallurgy occupy one-third of the volumes of Forbes, *Studies,* viz. VII–IX, separate chapters being devoted to each of the metals in common use. Within the limited space at our disposal we shall perforce concentrate on three only of the metals, namely gold, copper/-bronze, and iron/steel, prefacing our treatment with a brief review of the major processes and the prime techniques, which include hammering, tempering, quenching, annealing, as well as smelting and cupellation. These processes can best be grasped, not by defining them in the abstract, but by following through the development of a single metal, namely copper. In gold production the ore is collected and crushed; the gold particles are then separated from the enclosing material by washing or panning, and then the separated parts are melted together into a lump. The processes are tedious but uncomplicated: at no point do we encounter what Forbes describes as 'the most important discovery in metallurgy, namely the 'working' of the actual ore to produce the metal' (*Studies* VIII, 24). In the case of copper five phases of technical development are to be distinguished, some of them complicated: (1) shaping lumps of naturally occurring copper; this primitive operation requires techniques no more complicated than those used by early man in working with bone or stone; (2) annealing natural copper. Here the application of heat serves to make the metal much more malleable and easier to shape; it can also be hardened without becoming brittle; similar results were obtained, as we shall see, in the manufacture of wrought iron and steel; (3) smelting oxides and carbonates. Copper ores occur in a variety of chemical and physical combinations, but the separation of the metal from even the more tractable compounds (oxides and carbonates) requires greater heat than that of an

124

124 Ore-washing table at Agrileza, Laurion (second half of 5th century BC) showing washing table sluice and enclosing wall (see reconstruction, *ill. 25*).

open fire, and the required temperature can only be reached in a furnace with an induced draught. It now seems likely that this very important step towards controlling metallurgical processes was a transfer from the making of kiln-fired pottery, and more specifically from the use of copper in the process of applying a glaze to the surface;[184] (4) melting copper. The reduction process of Stage 3 requires temperatures of 700°–800°C, but *125* nearly 1100°C are needed to reach melting point. The melting of copper marks a new and important stage in metal technology, leading to the process of casting, and so to the fashioning of new shapes for implements and weapons; (5) smelting sulphide ores. The most abundantly occurring copper ores are the unweathered primary deposits which contain sulphur. Ridding these ores of their sulphur is a more difficult business: the ore must first be 'roasted', that is, subjected to moderate firing under open, oxidizing conditions; the exploitation of these ores came late in the history of copper technology; the first four stages had been reached long before classical times, but from the scanty evidence it would seem that Stage 5 had been reached by the Late Bronze Age, if not earlier (see Forbes. *Studies* VIII 30).

Improvements in the classical period

The main contribution of the Greeks undoubtedly lay in the working and casting of bronze. The predominance of this alloy of copper and tin is reflected in the language, the single word *chalkos* being used indiscriminately for both copper and bronze. It has been suggested that the well-known supremacy of Corinthian bronzes over those made elsewhere in Greece may have originated in an accident, when a Corinthian smith made the discovery that bronze containing a high percentage of tin could be forged in the cold state, the sample being raised to a given temperature and held there for a given time, then quenched in the famous spring of Peirene, the success of the experiment being erroneously ascribed to the sacred water instead of to the method employed. This unfounded assumption that bronze can be hardened by quenching is found in many classical texts from Aeschylus (*Agamemnon* 621) onwards.[185] Perhaps the story was circulated to keep an important trade secret.

The Roman contribution to the development of copper technology lay in the large-scale exploitation of the more difficult sulphide ores. There is some evidence for the view that the impetus was provided by the exhaustion of the simpler ores. To produce pure copper from sulphide ore requires no less than five treatments: first, roasting to remove the arsenic and antimony compounds, together with most of the sulphur; next, smelting with charcoal in the furnace, which produces a mixture known as 'copper matte', containing only smaller impurities; the third stage is a second smelting with charcoal and certain fluxes which remove the remaining impurities, forming a slag which covers the product and prevents oxidization of the metal. At this stage the copper is only 65 per

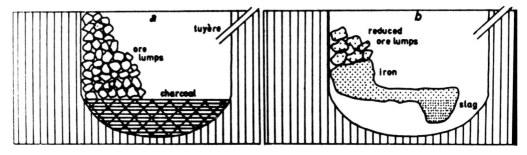

125 Bowl furnace (experimental) constructed and used by Wynne & Tylecote (1958) in the form of a small bowl or crucible, using a small *tuyère* at various angles, reaching a temperature of 1150 °C.

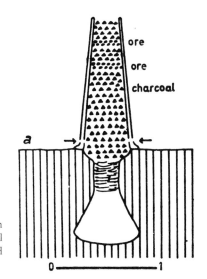
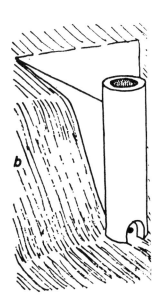

126 Reconstruction of a Roman shaft furnace used for experimental iron-making by Tylecote 1960 and more recently by Cleere 1971.

cent–70 per cent pure, and must be resmelted with charcoal, fluxes, and a blast of air. The product of this fourth stage, known nowadays as 'blister copper' or 'black copper', is now about 95 per cent pure. This is the form in which it is usually traded and passed on to the coppersmith for the fifth and final stage, that of refining.[186] Analysis of copper ingots from Wales indicates 'skilful smelting of pure oxidized ores' (Healy, 160).

Associated with the Roman period are two important technical developments, each a by-product of copper production. The first was the use of the zinc oxide which condensed on the roofs of furnaces during the melting of the ore to produce a brass of high quality, the second a method of extracting gold from copper during smelting, by the process known as 'liquidation', which makes its first appearance during the first century BC.

Refining of gold:
a Roman technical advance

As we saw earlier (above, p. 113) gold can be released from the matrix in which it lies embedded, either by washing away the deposits ('placer gold') or by crushing the conglomerate. From this stage to the refined product the process is much less complicated than in the case of copper, since the structure of the metal makes it

possible to remove the impurities by heating in a special type of porous pot, known as a 'cupel' (hence the term 'cupellation' for this part of the process), in which the base metals are absorbed by oxidation into the walls of the crucible. In the case of a naturally occurring lead-gold alloy the ore is heated in air on a bone-ash hearth: the oxygen causes the lead to change into litharge, which is removed partly by blowing from the surface of the molten metal, partly by absorption into the bone-ash, the gold remaining in the cupel. An improved process, in which mercury is used as the extracting agent, appears to be a Roman invention. The method used in this process was to crush the richer gold ore, then mix it with mercury, which dissolved the gold, forming an amalgam. This product was next filtered through leather and distilled: the gold was left behind in the still, and the valuable mercury recovered for further use. This process is described by Vitruvius (*De Arch.* 7.8.4), as used for recovering the gold from gold-embroidered cloth,

When gold has been woven into a garment, and the garment, being worn out with age, is no longer fit for wear, the pieces of cloth are placed in earthenware pots and burnt over a fire. The ashes are then thrown into water and quicksilver is added, which attracts all the particles of gold and combines with them. The water is then poured off, and the residue emptied into a cloth, and squeezed in the hands; whereupon the quicksilver, being a liquid, escapes through the loose texture of the cloth, while the gold, which has been concentrated by the squeezing, is found inside in a pure state.

This important development is evidently connected with the discovery, first described by Theophrastus (*De Lapid.* 60), of a simple chemical method of recovering mercury from its ore (cinnabar) by roasting in a furnace fitted with a device for condensing the vapour; the process was first used in Spain around the end of the first century BC.[187]

Lead and silver

The production of silver must be considered in close connection with that of lead, since about half the silver produced today comes from lead ores; in antiquity the proportion was higher. It has been inferred from samples of lead mined and processed in Britain (a major source of the metal from soon after its conquest in the first century AD), that its silver content, unlike that of the Laurion galena, which yielded up to 60 oz to the ton, was normally insufficient to make recovery of the silver worth while. Lead, on the other hand, held an important place in the Roman economy, being used extensively for water pipes, especially in the ubiquitous bath establishments, for the lining of cisterns, and the sheathing of the hulls of ships, as well as being alloyed with other metals in the manufacture of instruments.

Production of lead[188]

To produce lead, the sulphide ore, known as *galena,* must first be roasted to convert it into lead oxide. The oxide can then be reduced, and the lead released by heating in an ordinary domestic fire (the melting point of lead is a mere 327°C), the oxygen combining with carbon in the coals. In Britain, where mining was organized soon after the invasion of AD 43, beginning in the rich Mendip area, and fanning out later into Wales and the Midlands, the fact that the simple techniques of open-cast working and open-hearth processing could be suc-

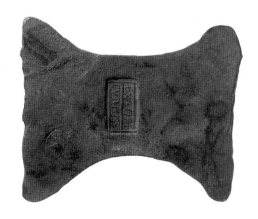

127 Roman silver ingot, 4th century AD, from Kent.

cessfully applied there, may well account for the rapid spread of the industry. Pliny's *Historia Naturalis* must have been completed by AD 77, yet his only reference to British mining is a statement that lead is so plentiful in the island that production has to be restricted by legislation (*HN.* 34.164). Unfortunately he offers no reasons for this move. Both open-hearths and furnaces are attested in Britain, 'in the open-hearth furnace the lead could be ladled out after the removal of slag and residual charcoal. In larger furnaces it was tapped from a hole in the side, or from a bowl-shaped hollow' (Tylecote 1962, *125* 77 f.) The well-known 'standard' lead pigs bear- *126* ing imperial stamp marks, will have been produced in large furnaces, from which the lead was run off into clay moulds by a continuous process. The Mendip region reveals a history of prolonged and continuous production extending over some three centuries, associated with a mining settlement of considerable size.[189] The waste heaps have yielded numerous specimens of the famous pigs, the inscriptions on which give valuable information on the organization and development of the lead industry.

Extraction of silver

The lead sulphide ore as extracted from the mine was first roasted to get rid of some of the sulphur, and then heated to a higher temperature, which further reduced the sulphur content, allowing the lead-silver alloy to form at the bottom of the furnace, where the charcoal prevented it from being oxidized. To remove the lead, the alloy was then melted in a porous crucible, and exposed to a blast of air, which oxidized and removed the lead. The extraction was not all that efficient: some Laurion slags have been found to contain 10 per cent lead. As we have seen, the silver content of galena varies greatly: thus, while in Britain recovery of the low silver content was not attempted, silver was later profitably recovered from earlier Laurion slags, as Strabo tells us:

The silver mines of Attica were originally productive but they declined. But those who worked them later, after getting poor results from their mining, remelted the old refuse or slag, and were still able to extract pure silver, since the workers of earlier periods had been unskilful in heating the ore in the furnaces. (9.1.23)

Production of iron[190]

'Strange to tell, when the ore is melted, the iron flows like water, and breaks up when solidified into spongy masses' (*H.N.* 34.). This sentence, which rounds off Pliny's somewhat confused and scrappy account of the varieties of iron and of variations in furnace practice, provides a good introduction to the complex problems of iron-making in classical times. The melting point of iron is very high ($1540°C$); if it occurred as a pure oxide it could be reduced at the comparatively low temperature of $800°C$. Unfortunately iron ores contain, in addition, considerable quantities of useless minerals which combine during smelting to form slag (see above, p. 121, f.s.v. 'copper making'). The smelting temperature is also high at $1150°C$, with the result that iron is produced in the form of a 'bloom' (Pliny's 'spongy mass' is an apt *126* description of it), from which the slag drains away to some extent, leaving a residue to be removed by a tedious succession of hammerings and reheatings. No wonder the resultant product is given the name of 'wrought' iron! This forging process was speeded up in the Middle Ages after the introduction of water-driven hammers, but this valuable extension of

water power, like that of the stamping mill, was not achieved in Roman times (see Ch. 5). Iron-making can also be made easier by the use of suitable fluxes, such as lime, which help to break up the slag. Did Greek and Roman smelters use fluxes? The question is not easily settled, since (and this is surprising) there is very little evidence to hand from the vast number of known sites about the constituents of the slag found *in situ*. Some literary references to so-called 'fire-fighting stones' *(pyromachos lithos)* have been taken to refer to the use of a fluxing agent,[191] but the suggestion has been vigorously denied. Analysis of residual slags might help to resolve the question.

From wrought iron to steel

In the process of smelting with charcoal the ore absorbs carbon; in an ancient furnace the amount absorbed will have varied according to where the ore was placed in the furnace, the ore in the upper portion absorbing more than that situated near to the source of the draught. This accounts for the observed variations in the carbon content of blooms (for recent experiments see below, p. 126). At the forging stage the smith, stage engaged in making tools or weapons requiring a higher carbon content for cutting edges, and a lower for welding, will have been able to take advantage of these facts in placing his partly forged implement or sword in the hearth. It is difficult, except on some such assumption, to account for the remarkable distribution of areas of high and low carbon content in some surviving artefacts (e.g. the Great *128* Chesterford scythes, below, p. 126). *125*
126

Hardening methods

'Iron that has been heated deteriorates, unless it is hardened by hammering; iron cannot be forged when red-hot, nor indeed until it has reached the threshold of white-heat.' Thus Pliny (*HN*. 34.149), with a good sense of observation, but, of course, with no knowledge of the structure of the metal, still less of the changes which cause it to deteriorate. What gives steel its hardness is the presence of the compound iron carbide or cementite. The latter, however, is unstable; if the iron is allowed to cool slowly, it loses its cementite. Hence the method of quenching, that is, plunging the white-hot steel into water, which secures retention of the cementite, and, therefore, of the hardness at low temperatures. Thus treated, the steel is still too hard and brittle for many purposes, e.g. for a sword or scythe-blade. To obtain the necessary flexibility the steel was tempered, that is, hammered, reheated and quenched a second time; or annealed, that is, reheated and cooled. Tempering by quenching was, of course, well known to metallurgy long before the Romans; when Odysseus and his companions put out the eye of the Cyclops Polyphemus, the sound made by the sizzling eyeball is compared with a familiar scene from everyday life:

Just as when a smith plunges into cold water a mighty axe-head or an adze to temper it – for this is what gives strength to the iron – and it hisses violently, just so did his eye sizzle around the olive spike. (Homer, *Odyssey* 9.391 ff.).

Until recently, however, the extent of knowledge possessed by Greek and Roman smiths

has been largely a matter of speculation; were the Romans, for example, able to distinguish between areas of high and low carbon content? Did they understand the process of hardening by quenching and tempering by reheating? Detailed examination of a Romano-British scythe-blade in a hoard of iron implements found at Great Chesterford near Cambridge established the following points: (1) the cutting edge appears to have been formed from a number of pieces of bar iron of high carbon content folded and joined together; (2) the backing strip and flange were similarly formed from pieces of lower carbon content; (3) the grain formation at the areas of junction shows that the lower carbon portions had been lapped over the higher; some sort of flux may have been used to unite the two parts.[192] It would be hazardous to draw general conclusions from a single investigation and to assert, without examining many more samples, that Romano-British smiths worked successfully without the aid of quenching and tempering, but of the quality of the product, as revealed by the intensive studies carried out by the Iron and Steel Research Institute, there can be no doubt.[193] Nor should we regard these scythe-blades as examples of isolated virtuosity; some years ago the discovery and drainage of the Wallbrook, an ancient tributary of the river Thames, produced a varied assortment of Roman tools, many of them doubtless accidentally dropped from the quayside or maybe from lighters moored alongside. Among them was a rebate chisel, remarkably well preserved as it lay in the tannic acid solution leached out from the bark of the hollowed-out tree-trunks which had been used to line the river bed. The finder, a carpenter by trade, was intrigued by its fine state of preservation, and was given permission to take it home. He returned with the chisel the following morning, reporting that a good cleaning, followed by sharpening and honing, was all that had been needed to restore it, making the instrument capable of doing once more the job for which its original maker had intended it almost two thousand years before.

128 Cross-section of Romano-British scythe blade from Great Chesterford, showing to left, flattened backstrip; at upper right, blade section enclosed and forged into central section.

10 Land Transport

In surveying the technical aspects of land transport, among the numerous questions to be answered three are of outstanding importance: first, what sources of power were available; secondly, how was the power applied (by load-carrying men or pack-animals, or by tractive effort, as with animal-drawn vehicles), and finally, what was the role of land transport in the economy relative to transport by sea, river or canal?

The answer to the first question is that for the movement of goods by land the civilizations of antiquity were limited to the muscle-power of men and animals. In looking at the various technical devices employed to develop the use of these basic sources of energy, we must remember that techniques are closely related, and indeed interdependent. The pack-saddle competes with the harness, bovine with equine traction. River and coastwise transport restrict the role of animals,[195] and where adverse conditions of terrain are found in combination with a low level of technology, as in the tropical rain forests, head-loading and porterage dominate the scene.

In looking at the second question we must not neglect the play of economic forces in promoting some forms of transport and restricting others; 'in proportion as a particular technique is better adapted to geographical conditions and is able to more goods at a lower price, it pushes other methods into the background'.[196] The answer to the third question will emerge in the course of the following discussion.

Techniques of porterage

In simple, undeveloped communities the cheapest and most ubiquitous form of land transport is a man or woman with a load carried on the head, shoulders or back. The simplest device for increasing the load, a pole across the shoulder, is well illustrated on a bas-relief now in Munich,[197] showing a peasant farmer arriving at the town gate to market his produce. In front is an ox, with a pair of lambs slung pannier-wise across its back; behind walks the farmer, with a basket of fruit in his right hand, and a pole across his left shoulder, from the end of which is suspended a pig. To this common category belongs the well-known load-carrying pole attached to the legionary Roman soldier's back, invented, it was said, by the general Gaius Marius (hence their nickname of 'Marius' mules').[198] Very common too was the neck-yoke (Gr. *anaphoron,* Lat. *iugum*), with a basket (Lat. *sirpiculus*) suspended at either end, like the milkmaid's yoke (Varro, *De Re Rustica* 2.3.9).

The absence of the wheelbarrow (a mediaeval invention), and of the porter's trolley is significant: the former is a fairly simple extension of the lever, and the latter a miniature wagon; where there is an abundance of cheap man-power, load-easing devices do not develop. Loads too heavy or too bulky for one man to handle are divided between two (e.g. a wine-amphora, weighing up to 55 kg when full, on a pole, and a load of manure on a hurdle).[199] Human load-carriers have a physical advantage over their animal competitors: a porter cannot cope with loads exceeding about 25 kg (55 lb) (except over very short distances), which is less than a quarter of what an average-sized panniered mule could manage, but he is much more versatile; and has the added advantage that he 'is, so to speak, self-loading and self-unloading'.[200] This adaptability is clearly illustrated in surviving Roman monuments, which show

129 Harbour work being carried out by six-oared tug *(navis codicaria)*, Ostia Antica, wall plaque from a mausoleum, Isola Sacra, 3rd century AD.

dockers off-loading and carrying ashore cargoes consisting of grain (in sacks), wine and oil (in *127* amphorae) and metal (in ingots).[201] In the early Empire Rome's annual importation of grain from Egypt, Sicily and Africa involved what was by far the most massive handling operation known to the classical world, viz., off-loading from sea-going freighters (see *ill. 129*), storing at the port of Ostia, and reloading into river barges for the final stage up the Tiber around half a million tons of grain, in sacks weighing 28 kg (60 lb) apiece. This formidable operation will have required some 17 million sackloads, all carried on the backs of *saccarii* without any mechanical aids.[202]

Where human muscle-power is cheap and abundant, it continues to prevail over more sophisticated sources and techniques; indeed its use persists in many parts of Africa and Asia today, alongside advanced forms of transport based on modern standards of highway and vehicle construction.[203]

Animals in land transport

The use of animals in transport falls into two main categories: the first consists of animals as load-bearers (e.g. the pack-horse which played an important role in the economic life of north-west Europe until the beginning of the nineteenth century), and the second, that of animals as loadpullers (e.g. the yoked draught oxen of southern Europe).

Pack-animals

Load-carrying mules and donkeys, equipped *132* with pack-saddle or panniers, played a far more important role in land transport than animal-drawn carts. There are two reasons for this: first, the high proportion of broken terrain to be found in most parts of the Mediterranean region lengthens the odds against any type of wheeled vehicle; in lands of harsh and changing relief, where deep gorges have to be crossed, the pack-animal is far more at home. Writing of his *Wanderjahre* in 1911–12 Arnold Toynbee reckons that he must have walked somewhere between 2,000 and 3,000 miles in the time; 'in walking,' he writes, 'I was following in the footsteps of the ancient Greeks themselves ... In the Greece of Philippides the runner and Pausanias the tourists' guide, roads capable of accommodating wheeled traffic were as rare as they still were in the Greece of 1911–12.'[204] The second reason is economic: where small-sized agricultural units are the norm, the pack-animal makes fewer demands on limited resources than a good draught-animal.[205] That this form of transport was highly competitive can be seen from official Roman sources: the regulations of the Theodosian Code give a loading maximum of 490 kg (1075 lb) on the state-controlled service for goods wagons (*Cod. Theod.* 8.5.17), a load which could easily be carried by four panniered mules. Supporting evidence has recently appeared in a newly discovered inscription from Salagassus in Asia Minor.[206]

In assessing the part played by pack-animals in land transport we must look next both at the range and weight of goods they carried, and the ways in which they were loaded: the questions are interconnected. The most primitive method, still prevalent in many parts of the world, is to put the load on the back of a donkey and rope it on. Heavier and bulkier burdens can be carried if the load is suspended on either flank, but this demands some form of saddle to prevent chafing, and an even distribution of the load. Two types of pack-saddle were used in classical times: (a) a wooden frame, covered with leather or cloth, with rings to hold the loads; (b) a covering of soft but durable material moulded to the animal's back, and opening out into a pair of capacious bags or panniers, usually woven from soft basketry.[207] The weight limits for each category of animal can only be roughly estimated, since there are no statistics, but on the basis of more recent figures it would seem that donkeys will have been able to carry loads of 70–90 kg (150–200 lb) and mules between 90 kg and 136 kg (200 lb 300 lb). As for the contents of the loads, literary references are rare but instructive, the most important being a passage in Varro's *De Re Rustica* (the author was himself the owner of several large stud-farms for the breeding of mules), 'the trains' (of pack-asses) 'are usually formed by the traders, as, for example, those who bring down to the sea from the region of Brindisi or Apulia oil or wine or grain or other products on panniered asses' (2.6.5).

Table on Land Transport

1 *Traction* (1 h. p. = 33,000 foot lb)

	Units
Horse walking at 2½ mph	22,000
Horse pulling cart 2½ mph	26,000
Pony (10,000 lb)	17,000
Ox	16,000
Mule	11,000
Ass	5,030
Man pushing or pulling horizontally	3,130
Man turning capstan	2,600
Man raising weights by pulley	1,560
Man raising weights by hand	1,480
Man wheeling barrow by ramp	520
Man lifting earth with spade 5¼ feet (1.6 m)	470

2 *Friction*—force required to carry 1 ton on level ground

Railroad	8 lb per ton
Macadam	44–67 lb per ton
Gravel	150 lb per ton
Turnpike road (hard and dry)	68 lb per ton
Turnpike road (heavily gravelled)	320 lb per ton
Loose, sandy road	457 lb per ton

3 *Loads* — porters, pack-animals draught animals

Porters

Man with head load	max 50 lb (24 kg)
Man with back load	max 100 lb (48 kg)
Shoulder load, e. g. Nigerian water carrier	80 lb (36 kg)
European soldier's pack	50 lb (24 kg)
Indian coolie's load	40 lb (18 kg)

Note: Chris Bonington reports (1973) porters on Everest 1972 carrying loads of 16 kg (35 lb) through the Ice Fall (from Base to Camp 7), and 14 kg (31 lb) from Camp 2 up to Camp 6; at lower altitudes normal loading was 36 kg (79 lb).

Pack-animals

Donkey with panniers	150 lb and even 200 lb (70–90 kg)
Mule with panniers	300 lb (136 kg)
Horse with panniers	400 lb (182 kg)
Bullock	400 lb (182 kg)

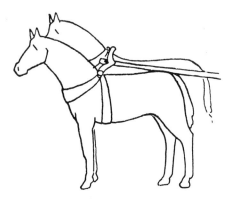

130 Sketch of breast-strap and girth harness, suitable for light chariots: the girth prevents the breast-strap from riding up and affecting the animal's breathing.

Animals as load-pullers

In moving goods from place to place by overland routes, animals may be used as carriers or as haulers. Mules and donkeys appear in both roles, oxen more frequently in the second; but they were used as carriers in classical times, as we have seen in the 'Going to market' frieze in Munich (above, p. 127). Information from written sources is scarce and very inadequate, and we must inevitably rely on a combination of pictorial evidence, which is not inconsiderable, and casual literary references. The agricultural writers have little to tell us, but some of their passing references are important; thus Cato's inventories of stock and equipment include, as we should expect, oxen (for ploughing), and donkeys (for pack-transport, cartage and driving the corn-mill), but neither mules

131 Bas-relief from Langres, northern France, showing a four-wheeled carriage, pulled by four heavy horses driven four-in-hand, 3rd century AD.

nor horses; he also provides (a rarity in ancient writers) precise details of the cost of transporting a complete olive-crusher by ox-wagon from the place of manufacture to the farm.[208]

Role of the ox-wagon

Except in very mountainous districts, where the hoe took the place of the ox-drawn plough, or rather, was not displaced by it (White 1967, 38), his plough-oxen were, to the Greek or Roman farmer, his most important physical asset; when not needed for ploughing, they were employed for heavy haulage, and in this department the ox has a number of advantages over the horse or the mule. The first point of difference is in the manner in which the power is applied in hauling the load; the ox applies his power head down, and the drive is from his powerful shoulders; the best harness for him is the yoke resting on the neck, connected to the vehicle by a centrally placed pole. The horse, on the other hand, exerts the greater part of his tractive effort from the hind legs; the most suitable arrangement for him is a collar resting on the shoulders; but this took a long time to evolve; the throat-and-girth harness, which was introduced as a replacement for the anatomically unsuitable yoke, placed such limitations on the load-pulling capacity of the horse as to rule him out of commercial transport. In addition, horses were both more expensive to rear and maintain, and less adaptable to varying conditions than either oxen or mules, and are, therefore, excluded from consideration. In pulling a loaded wagon along a surface, the ideal conditions for maximum efficiency in propulsion would be achived if the power were applied horizontally, and with no friction between wheels and road-bed. The higher the point at which the power is applied, the greater the effort needed to pull a given load; in technical terms, the lower the 'angle of draught', the greater the efficiency.[209] The lower height of his yoke above the ground compared with that of the neck-collar of a horse gives the ox a further advantage over his rival; slow but sure, his forward thrust on the yoke is about $1\frac{1}{2}$ times his body weight. Finally, he has an advantage of quite another kind, which compensates for the slowness of his pace; his digestive system enables him to convert and utilize an appreciably higher proportion of the protein in his diet than can a horse or mule.[210]

Precise information on the extent to which oxen were used in road transport is limited in amount; single pairs are shown drawing loaded goods wagons on a number of surviving monuments. The use of the public roads for conveying goods in wheeled vehicles was restricted by two factors: its slowness and its high cost compared with transport by water. As to speed, we have some interesting data relating to specific journeys over known distances made by passenger vehicles drawn by horses and mules, which suggest that an average daily run in a *carpentum* will have been some 70 km or 44 miles (*Amm. Marc.* 29.6). We have no comparative data for ox-drawn wagons, but the slower gait of the ox might suggest a daily run around 40 km or 25 miles. Lionel Casson (1974, 188) proposes the following averages for road travel: on foot (on the level) 15–20 m.p.d.; in a carriage, 25–30; if hard pushed, 40–45 m-p.d (but not for a prolonged journey).

By land or by water?

How do the two forms of transport compare? The main advantage of sea transport lies in the fact that a ship, given fair weather, can run many days without stopping; hence a Roman merchantman, sailing at (say) three knots, might average 72 m.p.d., or almost three times the suggested average for an ox-wagon. There was, in addition, a considerable saving in cost, which gave long-distance transport by sea a decisive advantage. Using the evidence of Diocletian's *Edictum de pretiis*, A. H. M. Jones (*LRE*, vol. 2, 841) calculated that a journey of 300 miles with a wagon-load of wheat would double its price, and that it was, therefore, cheaper to take grain from one end of the Mediterranean to the other than to carry it a mere 75

miles by road. This combination of slow speed and high coast is held to constitute a major weakness in the economy of the Graeco-Roman world: it has recently been called 'the most decisive failure of ancient technology'.[211]

If the slow and costly animal-drawn wagon had been the only available means of moving goods by land, it is evident that coastal centres of population would have enjoyed the benefit of cheap communications with others similarly situated, whether in relation to the importation of goods that they did not produce, or for the export of surplus primary products, such as grain, oil or wine, or of the products of their industries. Inland centres, on the other hand, will have been severely restricted in both directions, the effects of the differential costs being more crippling the further the distance between the centre and the sea. This is the picture presented in the overall study of the subject by M. I. Finley (1973).[212] But are these assumptions correct? Was the ox-wagon the sole means employed for the conveyance of goods by land? What share did the pack-animals referred to earlier in this chapter have in the business, and what sort of materials and products were involved in inter-city and inter-regional trade? Three passages may be cited in this connection. The first is from Varro's *De Re Rustica* (I.16.2–3) where the writer is discussing conditions in the neighbourhood of the farm:

Again, if there are towns or villages in the neighbourhood . . . from which you can purchase at a reasonable price what you need for your farm, and to which you can sell your surplus, such as props or poles or reeds, the farm will yield a bigger profit than if you have to fetch them from a distance.

In the second passage Varro (11.6.5) refers to transport by panniered donkeys as a regular traffic in southern Italy, as it still is today. 'Trains' (of donkeys) are usually formed by the traders, as, for example, those who load oil or wine or grain and other products from the Brindisi district down to the sea in Apulia on panniered donkeys.

The third occurs in a letter from the Younger Pliny to the emperor Trajan, in which the writer invites that emperor to support a project to build a canal linking a lake with the sea, and thus reduce transport costs on a major commercial route:

at present [he writes] marble, produce and building timber are transported cheaply and with little effort across the lake and on to the highway, but from there they have to be taken to the sea by cart with much labour and expense. (*Letters* 10.41.2)

Varro's references are to the small-scale but frequent interchange of commodities between farm and town and to the export of grain, etc., Pliny's to materials and products which are known to have been shipped considerable distances from the source of supply.[213] As might be expected, many of the best known quarries lay close to the sea or to a navigable river. But some heavy equipment had to be moved by road, and not all centres of population were on the sea coast; Athens lay seven miles from her port of Piraeus, and Corinth and Argos nearly the same distance from theirs. The building inscriptions from Eleusis give valuable information on the method employed to move the heavy blocks and drums of Pentelic marble from the famous quarries to the temple site 22 miles (35 km) away, and the time taken to complete the job. The loading requirements were met, in the absence of any better form of traction, by multiplying the available units; as many as 37 pairs of oxen were attached in file.[214] But the ox was not the only beast of burden available; there were mules and donkeys, and in certain areas, camels. Pack-animals are significantly included, along with wagons, ships, barrels and skins in the jurist Ulpian's list of essential equipment used in the export of produce from the farm (*Digest* 33.7.12.1). While it would be foolish to assume a priority of pack-animals over animal-hauled transport on the basis of the order in which they are mentioned in the *Digest* list, one may resonably infer that they were at least of equal importance in relation to the handling of agricultural produce. In commercial transport generally, the pack-mule had a prominent role to play, as it still has in the more mountainous parts of the Mediter-

ranean region, such as Spain, Sicily and Yugoslavia. If pack-mules carried similar loads to those carried in the above areas a troop of twenty mules would dispose of a total load about $2\frac{1}{2}$ tonnes, which would be equivalent to five wagon loads. Pack-animals have numerous advantages over draught-animals, provided they are not over-loaded (in southern Europe they invariably are!). They can carry their loads over roads rendered impassable for wheeled transport by storm or inundation, and over mountain trails which in antiquity were often the sole means of communication between populated centres in many Mediterranean countries. The following comments by a former veterinarian, now turned farmer and stock-breeder, make clear, with little need of further elaboration, why mules were preferred to horses: 'a team of mules can pull a heavier load than a similar team of horses. The mule is far stronger than the horse—he is quite indifferent to variations in temperature; he will perform his task under adverse conditions in which a horse would simply give up. He is never sick or sorry; he works longer hours, and all this on food which is far less expensive and easier to find.'[215] (Landels 1978, 172) adds two very important points to the list: (1) the proverbial surefootedness of the mule on rocky terrain; (2) its slow speed (a little more than 3 mph) is compensated for by the fact that it needs little sleep (4–5 hours in 24), and can, therefore, clock around 80 km (50 miles) in a day.

Role of wheeled transport

The literary evidence on the types of vehicle used and the different methods of harnessing is very meagre indeed. Surviving identifiable parts of vehicles are also scarce, and there have been few attempts to reconstruct wagons from surviving parts.[216] By far the best evidence in the way of representations comes from sculptured reliefs and mosaics of the Roman period, passenger vehicles being naturally more plentiful in the record than commercial vehicles.

132 Bas-relief from Langres, northern France, depicting a heavy four-wheeler carrying a capacious wine barrel and drawn by a pair of powerful mules: note the heavy chassis and platform, and the size of the barrel for bulk transport.

Each of the two classes of evidence presents its own special difficulties for the investigator. Apart from funeral chariots buried with their owners, the surviving parts consist in the main of the metal portions, and in the absence of sizeable parts of the wagon frame, reconstruction becomes highly conjectural. Investigation of the pictorial record is hampered in a number of ways; the harnessing is often sketchily presented, and the side-profile view, which is almost universal, means that the wagon frame cannot be clearly seen, so that little light can be thrown from this source on the important question whether the front axles of four-wheeled vehicles were fixed or pivoted. Nor does the archaelogical evidence supply any cut-and-dried solutions to the numerous problems connected with harnessing. The evidence here is disproportionate between the Greek and the Roman representations; on the Greek side war- and racing chariots predominate, and commercial vehicles are rare, while the Roman record is abundant varied for the latter class. Especially prominent are the Gallo-Roman sculptured reliefs, which furnish a fine array of two- and four-wheeled vehicles, both passenger and commercial; in particular, they provide important evidence of experiment in the harnessing of horses and mules. Access to this material has been made easier by the recent publication of a comprehensive range of photographs, which fill one complete volume of a two-volume work on the horse in Graeco-Roman antiquity.[217] Appendix 10 contains a list, with annotations, of the more important of these representations.

Evolution of wheeled vehicles

It is usually held that the animal-drawn wagon has evolved from the wheel-less sledge through the two-wheeler, and that in early wagons the wheels revolved with the axle which will have found it difficult to negotiate corners. But a different line of evolution is indicated by an early rock-engraving from Scandinavia, showing a four-wheeled vehicle frame which appears to have been made by attaching one triangular two-wheeled frame to another, thus forming a kind of 'trailer'. We thus have evidence that suggests two lines of evolution; the rigid four-wheeler with no swivel developing from the sledge, and the four-wheeler with a pivoted front axle from the two-wheeled chariot; articulated vehicles of the second type can be traced in the record from the prehistoric engraving mentioned earlier, through bronze votive models of the Roman period down to surviving specimens in contemporary Russia (the *telega*), and elsewhere in Europe.[218]

A much debated question is whether four-

134

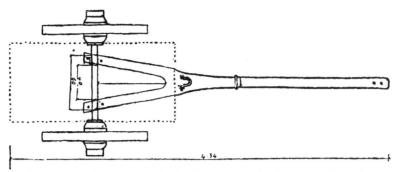

133 Diagram of primitive two-wheeler from the Landes district, southern France, showing central pole terminating in a forked section.

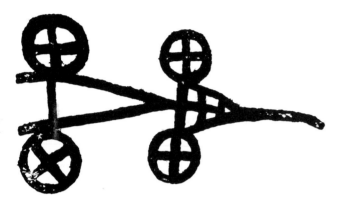

134 Plan of the Dejbjerg wagon.

wheeled vehicles of the classical period had pivoted front axles; the evidence is not conclusive either way. In representations of four-wheelers the front pair of wheels usually has the same diameter as the rear pair, and is usually shown as elevated above the platform or wagon-frame; there are, however, examples from the Roman period which show the frame rising clear of the wheels. On the strength of the first two points above it is argued that the front axle could not turn, or it would have brought the wheels into collision with the frame. But is this contention sound? If the front wheels are set slightly wider than the frame, the wagon would be able to turn on a swivel, as anyone with experience of a home-made soapbox racing car can testify. In an earlier discussion of the problem E. M. Jope (*HT* II, 539) asserted that four-wheelers could not have travelled on rut-roads (see Ch.8, p. 92) unless

equipped with pivoted front axles, and noted that none of the Greek terms for road vehicles indicated the number of wheels. Later he declared that there was no evidence for a pivoted front axle in Roman times. The gist of these somewhat confused statements appears to be that (a) there were no four-wheelers on the road in ancient Greece, (b) Roman four-wheelers had a fixed front axle, and were, therefore, very difficult to manage. Against the second statement we have evidence for the use of pivoted front axles long before the Romans (examples are known from Persia, Celtic Europe and, more recently, as reported by Piggott (1968), from the Caucasus region). There is a four-wheeler on a sarcophagus of the fourth century AD, now in Stockholm (Vigneron 1968, pl. 64), which appears to be turning to the right with the front axle pivoted. So we have a difficult problem! Swivelling axles had been invented long before the Romans, but we cannot prove that the Romans used them. Some light may be shed on the problem by the new fragments of the *Edictum de pretiis* of AD 301, found at Aphrodisias.[219] Among the parts of wagons listed is a *columella*. The diminutive of *columen* (a vertical pillar) is used in another technical context of the pivot fitted to the nave of the Catonian oil-mill *(trapetum)*, on which the upper stone revolved.[220] It is difficult to think of any other use for such a pin, if it is not be identified as an axle pivot. It was certainly an important component, and replaceable! One might add that it is difficult to fit a pivotless wagon into a

63

134

number of well-known literary contexts. How, for example, could a heavy four-wheeled lorry *(serracum)*, with its load of swaying logs so vividly described by Juvenal *(Satire* 3.255) have got throught the winding streets if it could only negotiate bends by skidding round them?

Wheel design

The fact that the wheels are usually clearly visible on the monuments depicting wagons, and that entire wheels as well as portions of them have survived make it possible to study the various types and their evolution in some
135 detail. Wheels vary in design from the solid type, usually made up from segments, through an intermediate type, in which the rim was held in position by a single, heavy transverse bar, supported by a pair of laterals at right angles to the bar, to the familiar spoked type, with multiple felloes into which the spokes were mortised. The primitive solid wheel or drum (Lat. *tympanum)* continued to be used, especially on the farm, throughout the classical period, while the spoked wheel underwent important modifications which enabled it to carry the loads sometimes required of it in commercial transport. The most important of these developments were the separation of the spokes from the felloes (see below) and the growth in size and strength
136 of the nave or hub, which enabled wheels to support heavier loads without deformation.

Wheel construction and mounting

Solid wheels (type a) were probably fixed to a rotating axle, as in a railway truck, while those fitted with cross-bars (b) or spokes (c) were evidently provided with naves or hubs rotating on a fixed axle (cf. the 'stub-axle' of a motor car), with a 'linchpin' passing through the stub. The most common type of wagon, as shown on bas-reliefs and wall paintings, was of the tumbril type, fitted with solid wheels and a revolving axle; hence the epithet *stridens,* 'screeching', commonly applied to a Roman cart on the move.[221] Problems of wear and lubrication are discussed below. The main problems involved in the construction of types (b) and (c) is that of securing rigidity of the bars or spokes, and a tight fit of the rim. Cross-bar wheels were fitted with two semi-circular felloes divided by the cross-bar: 'the hub was formed by strengthening the cross-bar at the central hole with a plate on the outside, and perhaps one on the inside, which involves simpler joinery than a radially spoked wheel' (Jope, *HT* II, 549, who suggests that the directions given by the Greek poet Hesiod *(c.* 700 BC) for cutting the timber for cart wheels may be referring to the cross-bar wheel). Representations of two-wheel chariots of the familiar Greek type have four, six or eight spokes splayed out at the ends for attaching to the rim. The method of attaching spokes to rim is not always easy to decipher, since no com-

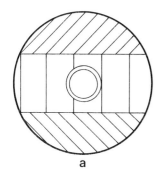 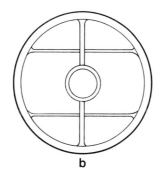 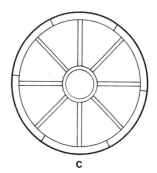

| a | b | c |

135 *a,* 3-piece solid wheel, still to be found in some under-developed regions; *b,* 'steering wheel' type, with single bar containing the hub, and two cross-members; *c,* multi-spoked wheel with 8 spokes and 6 felloes.

136 Wagon wheels with one-piece felloes from Saalburg, Germany.

137 Heavy Roman cartwheel from Newstead *(Trimontium)* Scotland, with thick spokes, 5 broad felloes and an enormous hub.

plete example survives. Early Egyptian and Celtic chariots had their felloes in one piece (Childe, *HT* I, fig. 525; Jope, *HT* II, fig. 482), and one-piece felloes have survived from Roman imperial times. The *Edictum de pretiis* distinguishes between four—wheeled vehicles with multiple felloes and those with one-piece felloes (see *JRS* 63 (1973), 101, 105). Surviving wheels of the Roman period show lathe-turned spokes with a square tenon for the nave and a round one for the felloe, displaying the art of the wheelwright at a high level. In the designing of wheeled vehicles the contribution of Celtic tradition in design and Celtic skill in execution was outstanding; it has often been pointed out that virtually all the Latin names for animal-drawn vehicles (e.g. *cisium,* a light two-wheeler (the 'chaise'), *petorritum* and *rhaeda,* heavier four-wheeled coaches), are of Celtic origin. Wheels fitted with the one-piece felloe, which, as the *Edictum de pretiis* informs us (15.32.35), were standard equipment for the most expen-

sive type of long-distance passenger vehicles, required a high standard of skill in the fitting as well as in the construction; the selected piece of ash had to be bent into a circular form, and the tensioning of the spokes was a difficult operation. Heavier loads need stronger wheels; they also put a heavier strain on the hub or nave. Surviving wheels show this development in design; the two almost complete wheels recovered at the Roman fort at Newstead in *137* Scotland have thick naves 40 cm (15½ in) deep; the same massive design can be seen in a similar series from the fort of Saalburg in Roman Germany. A nave belonging to a four-wheeler of the first century BC from Dejbjerg in Denmark *134* contains channels inside it which have been interpreted as part of a 'roller-bearing' system with wooden rods turning between nave and axle; this, if correct, would represent a major technical advance, but the interpretation is very doubtful.[222]

Rims and tyres

The parts of a Celtic chariot[223] found in the island of Anglesey, and dated to the early Iron Age, include tyres of high quality steel. The absence of nail-holes proves that they were heated and shrunk on before cooling by quenching, a process that has persisted right down to modern times. Iron tyres (Lat. *canthi*) are included in the newly found portion of the *Edictum de pretiis* from Aphrodisias, and were evidently standard on heavy vehicles in the fourth century AD, if not earlier. The two Newstead wheels were fitted with tyres 1 cm (3/8 in) thick, and 4½ cm (1¾ in) wide.

Braking

Good brakes are important, but the little evidence we have does not suggest that braking was satisfactory. A passage in the *Digest*[224] dealing with compensation for a slave run over by a wagon drawn by hinnies that ran backwards down a steep street in Rome may be taken as reflecting a common occurrence; it is significant that no reference is made either to brake failure or to any attempt to apply brakes; the

138 Bas-relief from Klagenfurt, Austria, depicting a covered travelling coach drawn by two horses; no details shown of the harnessing.

load was evidently too heavy for the light draught animals, and the muleteers, who were putting their shoulders to the wagon, are described as clearing out of its path to avoid being crushed between it and another wagon coming up behind.

Lubrication

Here again, we are inadequately informed; but the fact that Latin has a technical term for axle-grease *(axungia)*[225] provides in itself sufficient evidence of its use for lubricating axles! Lack of evidence of its use need not drive us to conclude with Landels (1978, 181) that water was the most commonly used lubricant (and a very ineffective one too!).

Harnessing techniques for draught-animals

Many different methods have been devised for attaching a draught-animal to a loaded vehicle, and many modifications and improvements have been developed since the first appearance of animal-drawn transport. All stem from one or the other of two basic devices—the yoke and the collar. The yoke consists essentially of a narrow piece of planking shaped so as to fit over the necks of a pair of animals; to this is attached either a second piece of wood similarly shaped, the whole device enclosing the animals' necks, or a pair of thongs that pass under their necks and serve the same purpose. Variations include the horn yoke, designed to fit over the animal's horns, and the withers yoke, designed to fit over the withers.

The neck-yoke is the form best suited to the anatomy of the ox, and to the way its power is applied to the load (see above, p. 130), since the thrust can be taken from the withers. To keep the yoke in position, all that is needed is a simple 'throat-and-girth' harness, the girth to hold the yoke down, and the throat strap to prevent the yoke sliding backwards. The yoke was attached to a pole which passed between the pair of oxen, and was fitted to the wagon frame at an appropriate angle such as to cause the animals to push somewhat upwards from the horizontal.

The collar can be either round, designed to fit the throat of the animal, or tapering from a round head to a pointed base so as to rest on the

139 Bas-relief (inco mp lete) showing a shafted two-wheeler with massive wheels and a basketry body drawn by two mules. The shafts turn upwards to meet the high collar.

140 Bas-relief (incomplete) showing a pair of draught mules with high collars pulling a shafted vehicle, fitted with upturned shafts. They are about to pass a milestone on a public road.

animal's shoulders.[226] The collar is the more suitable attachment for equines (horses, mules or donkeys) when used as draught-animals. There is some evidence that, in the Mediterranean area at least, the yoke preceded the collar, so that, when equines were introduced as draught-animals, they were provided with the same throat-and-girth harness. Unfortunately, the round neck-collar has a tendency to ride up the animal's throat, causing, if the load is heavy, some interference with the breathing due to pressure on the windpipe, which, in the case of equines, lies close to the anterior surface of the neck. Several modifications designed to overcome this problem are known from antiquity; they include, in addition to the restraining girth mentioned above, a harness which passes completely round the body at the level of the breastbone, to form what is known as the 'breast-strap' harness. Which of these varieties of har-

141 High-sided travelling carriage, with panelled body and light eight-spoked wheels drawn by a pair of powerful mules. The vehicle has a central pole, and the harness is unusually elaborate. It includes side traces attached to twin-strap collars, broad girths dividing to pass under the belly, hoof covers and elaborate head-pieces.

ness were used by the Greeks and Romans, and to what extent? The literary evidence, as we have noticed, is very scanty. For example, Xenophon's treatise *On Horsemanship* has a great deal of information on the harnessing of riding horses, but nothing about draught-animals, whose activities were of no concern to the country gentleman of his day! We have lists of vehicles of different types, but no details of the ways in which they were harnessed. We, therefore, have to rely heavily on the archaeological evidence, which is considerable in quantity, but unevenly distributed. Most surviving representations of road vehicles depict racing, hunting or fighting chariots; ordinary vehicles, whether freight- or passenger-carrying, are rare. The Homeric war chariot, with its yoke, pole and breast-strap, belongs unmistakeably to the *131* Mediterranean tradition, while the Gallo-Ro-*132* man reliefs, which depict both passenger and

goods vehicles of several types, exemplify a very distinct, non-Mediterranean system, of collar and shafts. It seems that, while the yoke and pole method was first employed to harness oxen, and then transferred to equines, the collar and shafts arrangement grew up in areas where equines predominated, that is in northern Europe. Henri Polge, in a lengthy discussion of these harnessing problems (1967, 20 ff.) suggests that imperial Roman Gaul, which provides the most important representations of shafted vehicles, was the meeting-point of the two systems. Hence we find in this region representations of two interesting types of hybrids—horses wearing yokes instead of collars, and oxen with yokes and shafts, in place of the usual poles.

It is evident that the Roman period witnessed, in the north-western provinces at least, a good deal of experiment with harnessing systems (for details see Appendix 10), but no real invention or innovation capable of overcoming the inherent inefficiency of their equine harnessing, which was one of the factors that restricted the load per unit of draught. Harnessing was not the only source of difficulty; Polge (1967, 28–54) advanced a number of other reasons to account for Roman difficulties with heavy transport. His list includes traction difficulties, including slipping, with heavy loads on stone-paved sections of road; use of steep gradients without either efficient bits to control the animals or adequate braking to hold heavily loaded vehicles. In addition they had neither 'dished' wheels (to cope with the cambered sections of road) nor pivoted front axles (for turning corners)[227] nor ball or roller bearings (to reduce friction between hub and axle).

It is not difficult to see why road transport continued to be dominated by the slow-paced ox, and why pack-animals continued to play an important role. We can also see in full perspective the effect of the introduction of the horse-collar that rested on the shoulder; gradually, not immediately, as can be seen from the mediaeval ploughing scenes, the horse began to take over the ploughing of richer but heavier soils that could previously only be subdued by increasing the number of oxen straining at the yoke.

But the horse-collar brought about no similar revolution on the roads of Europe; indeed, the Middle Ages were a period of decline, not of advance, in this area of technology. The key to the situation is the decline and fall of the road system, of that network of imperial highways, furnished with bridges, tunnels and storm-water culverts, that achieved a level of technical achievement not to be reached again for many centuries to come. Here, to illustrate the effects of that decline right in the heartland of the Roman Empire, is our concluding reference, part of an account, based on contemporary records, of travel conditions in central Italy during the fourteenth century:

The journey from Florence to Naples, which merchants sometimes completed in eleven or twelve days by riding from morning to night and travelling via Terni, Aquila, Sulmone and Teano, bristled with difficulties and dangers. The only way across the Abruzzi was by rough country track, and the borders of Campania were infested by brigands. The shortest route from Rome to Naples, the road through Terracina, [the *Appia Antica*, once the *Queen of Roads*] had such an evil reputation that only troops and public officials dared to use it, *while all trade went round by sea.*

11 Ships and Water Transport

This chapter embraces a wide range of technical problems, including several of prime importance to a proper assessment of the economy of the classical world. Chief among them are those which relate to the design, equipment, carrying capacity, manning and manoeuvrability of sea-going merchant vessels. Until recently the available evidence has been more or less confined to representations of ships on stone reliefs, vase-paintings, frescoes and mosaics, together with accounts of specific voyages, supplemented by casual literary references to the design and handling of ships at sea or in harbour. Neither ship-building manuals nor sailing directions have survived. But important new information has become increasingly available with the advent and expansion of aqualung diving, while the looting of wrecks for treasure has at least in part given way to systematic exploration,[228] and to the growth of underwater archaeology as a firmly based scientific discipline. We are now learning many new facts about the design and construction of ancient Mediterranean ships, as well as about the cargoes they carried, and the routes they followed. In addition, comparative studies have made it possible to discern technical advances of importance in relation to the design and equipment of merchant ships.[229]

Discussion of these problems has been somewhat hampered by a number of what may be called 'standard' assumptions, which can be shown to be erroneous, either wholly or in part. One of these is concerned with the relative importance of oars and sails as propellants. There is an important distinction here between fighting ships and cargo vessels. Both categories were equipped with oars and sails; for example, the best known of all ancient warships, the

Greek trireme, was fitted with a large square sail. When not stripped for battle, she cruised with a full spread of canvas, but when preparing for action she left all her gear, including the sail, ashore. Oared cargo vessels, while not unknown, were, for obvious economic reasons, rare. Crews of rowers must occupy space at the expense of cargo; the converted triremes used in the fourth century BC as horse transports were rowed by sixty oarsmen only, the remaining 170 being withdrawn to make room for thirty horses.

Design and operation of oared fighting ships

These, like all other vessels of classical times, were of timber construction. Before 1971, when a war galley, now indentified as Carthaginian and dated to the third century BC, was found lying in shallow water off the north-west coast of Sicily, no identifiable remains of an ancient warship had been discovered, so that the details of design and manning (in particular the placement of the rowers in a trireme) have been left to conjecture, based largely on surviving representations. Prior to the invention of gunpowder, and for a long time afterwards, the offensive capabilities of fighting ships were very restricted; they could put an enemy vessel out of action either by setting it on fire, or by piercing the hull below the water-line, or by boarding. Apart from the use of flame-throwers operated by air-pumps (both were Hellenistic developments (Tarn 1930, 116)), ancient warships lacked effective means of attack from a distance.[230] The Greek solution to this problem was to design warships capable of destroying or disabling an opponent by ramming, and advances in warship design were aimed at achieving the impact speed necessary for successful ramming without loss of stability. Oared ships enjoyed an absolute pre-eminence in naval warfare, and that for an obvious reason: ramming by a sailing ship, or indeed any form of collision between sailing ships would certainly dismast the attacking vessel. The fighting trireme, as developed during the fifth century BC, was essentially a projectile. In the heyday of the maritime empire of Athens, that city's admirals developed, through rigorous training of their crews, the tactical manoeuvre known as the *diekplous* or 'break-through and ram'; the attacking triremes peeled off from a line-ahead formation and, effecting a tight turn, aimed their heavily weighted bronze beaks so as to pierce the enemy vessel's hull just below the water-line. Impact theory[231] indicates that unless the attacker attained the critical speed of about ten knots at the moment of impact, it would crumple, while the enemy vessel would escape almost unscathed! Before the invention of the trireme the standard Greek warship was a single-banked vessel with a crew of fifty rowers (twenty-five a side), called a Pentekontor or 'fif-tier'. To produce the necessary increase in momentum meant more rowers, and more rowing power without an increase in size and weight of warship meant placing oars and oarsmen on top of one another. Unfortunately this redistribution raises the centre of gravity, and threatens the stability of a vessel with a displacement of only 1 m (3.75 ft). The inventors of the trireme solved this problem by means of an ingenious arrangement of the rowers which has only recently, after much patient research, been satisfactorily explained.[232] The dimensions of these remarkable fighting machines are significant. With an overall length of about 35 m (115 ft), a beam of 3.7 m (12 ft), a free-board of 1.4 m (4½ ft), a height above the water-line of 2.5 m (8½ ft), and a complement of 200, they had neither space nor weight to spare for cargo or passengers. They could, it seems, move fast under sail, reaching a maximum of perhaps fourteen knots under the most favourable

weather conditions. But when dressed for battle the trireme was nothing else but a stripped-down water-borne projectile, propelled by human muscle-power. The later Greek period saw many experiments in multiple-banked fighting ships, including some freaks ('thirtiers' and 'fortiers'), in which the possible arrangements of oars and rowers have provided a field for inconclusive speculation. But two of the more modest Greek developments, the 'fourer' *(tetreres, quadriremis),* and 'fiver' *(penteres, quinqueremis),* were adopted and used with some success by the Romans in their wars with Carthage and other Mediterranean naval powers. We know almost nothing about the structure of these vessels, and there has been much argument. Athenian 'fivers' used the same ship-sheds as their triremes, which means they were no broader in the beam, which in turn disposes of the possibility that they had more than three banks of oars. It has recently been suggested that the only difference between these vessels and the trireme lay in the provision of extra rowers to the oar, the quinquereme having two to the oar on each of the two upper banks, together with some additional strengthening of the hull.[233] This theory provides solutions to most of the problems, including the technical difficulty presented by the extreme distance between the two top banks of a 'fiver' and the water, requiring the use of oars far too heavy for one man to handle. The 'fourer' may have been only two-banked with two men to each oar, which will have greatly increased the effectiveness of the ramming, as well as permitting the use of a broader blade. As for the performance of the quinquereme, we hear of an engagement in the Straits of Gibraltar during the Second Punic War, in 206 BC, when, the triremes of neither fleet being able to stay on course because of the strong currents and tide-rips, a Roman quinquereme 'whether because her weight kept her more steadily on course, or because her multiple banks made her steer more easily, cut through the swirling waters of the Strait, sank two triremes and, driving at full speed past a third, completely swept away the

142

oars from one side of a fourth' (Livy, 28.30). In set-piece naval battles, however, the Romans contrived to make up for their lack of technical skill by adopting boarding instead of ramming methods, using a boarding-bridge (the *corvus* or 'crow'), which was designed to fix itself in the deck of the enemy ship, so locking the two vessels together and providing a gangway for the boarders.[234] Boarding was also made easier by the extension to the gunwales of the narrow raised decking, which took place, according to Plutarch (*Cimon*, 12.2) in 467 BC, leading soon afterwards to the production of the 'fully fenced in' trireme or cataphract. This important step made it possible later on to mount siege-engines on the heavier quinqueremes, but not with any measurable succes.[235]

Design and operation of cargo ships

Here it is the sails that predominate: most of the technical problems that arise are concerned with their design, arrangement and handling in relation to the tasks which sea-going vessels were required to perform. This brings us to a set of assumptions about the performance of ancient merchantmen under varying conditions of wind and weather. Most writers on the subject declare that transport of goods by sea, though a good deal faster, under favourable conditions, than by land, was subject to severe disruption through contrary winds and storms. The main reason given for this poor performance is that ancient sailing ships could not make progress without the aid of following winds; they could not, in the common nautical phrase 'beat against the wind'. In order to test

the validity of this assumption we must examine closely what happens when contrary winds are encountered. When a sailing ship meets a contrary wind, it will set its sails as close to the wind as possible (the phrase is also proverbial!), and proceed on its course by 'tacking', that is, by reaching as far as it can diagonally to the direct line to its objective 'as the crow flies', then 'luffing', that is, altering the angle of sailing so as to catch the wind on the opposite quarter, and proceeding as far as it can in that direction. The result of these changes is a zig-zag course, which may well carry the vessel twice as far as the direct line to the distant port, and take twice as long as it would have done if the wind had stayed obligingly astern throughout the voyage. Since the orthodox view of the matter is based on the drawing of a distinction between a number of different types of sail and of methods of setting them, we shall have to look closely at these systems. Broadly, arrangements of the sails fall into two categories: square-rig in which a single square sail (or multiples of it) are arranged athwart the line of the hull, and fore-and-aft rig, in which two or more sails are set in line with the hull. Greek and Roman ships, it is argued, could not sail except with a following wind, because the square rig prevented the adjustments of sail required to perform the tacking operations described above; no genuine improvement in navigation was possible before the invention of the fore-and-aft rig in mediaeval times.[236] But we now have irrefutable proof that at least three types of fore-and-aft rig were known in Roman times—the sprit-sail and two varieties of the lateen sail. Ships fitted with these rigs were certainly able to sail closer to the wind and there- *143*
144

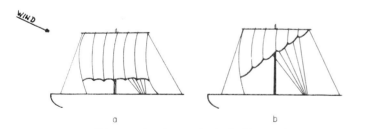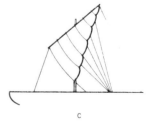

142 Diagram showing method of sailing against the wind with the sail brailed up.

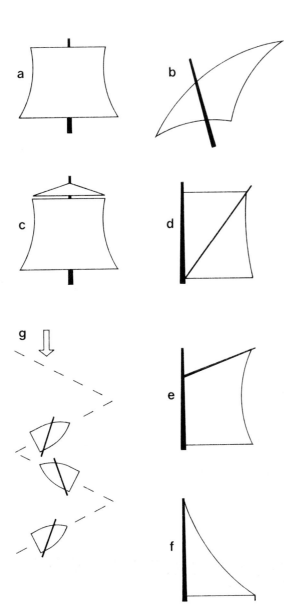

managed than the later square-riggers for which we have specific information, they will not have been able to head more than seven points into the wind. In ordinary language, this means that if steered in a northerly direction with a north wind blowing, they will have done no better than travel WNW on one leg and ENE on the other. This, if true, would make square-riggers useless against winds from an unfavourable quarter. Yet year after year square-rigged merchant ships made passage from the mouth of the Nile to Italian ports during the summer months, when NW winds prevail over most of the route! Ancient ships could certainly beat against the wind, but by how many points? The evidence comes from three main sources: first, recording of voyages on recognized trade routes; secondly, representations of ships under sail, showing details of equipment and handling methods while en voyage; lastly, casual literary references to navigation. The evidence on the first and last points has been sorted and classified,[237] but for the second category there is as yet no publication of the large number of surviving representations, which run to more than 600 items. David Rupp's recent (unpublished) study of more than 500 of them is important, but of limited value.[238] Study of the recorded voyages shows that the effect on average speeds of sailing under adverse wind conditions was very marked, and in some instances shattering. Thus the 'downwind' run from Pozzuoli to Alexandria (Casson, Table 1; n. 51) a distance of 1,000 nautical miles, was capable of being completed in 9 days at an average speed of 4.6 knots, while the 'upwind' run (sic!) to Marseilles (1500 n.m.) took 30 days, at an average of only 2.1 knots (Casson, Table 5; n. 86). It was on this difficult run that the 'Isis' (below, pp. 155ff.) was blown right off course and fetched up at the Piraeus! The picture that emerges from the study of the available evidence for the Roman period shows a water-borne commerce in which, while average speeds were slow, standards of seamanship and navigational know-how were sufficiently advanced to enable the development of regular

143 Diagram showing: *a,* square rig; *b,* lateen rig; *c,* square rig and topsail; *d,* sprit sail; *e,* gaff sail; *f,* leg-of-mutton sail; *g,* wind direction and line of sailing, maximum angle 80°.

fore make better progress against adverse winds than square-riggers. But what do we actually know about the capabilities of the square-riggers? Admittedly they were designed to operate with the wind on the stern quarter, 43 and the sail 'full and bye'. The extent to which they could tack to port or starboard was limited both by the shape of the sail and by the rig. Unless they were more efficiently trimmed and trade routes between the eastern and western ends of the Mediterranean, bulk cargoes of primary products such as corn, wine and oil, as well as heavy building materials and metals being conveyed long distances by well-built vessels capable of standing up to the elements. This picture is becoming clearer with the increasing volume of evidence emerging from underwater surveys and excavations.[239]

Size and carrying capacity of merchantmen

Underwater archaeology is now making possible more accurate estimates of the capacity of merchant ships than have hitherto been available. With the help of information from this source, together with that furnished by a number of official documents, it is reasonable to assume that (a) ships of 150 tons were common in Hellenistic times (from the end of the fourth century BC onwards); (b) ships in the range 350–500 tons were by no means rare (Casson 1971, 170–73). We know that the 'standard' grain-ships used on the regular run from Alexandria to Italy in imperial times were of 340 tons burden. Monsters like the ship built for King Hiero of Syracuse around 240 BC ran from 1700 to 1900 tons. The other famous 'superfreighter' was the 'Isis', for which we have the dimensions, but unfortunately not the capacity. There is plenty of room for guesswork here, since cubic capacity depends on shape, as well as on the basic dimensions of length, beam and height. A recent estimate puts it at 1,228 tons (Casson 1971, 184–88); if this is correct, we should observe that freighters of this tonnage do not appear in significant numbers before the end of the eighteenth century, when the size of East Indiamen rose from 800 to 1,200 tons. As we shall see later, these increases can be correlated with known improvements in design and equipment.

Design and building methods

Greek trading ships (called *holkades*, 'haulers'), had deep hulls and were broad in the beam,

144 Relief from tombstone of Alexander of Miletus (2nd century AD) showing lateen-rigged vessel.

145 Detail from tombstone of Naevoleia Tyche, shipper of Pompeii. Sailing vessel entering port, *c.* AD 50.

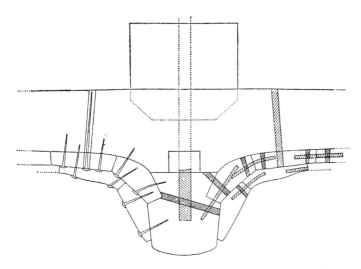

146 Diagram to show method of attaching garboard strakes to keel, floor timber and mast socket.

147 Shell-first technique of boat-building: a Swedish shipwright at work, 1929.

148 Model of Roman freighter under construction cut away to show the techniques employed.

which helped them sail close to the wind. The best evidence on the matter comes from a remarkably well-preserved specimen of the fourth century BC which was found a few years ago by a diver just outside the harbour of Kyrenia in Cyprus.[240] She may well have been the victim of a sudden squall. Her multifarious cargo marks her as the prototype of a modern tramp steamer, for she carried in her hold no less than ten different types of amphora. She was probably about 60 ft long, with a beam of 25 ft, a pronounced bow, and a blunt stern, which seems to have been fitted with a gallery, resembling those which became a standard feature of Roman merchantmen. Most of the information we have at present on ship-building methods comes from Roman shipwrecks. Not many ships have yet been so thoroughly excavated as to make it possible to reconstruct them, but the exhibits at the nautical museum at Albenga (Liguria), fruit of the immense labours of Fernand Benoît in this field, marks a big step forward in the nautical archaeology of the Mediterranean. From excavation, then, supplemented by numerous pictorial representations, the various stages in the construction of a Roman merchantman can be broadly determined (see Appendix 11). The first stage comprised the laying down of keel, sternposts and *146* stem. In the second stage the Roman shipwright proceeded to construct the vessel in the reverse order from that which has been normally employed for many centuries; Instead of shaping and setting up the frames or ribs, and then attaching to them the horizontal timbers, he made the hull according to curvatures laid *148* down by the designer, and then inserted the frames. The method used has been determined by piecing together a number of data, including a scene depicted on the funeral monument of the shipwright P. Longidienus, of Aquileia in north-east Italy, who worked in the late second or early third century AD. The relief shows him *149* shaping part of a frame for insertion into the completed shell. The horizontal timbers were fitted edge to edge by close-set mortise and *150* tenon joints, each being slotted into the plank

146

below it, as may be seen from timbers reco-
151 vered from wrecks; ill. 151 shows a plank
which has split lengthwise, exposing the construction. The exact method used in attaching
the lowest course to the keel is also known
from portions recovered from the sea-bed. The
third stage, that of inserting and fitting the
frames, was completed by the use of wooden
tree-nails, which were then pierced by bronze
spikes. Next came the attachment of the outer
strengthening timbers or 'wales', designed both
to give greater strength to the hull, and to pro-
152 tect it while berthing; these timbers encircled
the ship at gunwale level, and are a prominent
feature on reliefs and mosaics. The final stages
included smearing the seams with pitch or a
mixture of pitch and wax; sheathing, commonly with lead, over an underlay of tarred
fabric, and fitting an outer false keel to prevent
rubbing. Pliny (*HN.* 35.49) reports that even
merchant ships in his day were receiving a coat
of paint, in the form of melted wax, to which
the pigment had been added. Of more practical
use was the sheathing of the underwater surface

149 Funeral relief (late 2nd century AD) showing the
shipwright P. Longidienus preparing a rib for insertion into the finished hull. Ravenna, Museo Archaeo-
logico.

with lead as a protection against the destructive
activities of marine borers. The loss in most
cases of all but the floors of foundered ships[241]
means that we have to rely heavily on literary
and artistic sources, but the overall impression

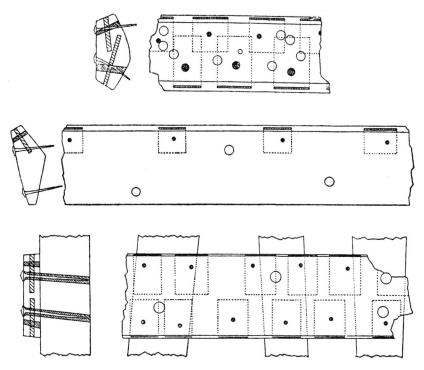

150 Diagram to show edge-to-edge attachment of planking by mortises and tenons.

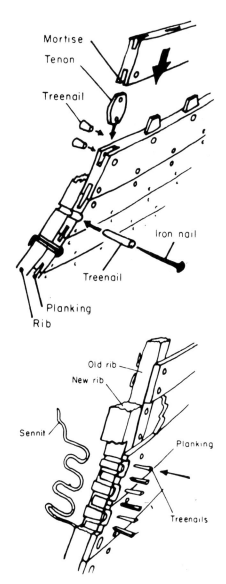

Mortise

Tenon

Treenail

Iron nail

Treenail

Planking
Rib

Old rib

New rib

Sennit

Planking

Treenails

151 Hull assembly showing mortising. Above, attaching planking to ribs. Below, attachment of a reinforcing frame, repairing with new ribs and sewing with sennits.

is one of massive strength: a finely carved freighter from Pompeii shows all her deck beams protruding through the hull; another, *43* from Beirut has them more widely spaced; the encircling wales were secured by massive through-bolts; massive too were the timbers that housed the great steering oars. For the hull planking, frames and keel Greek shipwrights are said to have preferred fir, cedar and pine

(Theophrastus, *Historia Plantarum (Enquiry into Plants 6.4.3).*[242] Some of the planks recovered from wrecks are 6 cm (2½ in) thick. Sailcloth was made from linen, as Pliny reminds his readers in a passionate outburst in which the deadly flax plant is condemned for luring men on to build bigger and better ships,

'so that even now we are not even content with sails larger than ships, but, though single trees are barely tall enough for the mighty yardarms that carry the sails, none the less they put on extra sails above the yards, and spread others besides at bow and stern, employing so many ways of challenging death (*HN.* 19.4-5).

The main points emphasized by Pliny can be illustrated from the monuments which show numerous variations in the number and the arrangement of masts and canvas. Two- and three-masters are common, and the following clearly recognizable types of freighter occur: (1) *a.* with mainsail only (Rougé (1966),1); *b.* with mainsail and angled *artemon* (Rougé, pl. *11* IIIa).

(2) *a.* with mainsail, mizzen and small square sail forward, on a mast with a strong forward *154* inclination, and forward of the prow; *b.* with mainsail and large foresail, inclined forward, like 2a, but aft of the prow (Rougé 1966, pl. IIIb)

(3) three-master (Rougé 1966, pl. IIb) ill. *153*

These and other questions about ship design, including the related topics of shape of hull, cargo capacity and navigational efficiency, merit close study; discussion based on the assumption that Roman freighters were all of Rougé's type 1 should be abandoned! Coastal vessels, which were only single-masted, were usually square-rigged, but some were fitted with the more flexible fore-and-aft rig, a sprit or a lateen (Casson 1971, 337, who illustrates the point with five sprit-riggers and three with lateens, figs 175-82).

Navigational efficiency: steering equipment and sails

The efficient operation of a sailing ship depends primarily on three factors, the stability of the

152 Housing and brackets for steering oar from ship of Odysseus. Museo Sperlonga, Italy.

hull, the quality of the steering equipment, and the manoeuvrability of the sails. Our knowledge of the steering system has been pieced together from close examination of the pictorial evidence. A prominent and curious feature of Roman sea-going ships is that on either quarter, at about the level of the upper deck, the run of the lateral timbers was carried aft so as to form a pair of trough-shaped projections, which were open at their after ends. The purpose of the two heavy beams which made up these projections was to keep the great steering oars in position. Throughout antiquity and right up to the invention of the single stern-post rudder in the thirteenth century AD, steering was effected by a pair of heavy, broad-bladed oars, suspended slantwise from each quarter, and operated from the deck by a tiller bar. They were not swept to port or starboard, but rotated.[243] Hence the strongly constructed housing, which also served to protect them when not in use. Both rudders were controlled by a single helmsman. In big vessels the distance from either gunwale to the centrally placed helmsman was too great for control by a single bar; the solution was to provide him with a pair of short

extensions connected to the tiller-bar by a sort of elbow-joint. A contemporary observer was moved to astonishment by the fact that the safety of a great ship was in the hands of a little old man holding a tiller that was no more than a stick (Lucian, *The Ship*, 5 —see below, pp. 155ff). There is no good evidence to support the widely held opinion that steering oars were inferior to the sternpost rudder, still less that the invention of the latter initiated a revolution in navigational technique! (see above, p. 143, n. 236). The arrangement and operation of the sails is less difficult to reconstruct; from the numerous representations of merchantmen at sea or in harbour we find that the dominant feature of the rig was the broad square sail, of such breadth that its main yard, when set athwart the ship, could be equipped for dropping rocks on enemy craft that approached it (Diodorus 13, 78 f.)! Corresponding in strength were the mast, and the various stays, lifts and shrouds that provided the stability and the adjustments needed during a voyage. The evolution of the rig began in the sixth century BC with the appearance of a foremast, with a forward rake, mounting a smaller version of the mainsail.

149

153 Mosaic from that Piazzale delle Corporazione, Ostia, depicting a threemaster freighter, with large foresail, small sail aft and cutwater. Above, the Claudian harbour entrance and lighthouse.

154 Mosaic from Sousse, Tunisia, depicting a Roman freighter fitted with large foresail; deck cabin astern and (unusually) oars as well as sails.

155 Relief (2nd century AD) showing freighter entering harbour at Ostia and then unloading at the quay. Museo Torlonia, Rome.

43 This feature, the *artemon,* appears in both a larger and a smaller form; and some larger vessels were fitted with a mizzen mast, carrying a fair-sized sail. All carried also a topsail in the shape of a shallow isosceles triangle, to give extra drive in the open sea. This was hoisted over the mainsail. The sails were fitted with a *155* series of vertical brailing ropes, enabling both the amount of canvas and the shape of the sail to be altered from the deck. Surviving representations show vessels being operated under varying conditions of sea and weather, including bowling along with a favourable wind and a full spread of canvas, lowering the sail down the mast to prevent the ship being 'blown down' in an adverse wind or sudden squall, raising a portion of the mainsail to cope with a stronger blow, or to reduce speed, as when *145* entering harbour at the journey's end.

Innovation and development: design of keels and anchors

Keel timbers and anchors have survived in greater numbers than any other remains of sunken ships. This has made detailed comparative study possible. Recent work on keels shows that some shipwrights made distinct advances in the direction of greater stability. As early as the fourth century BC an important innovation appears in the design of the keel
146 frames; half-ribs now alternate with the normal one-piece ribs.

This feature, which has been found in many wrecks, spanning several centuries, seems to form part of a development in shipbuilding technique aimed at strengthening the various keel members, while at the same time achieving greater flexibility against the buffetings of the sea, yet without abandoning the 'shell-first' method of construction.[244]

Effective anchors meant the difference between survival and total loss on a lee shore, and signs of both innovation and development are to be expected. The vast quantity of anchors recovered has led to a full classification of types from the rough chunks of stone of Homer's day

156 Keel of Roman freighter, viewed from above, showing close-set ribs beneath floor timbers.

to the advanced all-iron types with removable stocks that went out with the end of classical civilization, to return as a new (?) invention many centuries later.[245] Most anchors, however, were made from a mixture of materials, of which lead was the most common, whether cased in wood or forming the stock, with the remainder made of iron. There are two main problems in anchor design: first, how to get one of the arms to bury itself in the sea-floor: second, to get it to stay put. To achieve the first *158* aim, the stock was lashed at right angles to the arms, but this did not always happen; an important technical development in design has recently come to light; an anchor from a ship wrecked off the Lipari Islands was of asymmetrical design, which caused it to seesaw in the water, giving the arms a position perpendicular to the sea-bed, and thus enabling the arm to 'dig in' and hold the vessel. The Nemi barges (first century AD had fine examples of the still dominant lead and wood type, and of the iron that seems by the end of the next century to have taken over completely (Mercanti 1979, see figs 8 and 9), but much technical progress lay

157 Wooden anchor of Nemi barge with lead stock.

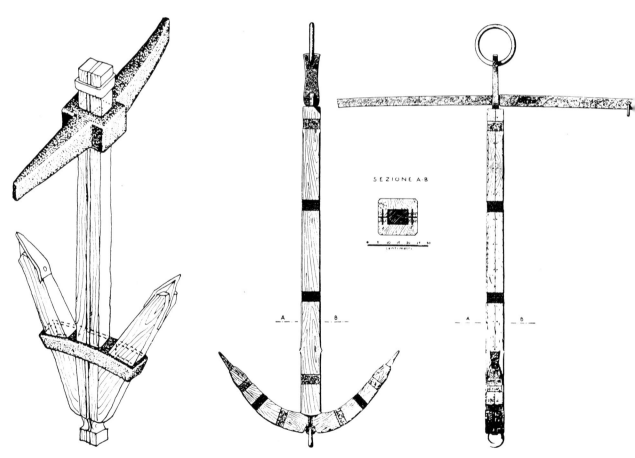

158 Reconstruction of an anchor with arms and shank secured by a lead collar.

159 Iron anchor with removable stock cased in wood. First half of 1st century AD.

SEZIONE A·B

152

ahead, as one may see by comparing the Nemi iron anchor with the vastly improved instrument used in Byzantine ships, such as the merchantman of Yassi Ada, which carried no less than eleven anchors with moveable stocks, although she was quite a small ship, not more than 22 m (70+ gt) long (Casson 1971, 256, n.130).

Cargoes and technical problems of loading and unloading

An increasing volume of information is now appearing from a variety of sources, which include the excavation of harbours and their installations (see Ch. 8, ff.), methods of handling and trans-shipment, and analysis of cargoes recovered from wrecks. We shall concentrate on the bulky cargoes of grain, oil and wine, and the heavy loadings of building stone and metals, which made up the bulk of the seaborne commerce of the classical world.

Grain

It has been estimated (Rickman 1980) that by the time of Nero (54–68 AD) shipments of Egyptian grain unloaded annually at Ostia, where the new harbour begun by Claudius (41–54) was fast outstripping the older Puteoli (Pozzuoli), amounted to around 100,000 tons. The total landed there from all major supply sources may have reached 300,000 tons. Divided into sackloads this means a total of more than ten million sacks to be shifted from sea-going freighters to warehouses, and thence to lighters for transport up the Tiber to Rome, the final operation calling for some 8,000 boatloads. The lighters *(naves codicariae)* which maintained a shuttle service between Ostia and Rome, were specially designed for the job, with stout masts set well forward, designed both to carry sail and to hold a tow-rope. The journey upstream was made for the most part, as on the Rhône and other northern rivers, with the aid of human muscle-power. Nero made effective use of the service by loading the empty lighters with the rubble cleared in vast quantities from the dis-

tricts devastated by the Great Fire of AD 64 (Tac., *Ann.* 15.43). From a series of inscriptions of the late fourth century BC, we can infer that a shipload of grain at that period amounted to 3,000 *medimni* or 120 tons. By contrast, the largest of the imperial grain-carriers carried well over 1,000 tons of grain (Casson 1971, 183–84; 187–89). Loading and unloading such quantities will have been a formidable task when the grain was bagged; but bulk shipment cannot have been uncommon, as we learn from a passage of the *Digest* (19.2.31) which refers to several consignments in a single vessel, 'each shipper's grain being kept apart by planks or partitions ... so that it could be identified'. Mixed cargoes are discussed below, s.v. 'stowage'.

Wine and oil

Containers for these important liquid cargoes have been brought up in vast quantities from the holds of wrecked carriers, and their shapes (slim and pointed, or squat and spherical) are very familiar. They were neither economical—an amphora holding 30 litres (6+ galls) might contain rather less than its weight of oil—nor easy to stow; the estimated cargo of 3,000 carried by the Grand Congloué will have weighed about 150 tons, and taken up a lot of space.[246]

Metals

The contribution of underwater archeology to our growing knowledge of sea-borne trade is now considerable. In the case of the standard ingot of lead—a cheap and easily worked metal with many different uses (see Table 13)—the volume of identified stamped ingots from Spain is large enough to provide a basis for extending our knowledge of mining in that area (full references in Parker 1971). Unloading and trans-shipment scenes related to both types of cargo are well known from mosaics and reliefs: a fine mosaic from Ostia shows a stevedore carrying an amphora along a narrow gangplank *160* linking a merchantman to a lighter (White 1975, pl. IXa). another, from Sousse in Tunisia,

160 Mosaic from Ostia (?) showing trans-shipment of cargo of amphorae (probably wine) from freighter (right) to river craft (left) with stepped mast.

9 shows men unloading metal bars from a small vessel that has been run ashore in shallow water; meanwhile, on the beach, officials weigh the bars in a large balance.

Stone

The impression gained from other sources that blocks, drums and columns of marble and other stone were shipped over long distances from quarry to building site seems to be confirmed by evidence from excavated wrecks. Out of a list of six vessels which foundered while engaged in the trade, the lightest cargo has been put at 120 tons, the heaviest at about 350

tons.[247] The dead weight of this type of cargo may have increased the risk of shipwreck in heavy weather, and there is also the possibility that an old ship might have been turned over to this sort of job after useful service in general trade.

Stowage

Heavy items carelessly stowed can break loose and cause severe damage; amphorae suffer from the additional disadvantage of being breakable, and therefore need planty of dunnage. Where enough of the hull of a wine- or oil-carrier has survived to show layers of amphorae still *in situ,* the stacking looks very efficient. 'Tiles were shipped stacked on edge in layers, and stacked so carefully that apparently a minimum of dunnage was needed' (Casson 1971, 199, with refs). In the holds of larger ships was a bailing machine, which took the form either of an Archimedean screw worked by tread-wheel or *12* of a set of force-pumps (a well-preserved set of four was found recently in a wrecked Roman merchantman (details in Landels 1978, 81 f.).

Speeds

Wind direction and velocity are so important in determining the speed of sailing ships that a voyage against contrary winds can take as much as three or four times as long as one with favourable winds. For example, recorded voyages on well-known routes, as used by the convoys of grain-ships from Alexandria to Rome, show that the outward voyage, mostly against the prevailing north-westerly summer winds, which the Greeks called the *Etesioi* or 'Yearlies', could take a minimum of one month, and a maximum of two, while the return trip (downwind nearly all the way) could be completed in two or three weeks. The upwind course followed by the big grain freighters was inevitably a roundabout one; thus St Paul, en route from the port of Caesarea in Palestine to Rome, started his journey on a short-haul freighter of Syrian registration bound for ports in the Roman Province of Asia. After calling at

154

Sidon, they steered round the lee of Cyprus 'because of the head-winds, then across the open sea off the coast of Cilicia and Pamphylia, and so reached Myra in Lycia' (*Acts* 27, 2–5). There the centurion in charge of Paul had the good luck to find in harbour a grain-carrier out of Alexandria bound for Italy. This vessel too was dogged by contrary winds, and finally fell victim to the dreaded Nor'Easter, being eventually obliged, in the full fury of the gale, to lower the great sail, and 'let her drive'. Neither Greek nor Roman freighters were built for speed, 'Even before a fine breeze from the right quarter their ships did no better than six knots' (Casson 1974, 152). Such statistics as we have on lengths of voyages cannot be used as bases for calculating speeds, for reasons which are clear from the earlier discussion of the ability of ancient merchantmen to sail into the wind; thus the nine days given by Pliny (*HN* 19.3) for the Alexandria–Rome run, would give, by simple division, an average speed of 4½–5 knots for the distance of 1,000 nautical miles, as the crow flies. But the actual distance covered will have been substantially greater, and proportioned to the sailing ability of the vessel!

We conclude this survey with two first–hand impressions, each from the pen of a Greek writer, but separated in time by a gap of two centuries. The first was written by the travelling lecturer and publicist Lucian of Samosata (120 till after 180 AD). We owe his account to an accident which was probably not uncommon. The vessel which gives the dialogue its title, the *Isis,* was a big freighter loaded with Egyptian grain bound for Ostia. After much buffeting by adverse weather she eventually turned up, a long way off course, at the port of Piraeus, once a great entrepot of trade in the heyday of the Athenian Empire, but now a quiet backwater. The whole city turned out to admire her, as Lucian explains:

Samippos: What a monster she was! All of 180 feet in length, so the ship's carpenter told me, with a beam more than a quarter of that, and a full forty-for feet from the deck to the bottom of her hold. And the height of the mast! And what a whopping great yard it carried; and what a forest of rigging held it up! And the way her stern rose in a gradual rake finishing up in a gilded goose-head, balanced for'ard by the flatter curve of her prow with its figure-heads of Isis on either side. The whole ship was a dream! The rest of the fittings, the red tops, even more breathtaking the anchors with their capstans and winches, and the cabins. She had a crew as big as an army, and enough grain aboard to feed the whole of Athens for a year! And the entire outfit depended for its safety on one little old man who turns those ruddy great steering oars, with a tiller that's nothing more than a broomstick! they pointed him out to me; woolly-haired little feller, half-bald, name of Hero, I gather.

Timolaos: He's a marvellous hand at his job, so the crew say; knows more about the sea than Proteus. Did they tell you how he brought them here, and what adventures they had? how they were rescued by a star?

Lykinos: No; you can tell us about that now.

Tim.: I got the story from the skipper—a nice intelligent chap to talk to. They set sail with a moderate breeze from Pharos, and sighted Akamas on the seventh day out. Then a westerly got up, and they were carried as far east as Sidon. On the way from there they ran into a terrific gale, and the tenth day brought them through the Straits to the Chelydonian Islands. There they jolly nearly went to the bottom... However, the Gods took pity on them, and showed them a fire that enabled them to make out the coast of Lydia. A bright star—either Castor or Pollux—appeared at the masthead, and guided the ship into the open sea on the port side. They were only just in time—she was heading straight for the cliff. Having once lost their proper course, they sailed on through the Aegean, beating up against the Etesians, until they anchored yesterday at Piraeus, which was the seventieth day of the voyage! If they'd kept Crete to starboard they'd have doubled Malea and got to Rome by now! (*The Ship,* 5)

The second piece comes from the *Letters* of the fourth-century bishop Synesius, and tells of a very much shorter voyage from the same port of Alexandria which also nearly ended in disaste. Unfortunately only a few sentences of this vivid description can be quoted here.

A Voyage from Alexandria

Hear my story then, and I will tell you first of all how our crew was made up. Our skipper was being crushed to death by bankruptcy; beside him we had a crew of twelve, making thirteen in all. More than half of them, including the skipper, were Jews, an uncouth race... The rest were a collection of peasants, who even as recently as last year had never handled an oar. Both lots had one thing in common: every man jack of them had some personal defect... We had taken on board more than fifty passengers, about a third of them women...

As soon as he'd doubled the temple of Poseidon, near you, he made straight for Taposiris, with all canvas spread; it looked as if he was bent on challenging Scylla, over whom we all used to shudder when we were children doing our school exercises. We spotted the manoeuvre just as the ship was nearing the reefs, and we all raised such a tremendous cry that he was forced to give up his attempt to battle with the rocks. All at once he swung away as though gripped by some new idea, and turned her head to the open water, struggling as best he could against a contrary sea.

Presently a fresh southerly breeze sprang up, and swept us along; soon we were out of sight of land and got into the track of the double-sailed cargo ships, whose business doesn't take them to our Libya: they are sailing on quite a different course... a gale got up from the north, and the violent wind soon raised mountainous seas. This gust falling suddenly on us, drove our sail back, turning it from convex to concave, and the vessel was all but capsized by the stern. With great difficulty, however, we managed to head her in... [As the gales persisted] What made death gape at our feet was the fact that the ship was running with all canvas spread, and there was no way of taking in sail; every time we tried this, we were thwarted by the ropes which were jammed in the pulley-blocks; and, what is more, we were secretly afraid that, even if we survived the gale, we should be approaching land at dead of night in this helpless state.

Day broke, however, before all this could happen, and never, I am certain, did we see the sun with greater joy. The wind lessened as the temperature became milder, and thus, as the moisture evaporated, we were able to work the rigging and handle the sails. We were unable, it is true, to replace our old sail with a new one (that was already in the hands of the pawnbroker), but we took it in like the swelling folds of a tunic, and inside four hours we, who had imagined ourselves already in the jaws of death, were disembarking in a remote desert place. (Synesius, *Letters,* 4)

The following is a vivid and presumably accurate description of the way in which the steering oars of a Roman ship operated in heavy seas. The passage quoted, though it comes in the form of a simile, bears the stamp of direct experience, even though the the writer was a bishop!

Jerome: *Letter* 100.14: (translation from Greek of a letter of Bp. Theophilus): a simile, underlining the point that in a period of great strain on the steering, with high winds and big waves approaching, danger can be averted by dividing the shock.

Similarly, the helmsmen, when they see a tremendous wave coming at them from the deep ... meet the foaming water and withstand its onrush by thrusting the prow at it, turning the steering-oars first one way, then the other, and as the gusts of wind and necessity dictate, by making the rudder-lines taut or slack. When the wave has fallen away, they release the straining bonds of the rudders on each side of the ship, so as to prepare for the oncoming wave during the moment's calm. When it comes, they tighten their grips on the heads of the tiller-bars, and turn the blades broadside on to the water, so that, by splitting the gusts, some to one side, some to the other, the strain becomes equal on both sides, and what could not be withstood coming all at once becomes more tolerable when divided.

12 Hydraulic Engineering

Hydraulic engineering is the grander name for a wide variety of processes connected with the control, storage and distribution of water, and for the application of techniques and devices to serve many different requirements, from the raising of water from rivers, lakes or man-made reservoirs for human consumption or irrigation to the channelling and distributing of water from a higher source to a lower. It is important for the history of hydraulic engineering that the task of removing water from the land requires much the same techniques as that of supplying water to areas where there is a natural deficiency. Like food, water is essential for survival, but for very long periods in his history man has relied on natural sources alone, early food-gatherers camping in the vicinity of rivers or natural springs. In dry areas the water table drops catastrophically in the dry season, and the springs fail. The historian Herodotus (*Hist.* 3.6) records how, to overcome the problem of water-shortage while crossing the desert, empty wine-jars were collected at Memphis, filled there with water, and then dumped at intervals along the desert tracks connecting Syria with Egypt (reminding us of the petrol dumps on the trans-Saharan tracks). The first and most obvious artificial device for improving this situation is the well.[248] These were always hand-dug in antiquity, and lined with stone, brick or timber. Some wells reached considerable depths (115 ft at Hermopolis in Egypt, and 250 ft at Lachish in Palestine), but most were less than 30 ft deep. The simplest method of extraction was by manpower with rope and bucket (the well-heads of many surviving specimens in north Africa are deeply scored from the rubbing of countless ropes), but improvements came early; first, the simple spindle converting

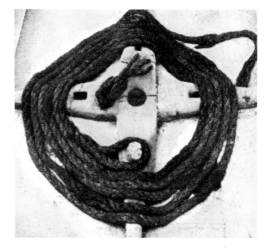

161 Windlass and ropes for a domestic well. Pompeii, Casa a Graticcio.

162 Well bucket from Newstead *(Trimontium)*, Scotland.

163 Egyptian wall-painting showing a gardener watering from a *shaduf* with twin buckets. From a tomb in Thebes.

the upward lift to the downward pull, then the addition of the windlass to provide the easier rotary motion. Mechanized methods also appeared in preclassical times; ox or camel-power on the rope, the *shaduf* or swipe (a swing-beam with a compensating weight, making some improvement in the use of muscle-power), the bucket-chain and finally the twin-cylinder force-pump, a very important develop-ment which is certainly Greek in origin. These and other water-raising devices are fully dis-cussed in Chapter 4, pp. 32ff.

163

23

4

Gravity tunnels

The earliest large aggregations of population in the first urban centres of the Near East obtained their supplies either directly from the great rivers of Mesopotamia or from multiple wells. But in Palestine, Syria and Mycenaean Greece fenced cities set on hills had to get their water from springs down in the valley beyond the defensive perimeter, so that in time of war the springs were protected by means of water tun-nels with access from the citadel. The remains of such works indicate some degree of skill in shaft-sinking through rock, and in the design

and construction of the conduits. In the more mountainous regions of Armenia and Iran more sophisticated techniques were developed to transport water to densely populated cities by means of long-distance water tunnels. Many of these tunnels survive in Iran, where they are still called by the ancient name of 'qanat';[249] they average 2 ft in width by 4 ft in height, and the gradients vary between 1:100 and 3:100.[250] Similar engineering works have been disco-vered in Greece, the oldest being a four-mile drainage tunnel constructed in Mycenaean times to drain off the waters of Lake Copais in Thessaly. The technical requirements for con-structing a qanat were simple enough, consist-ing of a vertical shaft to reach the source, and a lateral adit to bring the water out. More diffi-cult problems arose when the mountain was not a source of water to be tapped but an obstacle in the path of a proposed water conduit. The ear-liest example for which we have information on the actual construction comes from Jerusalem, and is dated to the eighth century BC. Threa-tened by Sennacherib, king of Assyria, the inha-bitants diverted the water from the spring of Siloam by underground conduit. A surviving inscription shows that gangs worked from both sides of the citadel, but there were wrong direc-tions as well as false starts, and the final length of the tunnel was 1,500 ft against a direct dis-tance of only 1,000.[251] About century later, Theagenes, tyrant of Megara, built the earliest known underground conduit system, but we know nothing about its construction.

More than a hundred years later still, about 530 BC, the two-way construction system was employed on the island of Samos to convey water through an intervening mountain. The tunnel itself survives; it has been well photo-graphed and described and meticulously sur-veyed, and can, therefore, be profitably dis-cussed in some detail. We begin with the account of Herodotus, who named it as one of the 'three greatest works *(ergasmata)* of all the Greeks'.

164

I have dwelt rather long on the history of the Samians because they possess the three greatest

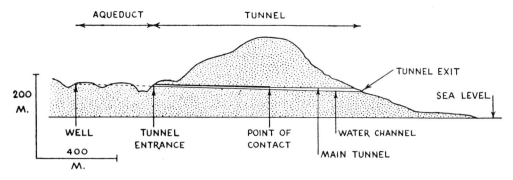

164 Diagramatic section of the aqueduct of Samos, probably 6th century BC. The tunnel was cut from both ends, but by miscalculation a large bend was required to enable the sections to meet.

works of the Greeks. One of these is a tunnel running right through the base of a nine-hundred-foot high mountain. The tunnel's length is seven stades [about four-fifths of a mile], its height and width both eight feet. Throughout its length another tunnel has been dug three feet wide and three feet deep, through which the water flowing in pipes is led into the city from an abundant spring. The builder *(architekton)* of the tunnel was the Megarian Eupalinos, son of Naustrophos. *(Hist* 3.60).

In spite of the attention given to it since its rediscovery nearly a century ago, most of the problems raised by Eupalinos' tunnel remain unsolved. For example, what calculations were made to plan the direction and slope? Once determined, how was the plan carried out? Why were the pipes laid in a second tunnel instead of being laid along the floor of the main tunnel? Fabricius (1884, 165–92), who took the first measurements, was able to prove that the tunnelling was carried out from both ends, with accurate planning and tolerably accurate execution. The evidence is unmistakeable:

The part of the tunnel dug from the south, after turning slightly to the right (east), comes to a dead end; the pick marks can still be seen on the blind head-wall that terminates it. Short of this wall the tunnel coming from the north, which has also turned east, breaks into the south tunnel through its western sidewall. In other words, a few feet before meeting, both tunnel halves turn somewhat to the east and thus meet again almost at a right angle. In addition, at the meeting place, the floor of the north tunnel is about even with the ceiling of the south tunnel. (Burns 1971, 173–74)

Kastenbein (1960), who re-measured the tunnels much more recently, using more accurate methods, and was able for the first time to establish the elevation, calculated that if the two tunnels had not turned east, they would have met head-on in a straight line, but with some difference of level. There are several possible reasons for the deviations: (a) distortion of sound underground, causing the foremen to misjudge the direction; (b) an error in calculation of the length, causing the diggers to start looking for each other; (c) intentional deviation, to avoid a possible overlap. We know some of the details of an error of precisely this nature in a similar project carried out in Morocco in the mid-second century AD, and the last-named theory may well be the correct one.

So how was the direction of the tunnel determined? The only possible clue is a passage in Hero's *Dioptra,* Chapter 15, dated to the first century AD, where the author poses the question of how to dig through a mountain in a straight line, given the two entrance points. After explaining that tunnels can be driven from either end or both, he provides a solution to the design problem based on similar triangles, and a series of horizontal sightings round the base of the mountain. Resemblances between the drawing of Hero's mountain, which appears as an illustration in a codex of his works dated to the eleventh or twelfth century AD, and the shape of the mountain at Samos have prompted speculation as to whether Hero had this particular tunnel in mind, and

159

169 whether the drawing can be used as evidence for the method used by Eupalinos. The tunnel could have been surveyed either by Hero's method or by a series of sighting poles over the mountain. The gap of 600 years that separates Hero from Eupalinos is no valid objection to the view that Hero used the Samos tunnel as his model. The surveying instruments and techniques he mentions are, with the one exception of the *dioptra,* part of a tradition that had remained virtually unchanged since ancient Egyptian times.

Next we come to the subsidiary tunnel. The water pipes for this were laid in a conduit running along the east wall of the tunnel. For a short distance from the north entrance it was dug as an open channel some 4 ft deep, but it soon took the form of a second tunnel running beneath the first, and connected with it by shafts at regular intervals. The usually accepted explanation for this feature is that of Fabricius, that it was made necessary because the calculations of gradient had been incorrect, giving insufficient flow, so that a second tunnel was needed. But there are similar double-tunnels in three populous cities, Akragas, Syracuse and Athens, each with the same connecting shafts. Assuming, then, that the double-tunnel system was the norm, what was its purpose? Burns (1971, 175) thinks that the primary reason was 'the difficulty of holding a constant slope of less than 1% in the depth of a mountain with primitive instruments' (on errors of gradient in surviving aqueducts see Usher (1954, 148). But it has recently been demonstrated that the main tunnel at Samos was horizontal, with a slight sag in the centre (Burns (1971, 179). The most sensible explanation is that the upper tunnel was deliberately made horizontal, and then the water conduit was constructed, in the same manner as the more ancient qanats (see above, p. 158, n. 249) by sinking shafts at regular intervals, and calculating the gradient from shaft to shaft, a job that could easily be carried out without the aid of sophisticated instruments such as *173* Hero's *Dioptra* (see Ch. 8, p. 101).[252] If this was the method adopted by Eupalinos, he was

following a tradition which remained unchanged over a long span of time, the drainage tunnels explored in recent years in southern Etruria being of identical construction.

There are two lessons to be learnt from the study of the Samos water-tunnel, and there is a *164* close connection between them. The first is that the instruments and procedures used in this area of hydraulic engineering were quite adequate for their purpose; the second is that empirical methods (often dismissed as primitive) were normal and effective here as they are found to be in the allied branch of water-raising (see above, p. 158). Under normal conditions the appropriate rate of flow from a spring of modest size can be adjusted on the job, as anyone familiar with water-leading problems on the small farm knows perfectly well. Even in the case of largers chemes, such as those in which a supply of water is carried to a city by gravity flow, the observed discrepancies are not significant. In the matter of tunnel alignment, however, small errors of calculation, and failure of the tunnellers to work to specification, could have serious effects, as the story of the Saldae water-tunnel demonstrates only too clearly.

A long inscription from the province of Mauretania in North Africa tells, with a great deal of amusing incidental detail, the sad tale of a major water-tunnelling scheme which went seriously astray (whether through inaccurate calculations or indifferent execution is not made clear). The problem of supplying the city of *Saldae* (modern Bougie) was rather like that presented to Eupalinos, but on a much larger scale, to construct a water-tunnel through a mountain lying behind the city to the north. The scheme was drawn up by one Nonius Datus, an experienced water engineer *(librator),* seconded for the work from the provincial military headquarters at Lambaesis. In spite of careful planning, things went badly astray, as we learn at some length from a contemporary inscription (*CIL* VIII. 2728—*ILS* 5795), the full text of which appears in Appendix 11. As the examination of Eupalinos' tunnel has shown, a very slight deviation by either of the two gangs of tunnellers from the set alignment could end in failure of the two parties to meet; but at Samos the supervision was evidently good, and the error was corrected in time; hence the slight 'kink' in the line. At Saldae Datus had apparently taken all the necessary measures to ensure success before returning to base; these included the choice of two gangs of troops, a detachment drawn from the 'marine infantry' *(classici milites)* and a second from an 'Alpine brigade' *(gaesates)*, who would be expected to show a competitive spirit and get the work finished in good time. But things did not go according to plan. On returning to the site after four years' absence Datus found a general air of gloom: the project was now regarded as a 'write-off', for the excavated length already exceeded the width of the mountain! It was presumably the contractors, and especially their site foremen who were at fault; one may imagine a concentration of attention on encouraging the tunnellers to keep up the competition and exceed their respective targets (which they did!), but with no thought for the vital matter of the alignments.

The writer of the letter concludes his report to the provincial governor by stating that he would certainly have asked for an extension to Datus' contract, to enable him to finish the job, had the latter not unfortunately fallen ill as a result of his exacting labours.

Siphons and bridges

Tunnelling was an expensive operation, and presumably was only resorted to when the fall from source to delivery point was too slight for the longer circuit around the mountain. When the water engineers were faced with the opposite condition—a gorge or depression instead of a hill, they could do one of three things: they could either dodge the problem completely by leading the channel round the depression, 'if the distance round is only moderate' (Vitruv. 6.5); or, if the valley was not too deep, they could bridge it with an arcaded aqueduct—a very popular solution, this, to judge from the

number of surviving bridges (see Appendix 12); or making use of the principle known as the 'inverted siphon', they could enclose the water in pipes, and convey it down the slope and up the other side without appreciable loss of flow. We shall deal first with the siphon method, as it involves a number of technical problems which have not always been satisfactorily explained.[253] The basic elements of the system are, in fact, easy to grasp, and are clearly spelt out by Vitruvius at the beginning of a passage which later on becomes very obscure:

If there is a regular fall from the source to the city, with no intervening hills high enough to interrupt it, but with depressions, then we must build *substructures* to bring it up to the level ... if the valleys are extensive, the course of the siphon will be directed down the slope, forming the *venter,* known to the Greeks as *koilia.*

So far so good:

on reaching the opposite side [of the valley],' 'the length of the *venter* reduces the velocity of the water as it swells up to rise to the top.

This incorrect statement on the behaviour of a siphon is followed by a very obscure passage (Appendix 12) on some device (the text is doubtful) for dealing with air-locks (?) or reducing pressure at the turns of the U-pipe (?). It is this passage which seems to have given rise to the popular and well-entrenched notion that siphons were rare, because the atmospheric pressure at the foot of the *venter* was too great for the pipes to withstand.[254] But the fact is that siphons were by no means uncommon (see Appendix 12); and the pressure argument cannot be sustained: experiments on the Lyons system (see below, p. 163) show that in this instance at any rate the internal pressures will have been well below danger level, assuming that lead pipes were used. Air-locks and sediment-formation do occur in siphons, and these may be the problems which Vitruvius (or his source) is discussing. An interesting suggestion is that possibly Vitruvius failed to grasp the meaning because his Greek was deficient.[255] We shall turn now to the archaeological evidence from four well-known sites, Pergamum and Aspendos in Anatolia, Lyons *(Lugdunum)* in France, and Lincoln *(Lindum)* in Britain. The technical problems at Pergamum were formidable, for the water had to be piped from a source 375 m above sea-level, whence it descended and, crossing two valleys separated by a saddle, climbed the successive terraces which form the spectacular citadel, the summit of which stood only 40 m below the level of the spring. The entire course of the pipeline has been traced, and identified from numerous surviving stone 'sleeves' into which the pipes were set. The latter could in theory have been either of earthenware or of lead; but the absence of any remains supports the view that they were of lead, which, in addition to its intrinsic value, had the advantage of being easily made into something else! It is possible that the original system built by Greek engineers had developed serious faults; when the Romans took over the former kingdom in 133 BC they replaced the closed-pipe system with an open conduit, the water being conveyed across the two valleys in open channels linked by a tunnel through the intervening saddle. The system seems to have been further modified by fixing the point of delivery well below the summit, water for the upper levels being presumably pumped up from a reservoir at the delivery point. The Pergamum supply line was not, as Forbes *(Studies,*

I, 161) suggests, an isolated experiment. Excavation has shown that the same method was used by engineers at Aspendos, where similar difficulties of terrain were encountered; furthermore, the excavations there have revealed a series of superimposed tanks designed to release accumulating air from the system, and thus eliminate 'drag' on the pipeline; what is clear is that measures were taken to deal with air-locks, pressure and sediment formation in the system; what is not certain is the extent to which the difficulties were overcome.

Aqueducts

This brings us naturally to a discussion of the aqueduct, by far the best known of all the ancient methods used for supplying water to cities and towns, because of the abundant and impressive remains of the arcaded portions still to be found *in situ* all over the Roman world, in areas as far apart as north-west Greece, central Spain, Algeria and metropolitan France. The surviving arcades, however, represent, in most cases, only a very small portion of the aqueduct. For most of their length from their sources in the mountains to the final distribution points in the cities the channels ran underground: in fact, of the seven major aqueducts supplying Rome in Imperial times, only just over 12 percent of the total length was carried above ground.

The Aqueducts of Rome

Name	Total length	Above ground	Underground	Date
Anio Vetus	63.6 km	0.32 km	63 km	272/69 BC
Aqua Marcia	91.7	10.25	81	144/140
Aqua Iulia	22.8	9.50	13	35/33
Aqua Alsietina	32.8	0.50	32.3	30/14
Aqua Virgo	20.8	1.03	19.8	21/19
A. Anio Nova	86.8	15.00	71.8	35/49 AD
Aqua Claudia	68.7	13.00	55.7	38/52
	387.2	49.60	336.6	

Note

A recent study of the aqueducts supplying Rome reveals a correlation between date of construction and height of delivery point; the later the date, the higher the delivery point, suggesting a possible connection with population growth.

165 Twin aqueducts carrying water across the River Louros, Epirus, Greece. A second (later) aqueduct was constructed because of insufficient head provided by original spring.

Two of the longest were built by the emperor Claudius in the mid-first century AD. The system that bears his name—the *Aqua Claudia*—ran for a total distance of 42 miles, the last eight being conveyed across the Campagna on stone arcading, long stretches of which can still be seen. Among the most interesting examples outside Italy are those at Segovia in Spain (built in 109 AD and still in *167* use), Carthage, where the main road from Tunis to the south passes directly under a fine stretch of the arcading, and Nimes in Provence, served by the exquisitely proportioned Pont du Gard, which carried the city's supply across the wide valley of the River Gard on three superimposed arcades. Although open channels with gravity flow were common in Roman times, there are some interesting examples of the use of the U-tube or inverted siphon, employed earlier by Greek engineers (see above, p. 161 ff.). The city of Arles in Provence was supplied with water which had to cross the River Rhone en route from its source. Here the pipes were laid in the river bed. Further up the river, at Lyons, the engineers had a very difficult access problem, because the water had to traverse three valleys, 'so deep that an open-channel system would have required arcades 215 to 300 feet high, while a system of piping only would have

developed pressures at the bottoms of the inverted siphons of nine 'to ten atmospheres'. So they used a combination of siphons and gravity channels, reducing the maximum height of the arcading to 60 ft, and lowering the pressure by carrying the pipes along the arcades. The sup-

166 Aqueducts of *ill. 165.* merge to carry water underground to Nikopolis. Entrance exposed by cutting for modern road.

163

167 Central arcaded portion of Roman aqueduct. Still in use until recently. Segovia, Spain.

ply was carried, not in a single pipe, but in nine pipes of smaller diameter. The reason for the choice is not clear; was it due to a mistaken notion that pressure could be reduced by this means? At all events Montauzan who surveyed the entire system, was in no doubt about the quality of the work; he commends the third siphon as 'an engineering work of the highest order. Of all Roman aqueducts this is perhaps the one that brings the greatest credit to the scientific knowledge and engineering skill of the Romans'. The supply line of which it formed a vital part, had a total length of *c.* 80 km (50 miles), and it included a thirty-arch bridge over the River Izeron, two siphons and a final arcaded section which delivered the supply to a reservoir for urban distribution.

Of equal interest is the closed-pipe system employed to bring water to the Roman colony of *Lindum* (Lincoln) in eastern England. Recent excavations[256] in this distant province have revealed a remarkable combination of a well-thought out solution to a difficult problem in hydraulics, and a high degree of practical engineering skill in carrying out the adopted scheme. The town stands on a limestone ridge, some 70 m above sea-level, which has very little water. With such a low water-table, the usual

wells were out of the question—they would have to go down to at least 14 m, much too deep for the manually operated force-pumps which would normally be employed, as at Silchester, *Londinium* (London) and many urban centres. The alternative, to bring in a supply from the nearest perennial spring, which lay some 2.5 km away, would involve a lift of more than 20 m. This could be achieved in one of two ways: either by raising the water to the full height at source, and thence by gravity-flow in open channels, or by means of a closed-pipe system, using a pump to force the water up what is now known as a 'rising main'. Excavation has shown that the second solution was preferred. The pipeline has been traced, several sections being discovered intact, but so far no trace of the pumping machinery has come to light. When the results were published in 1955, the only force-pumps known to the excavators were the small systems which could only deliver a maximum of 1500 l (*c.* 300 gals) per hour, but recent research has shown[257] that much more powerful pumps could be produced when required. What about the materials for the piping? The engineers were already familiar with a variety of materials, including bronze, lead, wood, earthenware and concrete. Bronze is expensive and difficult to work, wood rots and splits, while earthenware and unreinforced concrete cannot withstand much internal pressure. For closed supply systems Vitruvius (8.6.1) mentions that the pipes may be made of lead or earthenware, and gives a number of reasons why the latter material is to be preferred. In the first place, lead is dangerous to health, affecting both those who work with it, and those who drink from lead pipes, at least if the water is soft; hard water furs, and this reduces the rate of formation of the highly toxic oxide *(cerussa),* and consequently the risk of contamination This particular objection is repeated by Pliny (*HN* 31.57), and by Palladius (*Op. Agri.* 9.11.3), after whom the point was forgotten until Benjamin Franklin diagnosed lead poisoning fourteen centuries later.[258] Secondly, while earthenware pipes can be laid by a bricklayer, working with lead calls for special skills, both for installation (see below), and for maintenance. Finally, lead is much the more expensive of the two materials.

Manufacture of pipes

Earthenware pipes were easily made in moulds, the joints being sealed with putty.[259] The joints of the Pergamum pipes were supported by stone 'sleeves' (above, p. 162). Lead pipes of the Roman period were made, as Vitruvius explains, out of rectangular sheets and formed into a cylindrical or triangular shape (both occur in surviving systems) on wooden formers. Both lengths and sizes were standardized, the standard length being ten Roman feet (= 2.95 m or 9 ft 8½ in), while the standard sizes were based on the width of sheet used, not on the diameter (details in Landels 1978, 43 f., who includes a table of sizes, showing both the weights of lead required, and the diameters).

Pipes or conduits

A further argument against piped supplies is the difficulty involved in tracing leaks. Even today our city streets are suddenly made impassable by the bursting of underground water mains. Another serious problem in piped circuits was that of sediment formation. Settling tanks at the entrance would reduce, but not eliminate, the problem. In modern closed-pipe systems the danger of an obstruction at the most vulnerable point, the bottom of a U-bend, is avoided by providing a sludge-cock, by which the sludge can be removed; something of this sort was probably provided.[260]

For a number of reasons, then, Roman engineers preferred a gravity flow in open channels, with a slight and fairly constant down grade of about two or three feet to the mile. This estimated gradient (equivalent to 1:1750–1:2640) is twice that proposed by Pliny *(loc. cit.).* In fact, measurements taken on a number of surviving aqueducts in Roman France show wide variations, the gradient at Nimes being 1:3000 and that at Vienne 1:860. As to the choice between bridge and siphon, the determining factor may well have been that of instability at great heights, the height of the highest bridges (160 ft at Alcantara and Pont du Gard) being much lower than the depth of any of the major siphons (e.g. the 380 ft depth of the Tourillons siphon on the Lyons system, see Appendix 11). The delivery point would have to take account of the highest point in the area to be supplied. If we look at the relevant columns in the table, on *The Aqueducts of Rome* (p. 162), we find what appears to be a correlation between date of construction and height of delivery point: the later the date, the higher the delivery point. Does the sequence, if it really is a significant sequence, indicate a response by the engineers to an expansion of the population into the higher parts of the city?

Urban distribution

The survival of a major work on the aqueducts supplying Rome, published at the end of the first century AD by Rome's Controller of Water Supply, Sextus Julius Frontinus, provides a great deal of information about the technical arrangements for the distribution of water, the personnel employed in its various branches, and the practical problems encountered, including fraud and theft of water, and the degree of competence with which they were handled. There is no substitute for reading Frontinus: our task is to draw attention to some of the more important features of the system. The distribution system began at the end of the aqueduct with the passage of the water through one or more settling tanks to release the mud and pebbles, after which it was piped into a water *22* tower *(castellum),* from which it was made to flow into smaller tanks, from which again it was led off in lead pipes to fountains, baths, industrial establishments, and private users. The number of tanks into which each aqueduct flowed was proportionate to the volume of water coming through the particular aqueduct, the total in Frontinus' time being 247. The bottom levels of the tanks were tapped to provide three mains delivering water to fountains, public baths and official buildings, while ten higher pipes served the needs of private consumers (houses and industrial establishments), who on a gravity-flow system would be the first to suffer from any reduction or interruption. The

private users were charged on the calibration of the pipes they used, which were regularly inspected and stamped. Charges were crudely based on pipe-diameter, not on the rate of flow: the unit used in measuring water was a *calix* or standard nozzle. The standard *calix* was the *quinarius,* a length of bronze pipe $1\frac{1}{4}$ digits (= 0.728 in or 1.7 cm) in diameter and 12 digits (= 8.75 in or 21 cm) long, connecting the distributing tank to the user's pipeline. Users of large *calices* were charged in a rough proportion to the cross-sectional area of their nozzles ... These charges were made on the assumption that doubling the cross-sectional area would double the flow, whereas, in fact, it would more than double it. Roman engineers knew that pressure affected flow, but not knowing by how much, they worked only from the size of the aperture. The same methods were still in use up to the end of the nineteenth century. This major technical defect was due to lack of apparatus capable of measuring flow.[261]

From the information provided by Frontinus numerous attempts have been made to

168 The great aqueduct intersection outside Rome Showing (foreground left) the combined channels of Anio Novus and Claudia about to intersect the arcading; (centre right) carrying the Marcia, Tepula and Julia; background (right) a second intersection. Painting by Zeno Diemer. Deutsches Museum, Mainz.

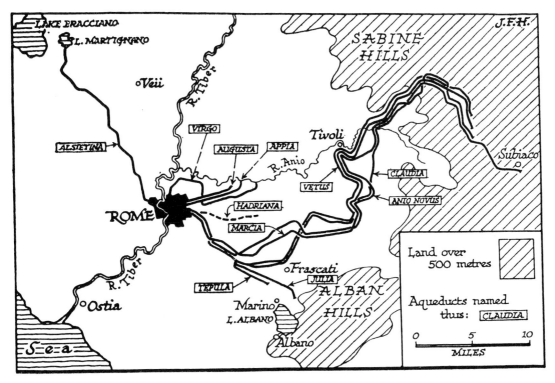

169 Map showing the routes taken by the principal aqueducts serving Rome.

estimate the total amount of water available to the citizens of Rome at the height of the empire. But the information he supplies is, as we have noted above, defective in one vital particular: when Frontius tells us that the total supply amounted to 14,018 *quinariae,* he is using, not the rate of flow, but the cross-sectional area of the conduits. There are formulae for calculating the rate of flow through various types of delivery-pipe, but the fact that 'modern estimates of the total volume ... run from 85 to 317 million gallons a day' (De Camp 1960, 201) suggests that some at least have been based on guess-work![262] Favourable and misleading comparisons have often been made by writers on Roman life, but there is no true basis of comparison between the metered demand system of a modern water authority and the constant-flow

170 Two channels crossing the Via Praenestina and the Via Labicane at Porta Maggiore, Rome. The arches carry the Aqna Claudia (below) and the Anio Novus (above) over the Via Casilina.

system of Roman times, where 40 percent of the supply went to public buildings, fountains and baths. Indeed, the only comparable feature is the fact that any Roman, like any Londoner, could get water from the public supply, even if he had to carry his own jugful to the top storey of his block of flats.

According to figures supplied by Frontinus (*De aquis urbis Romae* II, 78–87) 38.6 percent of the supply went to private consumers, whether for domestic or industrial purposes. The method of charging the consumer by standard measures had the inevitable consequence that dishonest citizens would attempt to beat the system. Frontinus devotes a great deal of space to the subject, noting with 'official horror' how the nozzles were illegally altered to get more than the user was paying for, and how he had found pipes running to irrigated farmlands, shops and even brothels! Out in the country-side, where supervision was difficult, we hear of water being surreptitiously withdrawn from the aqueducts by neighbouring farmers. Another perennial problem was that of cracks, leaks and seepage. Frontinus attributes losses on this score to faulty workmanship, but cracks in the conduits were also due to failure on the part of the construction engineers to make provision for the thermal expansion of the cement lin-ings: the expansion and contraction of a straight concrete channel several miles long, between a hot summer day and a cold winter night, would cause cracking of the cement. References to leaks are familiar in Roman litera-ture; e.g. the satirist Juvenal's account of the departure of his friend Umbricius en route for the Campanian coast, when he makes him halt at the Porta Capena on the Appian Way, which the poet calls 'soaking wet' *(madidam)*, from the seepage from the old arcading where the aqueduct crosses the road (*Sat.* 3.11) Finally for Rome, it may be noted that little was done to utilize the outflow from the baths for other pur-poses. In fact, the idea of passing on water from the public baths, laundries or other industrial establishments into the drainage system is only feasible where really large amounts of water are

involved; it is, therefore, not surprising that for most of the Republican period the famous Cloaca Maxima which started life as a humble ditch to drain the swampy ground between the seven hills, remained a storm-water drain. But the increasing size of the city led to the building of a number of large public latrines connected to the sewerage system, and flushed by water from the baths and fulling works.[263]

Purification of the water supply

Among the small communities of early Greece the emphasis in water supply was on finding a strong spring of wholesome water and keeping it free from contamination. But growth of population meant that the water, whether obtained from springs or wells, had to be stored in cisterns, and this gives rise to new problems of contamination; untreated water, whatever the source, when left stagnant, encourages the growth of algae and other organisms, rendering the water malodorous, unpalatable, and after a time, undrinkable. The need for filtration was recognized early, as was the need to test water for quality; filtering agents and settling tanks are mentioned by Vitruvius (8.6.15). Pliny's list of tests contains some worthless recommen-dations, and his observation that 'all water is the better for being boiled' suggests that boiling was a common way of making it safe for drinking. Boiled water is notably insi-pid, and it is, therefore, appropriate that Pliny should follow up his advice on how to treat bad water with a hymn of praise for the most famous water in the world, that of the *Aqua Marcia*.[264]

Irrigation

Uncommon in Greece and Italy, where the techniques of dry-farming prevailed, irrigation was extensively developed in the semi-arid zones of the north African littoral, notably in Algeria, and to a lesser extent in Tripolitania. In the former area there are impressive remains of dams, cisterns, and large-scale hydraulic sys- *90*

168

171 One of many surviving water storage cisterns in North Africa. One was used as headquarters for forward units of the British Eighth Army in World War II.

tems, and in the latter also the strong masonry dams across the many wadis provide powerful evidence of the attention given in this particularly arid region to water control. These remains have been exhaustively studied. In a recent contribution to the subject a hydraulic engineer, J. Birebent (1962), has drawn attention to the skill exhibited by Roman engineers in tapping underground water supplies in the Aures mountains, linking the aquifers held in successive layers of impermeable clays, and providing a supply both for urban and agricultural purposes from the same source.

Inscriptions provide a good deal of information concerning the methods employed in allocating irrigation water from a common source among neighbouring farmers. The most important of these was found at Lamesba in the central area of what is now Algeria, and gives details of the allocations made from a common source (the source itself is not described). The recipients are divided into two groups, consisting of those whose irrigated plots lay below and those that lay above, the level of the gravity flow. The members of the latter group must have been served by mechanical means (again the nature of this is not indicated), and they were compensated for the slower rate of flow by the longer periods allocated to them.

Water supply for industry

Urban industries received their supplies either directly from the central supply, or in certain cases from the overflow from baths and fountains; to the former category belong the cleaning and dyeing establishments; the latter include mills like those found in the basement of the Baths of Caracalla at Rome. Some buildings had their own reservoirs filled by water-wheels or force-pumps. A different organization of supplies was required for the mining industry.

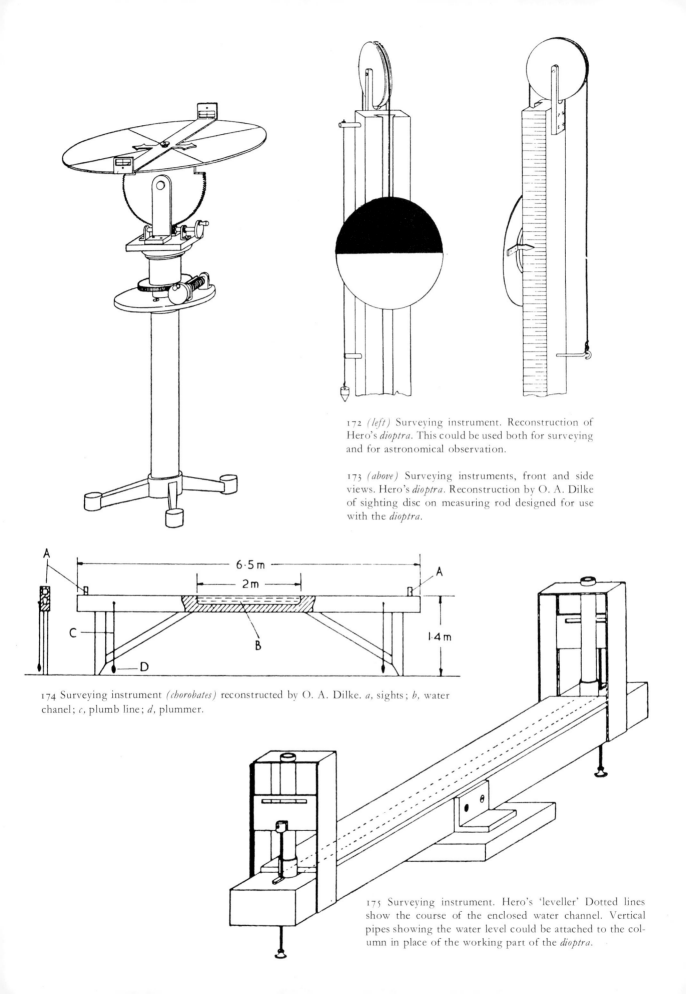

172 *(left)* Surveying instrument. Reconstruction of Hero's *dioptra*. This could be used both for surveying and for astronomical observation.

173 *(above)* Surveying instruments, front and side views. Hero's *dioptra*. Reconstruction by O. A. Dilke of sighting disc on measuring rod designed for use with the *dioptra*.

174 Surveying instrument *(chorobates)* reconstructed by O. A. Dilke. *a*, sights; *b*, water chanel; *c*, plumb line; *d*, plummer.

175 Surveying instrument. Hero's 'leveller' Dotted lines show the course of the enclosed water channel. Vertical pipes showing the water level could be attached to the column in place of the working part of the *dioptra*.

Of the three main types of metalliferous mining employed in classical antiquity, placer, alluvial and hard-rock extraction, the first two involved the use of water in considerable quantity. Recent investigations in the gold-mining areas of north-west Spain show ingenious combinations of hydraulic techniques, including aqueducts, reservoirs and a water-tunnel. The most interesting feature is the narrow water-tunnel on the Puerto del Palo mine, designed to tap the water from one valley and pass it along to another in order to furnish a sufficient head of water for mining operations carried on at a height of over 3,000 m, exploiting gold deposits too high up to be served by the tributaries of the Rio de Oro. The remains also include a well-preserved hushing tank, 55 m long by 5 m wide, with walls still standing to a height of 3.5 m (see Ch. 9, p. 117).

122

Instruments and techniques connected with water supply

What technical equipment did the planners and surveyors have at their disposal, and how far did practice measure up to theory? In addition to the cross-staff *(stella),* the plumb-line level *(libra, libella),* and ranging rods *(decempeda, pertica),* used by the land surveyors, the hydraulic engineer had available a number of levelling devices, the principal ones being an adjustable plane-table set on a vertical column, (the *dioptra*), and a horizontal water-level, the *chorobates,* designed rather like our spirit-level. The *dioptra* base could also be used for mounting a more sophisticated type of water level *(libra aquaria).* The *dioptra* and the *libra aquaria* are described in detail (see pp. 180ff.) by Hero of Alexandria.[265] The earliest account we have of these instruments (including the only surviving account of the *chorobates*) comes from Vitruvius' great work on architecture, dated *c.*19 BC (see p. 185ff.). Of course we do not know how long before this time they had come into use. This is how Vitruvius begins his chapter *(De Arch.* VIII) on the supply of water to country houses and towns:

172
173
174
175

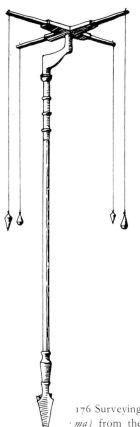

176 Surveying instrument *(groma)* from the restored model in the Museo Nazionale, Naples.

The first stage is levelling. This is done by *dioptras* or water-levels *(librae aquariae)* or by the *chorobates.* The last-named is the most accurate, since *dioptras* and levels are liable to error. The *chorobates* is a straight plank some twenty feet long. At its extremities it has legs of equal length fastened to it at right angles, and between the plank and the legs are cross-struts attached by tenons. These have perpendicular lines marked on them, and plumb-bobs hanging from the plank, one over each cross-strut. When the plank is in position, the perpendiculars, touching the corresponding marks evenly, show that the position is level. If there is interference by wind, causing movement and preventing the lines from certifying the level, the *chorobates* should have on its top side a channel five feet long, one inch wide and an inch and a half deep. Let water be poured in; if it touches the lips of the channel evenly, we shall know that a level has been established.

Hero's description of the water-level *(libra aquaria),* may be summarized as follows:

171

the hollow column carried a horizontal rod hollowed out to take a bronze tube. At either end this tube was turned up to take a small vertical tube of glass. Each of these tubes was enclosed in a frame along which a bronze plate fitted with a sighting-slip could slide, its movements being controlled with a screw. In taking levels, water was poured into the tube. Then two measuring poles, equipped with a plumb-bob and a disc, half painted black and half white *172* (see diagram, ill. 172), which could be raised or lowered by means of a cord over a pulley, were held by assistants, one placed at either side. These measuring-poles were calibrated, so that the readings could be called out to the surveyor, and the differences in height registered.

It seems astonishing that an instrument which would appear from Hero's account to be well designed for accurate determination of levels should be condemned by Vitruvius as un- *174* reliable.[266] The *chorobates* appears to have possessed two advantages over its rivals: first it was constructed on a four-legged frame, which is easy to set plumb, whereas the rival instruments were set on hollow cylinders, with the disadvantage that they would be more difficult to set.

The foot of the *dioptra,* 'probably a tripod or *172* table of wood' (Drachmann 1963, 198) is unfortunately not described, so that we can only conjecture how it was adjusted. But we must not assume that Vitruvius is mistaken in his estimate. The fact that it was 20 ft long would give the *chorobates* an advantage in determining gradients, since a small error in the sighting would make for less deviation in the actual construction than if a similar error occurred in the *dioptra.*

In this department of hydraulic engineering empirical methods still predominate today, and there is nothing particularly remarkable about the variations in the gradients of aqueducts. As we have seen above there were errors, some of them costly in terms of manpower; but it is no answer to the question of the level of efficiency attained in antiquity to assume, against Vitruvius, that the *dioptra* was regularly used for water-levelling, and then to assert, without supporting evidence, that 'when these instruments were used skilfully, it was possible to maintain a gradient of 1:2000 very closely' (Forbes, *HT* II, 672) who adds that 'we seldom hear complaints of careless and negligent work'!

Conclusions

The technical achievements of classical antiquity have been variously assessed. In his widely read and influential work, *Mediaeval Technology and Social Change* (1962), Lynn White established the now fashionable tendency to deny or at least to downgrade them. The attractive style and bold, sweeping assertions, supported by an impressive array of footnote references, not all of which are germane to the argument, have given this book an influence that far exceeds its intrinsic and undeniable merits. If the yardstick chosen is that of invention, the period of more than a thousand years that spans the gap between early Greek and late Roman civilization was, to say the least, not very productive. More cautious in his approach to the same

problems of technical development, Finley, in *The Ancient Economy,* also overrates the importance of the restricted category of invention (see Appendix 2). His discussion of manufacturing industry seems to have been too much influenced by the notorious bias of the Roman ruling class so carefully analysed and illustrated in the second chapter. In an area much given over to speculation, one may also speculate that the leaders among the inventive Celtic peoples of the western provinces will have been neither so hidebound in their attitudes nor so status-ridden as their counterparts in the Capital. Where the argumental basis is one of conjecture, and the literary record so full of gaps as to leave ample scope for speculation, the contribution of the archaeological record is a crucial one (see pp. 174 ff. below). The reader of these pages, having taken the measure of the innovations and developments of inventions reviewed in the chapters and tables, may be inclined to take a less unfavourable view of the standards reached by classical architects, builders and engineers, as well as by the farmers, the food processors and the men who built the ships that carried the products of their labours. In a review of John Landels' book, *Engineering in the Ancient World* (*JRS* 69 (1979), 203) Mark Hassall emphasizes the great importance of the archaeological record, citing in support of his plea two inventions of major importance which, according to the 'received doctrine', were neither discovered nor used by the Greeks and Romans, namely the crank and the shafted

12 vehicle. Yet one of the Archimedean screws found *in situ* at Alcaracegos 'was turned by a crank' (Healy 1978, 95), and 'shafts for carts', we are told (Landels, on p. 174), 'were first used in the ninth or tenth century', while actually they appear on relief sculptures from the Rhineland in the third! Outstanding in the record are those examples of applied scientific knowledge in which either a specific machine was produced by means of a combination of a number of prin-

177 ciples, as in the water-organ, or a large-scale system was put into operation, as in the gold

1, 122 mines of the Asturias; here the technical advances made in one area of technology, that of water-supply, were successfully transplanted, and put to use in a bold and imaginative effort for the removal of the overburden.

There is still a great deal of work to be done in this complex and difficult area of enquiry. As the first tentative experiments have demonstrated, new light can be thrown on ancient technical problems by co-operation—between historians and engineers, between linguistic scholars and scientists, and especially between archaeologists and historians. Many arguments advanced here are due to be undermined by more systematic research carried out with more refined instruments. This book is no more than a survey, and a starting-off point.

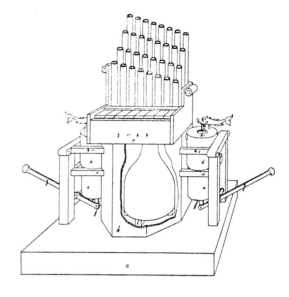

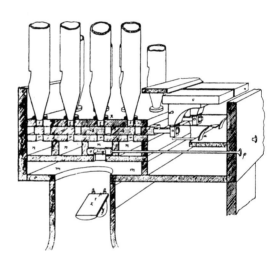

177 Reconstruction of Ktesibios' water organ. Front view (above) and back view (below). These instruments, which combine the use of water power and compressed air, with their keys connected to slides fitted with iron jacks and their pipes, reeds and valves are briefly described by Vitruvius (x. 9). The remains of the Budapest specimen bear out his comment that the instrument is carefully and exquisitely contrived in all respects.

Sources of Information

Primary sources

Surviving implements,
parts of machines, etc.

Examining the range of raw materials used in the manufacture of implements, utensils and mechanical devices which are the subject of our discussion, we find marked disparities in the distribution of the material, both in the quantities of surviving specimens of the implements or machines used in particular processes and in their geographical distribution. Of the factors affecting survival the most important is the durability of the material under varying conditions of pressure, temperature and moisture, stone being by far the most durable of naturally occurring materials and wood the most subject to decay. Of man-made materials used in antiquity, baked clay, though easily broken, is virtually indestructible, so that shattered pottery utensils can readily be restored. Leather, too, is extremely hardy, as witness the quantities of well-preserved footwear recovered from Roman military establishments. Wrought iron, on the other hand, while possessing a long life-expectancy, is very susceptible to erosion by rust in the presence of moisture; in addition, its laminated structure assists the process of decay. Woven or stranded fibres, which were used in a great variety of machines from cranes to catapults, have a poor chance of survival, except under arid conditions. The investigator invariably finds himself drawing on the abundant materials that have survived in the exceedingly dry climate of Egypt, but must resist the temptation to draw general conclusions about the type, design or function of the object under study.

Where an implement or device is made of several different materials, e.g. an iron scythe fitted with a wooden handle or a grain-rubber operated by means of a wooden lever, the task of reconstruction is far from easy. Thus we have thousands of surviving iron ploughshares, but not a single complete plough has survived, so that different evidence, both representational and literary, has to be brought into play. These other classes of evidence, as we shall see, provide their own crop of difficulties and hazards.

Again, while numerous press-rooms for oil and wine have been identified *in situ,* no ancient press has survived, but this particular drawback is cancelled out by the fact that Cato's *De Agri Cultura* contains a detailed specification and working instructions for setting up a lever-press for olives. The great advances that have taken place in recent years in the techniques and processes of conservation have vastly improved the survival rate of objects recovered from the soil or the sea-bed; buried timber can now be readily protected against the disintegrating effect of exposure to the atmosphere, and the reduction of iron artefacts to shapeless rust-corroding laminations should now be avoidable. Reconstructions based on surviving portions have not always been successful. Among the obvious examples of incorrect reconstruction may be cited the Catonian oil-mill *(trapetum)* at Pompeii, incorporating stones which do not belong to the same original mill, and the drainage water-wheel from the Rio Tinto mines, now in the British Museum.

The Gallo-Roman harvesting machines have been the subject of many absurd and fanciful reconstructions in the form of models, allegedly based on Palladius' description. The makers of the full-scale replica now *in situ* at Montauban,

Buzenol, not only omitted a vitally important feature, but used timber so heavy that the machine broke down at the first attempt. Apart from this fairly recent example, many dubious reconstructions have managed to find their way into most of the standard text-books, the authors of which have little or no direct knowledge of the sources and even less of the published refutations or corrections of error; 'received' dating and attribution are accepted without question as coming from 'reputable' authorities. Examples abound; here we cite a single well-known instance—the reconstruction of a sprocket-driven bucket hoist, propelled by an undershot water-wheel. In his full study, *Roman and Islamic Water-Lifting Machines* (Odense 1973), the author Dr Thorkild Schioler, states (fig. 43, 65) that the construction has no counterpart in practice, and that the attribution to Philon is doubtful. According to the text of Philon, the machine is not a hoist at all, but a water-wheel driven by falling water!

In a detailed review of M. Della Corte's article on the Groma (*Monum. Ant.* 28 (1922), 5–100) in *Class. Phil.* 21 (1926), 259–62, F. W. Kelsey pointed up the difficulties involved by reference to a contemporary experience of the problem:

The difficulty of reconstructing even a simple instrument from insufficient data may be illustrated from an experience in the late [i.e. First World] war. A part of a German periscope came into the hands of some American officers. They immediately attacked the problem of supplying the missing parts, but no two reconstructions of the parts were alike. Finally a German officer who had been taken prisoner showed how the instrument was made; in the light of his explanations, all but one of the reconstructions seemed absurd (cited by O. A. W. Dilke, *Aufstieg und Niedergang der Römischen Welt*, 2.1, 1974, 572–73).

Representations

Here again there is wide variation in the amount of surviving evidence, as well as in the availability of the material. Ideally, and especially where the object consists of a complex assemblage of parts, such as a ship or a wagon,

the investigator should personally examine all the available evidence; this entails a great deal of time and expense in travelling and photographing of monuments. In some cases they are difficult to locate, and even when located they often prove difficult to photograph adequately and consequently interpretation is far from easy. In recent years, however, there has been a welcome recognition of the fact that this class of evidence requires to be studied with at least as much diligence as the artefacts themselves. What is needed now is good publication of the material, with photographs of high quality, and in reasonably accessible sources. The nature of the problem is indicated clearly in a key chapter of Landels (1978), on cranes and hoists. The sources of information on these machines include a sculptured relief panel containing a detailed representation of a tread-wheel powered crane in action. Here we have an obvious example of the need for accurate and readily accessible illustrations, yet the best photographs, as Landels points out, are in the Italian periodical *Atti della Academia dei Lincei,* which is a very expensive publication and not readily accessible to English students. The need for something in the nature of a *Corpus Machinarum Antiquarum* (printing costs would· rule out a conventional series such as the *Corpus Vasorum*) could be met by the use of microfiche reproductions; ships and land transport vehicles are two strong candidates for inclusion in such a series.

Secondary sources

Inscriptions

These cover a wide area of technology, with building construction and water supply particularly well represented. Important inscriptions are fully discussed in the relevant chapters of Part II, but attention may be drawn to two outstanding examples, the masterly handling of Greek building inscriptions by Bundgaard (1968), and the valuable summary by Eric Birley of inscriptions connected with the building of Hadrian's Wall.

Papyri

This vast, but as yet largely untapped, store-house of information (the published material represents probably a mere fifth of the total of documents so far recovered) has been little explored, at least so far as the technical references are concerned. As might have been expected from documents of Egyptian origin, the published papyri contain numerous references to water-raising machinery. These documents, referring to day-to-day operations, are not only original sources of information, but they can be dated, some of them very closely—a matter of great importance in relation to the development and extension of techniques.

Thorkild Schioler's *Roman and Islamic Water-Lifting Machines* sheds much light on the complicated problems related to the design of one of these machines, the water-wheel. It contains an entire chapter on evidence from papyri and from excavations (pp. 111–67); twenty-four passages out of more than two hundred which refer to irrigation by machinery are presented in translation, with technical commentaries, and there is a valuable list of references to papyri not included in his study.

Literary sources

Before proceding to discuss the individual writers and their contributions, we must take notice of the fact that technical treatises (and the technical sections of other authors, e.g. Pliny) are particularly susceptible to the sort of transcriptional errors that make it difficult to understand precisely what the writer was trying to say. The mediaeval scribes, to whose labours we owe the texts as they have come down to us, lacked the necessary technical knowledge to comprehend what they were copying; the result is an abundance of the sort of errors that can now be identified and corrected, but all too many passages to which the harassed editor must attach the despairing comment: *locus corruptissimus necdum sanatus*. The text of Vitruvius, for example, abounds in instances of both

kinds; indeed, a perusal of several renderings of the same passage often leaves the bewildered reader doubting whether the translators were working from a more or less common text!

The obvious need here is for a well organized programme of translation of the major literary sources, with some guidelines laid down, which would stress, *inter alia,* the importance of consulting engineers and technologists on the numerous passages in which the text is obscure, and perhaps suspect, and the need for the translator to look for the most recent work on the particular subject, especially in languages other than his own. Many examples could be cited, from recent translations, of an editor quite unaware of recent discussions of difficult or obscure passages (e.g. Le Bonniec's edition of Pliny, *HN* 18 in the Budé series).

The Greek writers

It would be surprising if the ancient Greeks, who invented all the major literary forms, had not found it useful to compile scientific and technical treatises (including practical manuals). The inclusion in the epic tradition of detailed technical descriptions like that which depicts Odysseus as a skilled shipwright constructing the sturdy craft that is to carry him away from the island of Calypso, shows that such material was acceptable to the aristocratic audiences to which the Homeric bards addressed their songs. Hesiod's almanack and seasonal calendar for farmers, the *Works and Days,* contains directions for making a wooden plough (*WD* 430 ff.). It has been asserted that in the majority of ancient crafts, the traditional skills and methods were transmitted from master to apprentice by word of mouth, and did not find their way into technical manuals; but the range and volume of farm manuals surviving from the Roman period make it unlikely that the techniques of traditional craft industries were never committed to writing.

With this introduction, we now pass to a brief review of the technical writers whose works survive in whole or in part, with more extended treatment of the most important of them, Hero of Alexandria. We begin with the earliest surviving work, the *Mechanica*, which has come down to us among the so-called 'minor works' of the philosopher Aristotle. Most scholars who write about Aristotle ignore the work or deny his authorship of it; the usual argument advanced is somewhat circular: 'They say Aristotle could not have written the *Mechanica* because he was not interested in the subject, and we know that he was not interested in the subject because he never wrote about it' (Sprague de Camp 1960, 118). It has been fashionable to ascribe the work to the hand of a distinguished pupil of Aristotle, the physicist Straton.

After some verbose theorizing on the mystical properties of the circle, the author of the *Mechanica* passes naturally to a discussion of gear wheels, finding fascination in the fact that in a system of gears each gear wheel turns in a direction opposite to the one with which it meshes. This is the earliest surviving reference to gear wheels; one would like to know whether his wheels are toothed cogwheels or smooth wheels engaging by friction: he does not tell us. The transition from smooth to toothed wheels may well have taken a long time, with possibly an intervening stage when the rims were roughened to prevent wheelslip (see Sprague de Camp 1960, 120). His next topic is the law of the lever, introduced at the beginning of the work, but now seen in a number of applications, which include the balance, and the oars of a galley.

Turning from oared to sailing ships he explains, in his customary question and answer form, how a sailing ship can be made to steer a course against an adverse wind: The author then goes on to review in turn all the 'mechanical advantage' devices apart from the screw (viz. the roller, the wedge, the pulley, the sling, the horizontal capstan, and the vertical windlass), concluding with some observations on the remarkable weights that can be lifted by one man through a system of multiple pulleys. Many of the questions he puts are relevant to day-to-day situations, for example, the breaking

strains of pieces of timber of different shapes, which had obvious practical implications for the makers of torsion catapults. Altogether the author of this treatise shows himself to be, on the whole, a practical, down-to-earth writer, with a shrewd grasp of the fundamentals of his subject. Many of the questions he puts could not be answered within the limitations of the scientific knowledge of his time; the answers to some of them were not given until after the lapse of many centuries.

Our knowledge of the life and work of the first of the leading Hellenistic engineers, Ktesibios, is derived entirely from later writers. We have no firm dates for his career, and precious little in the way of biographical information of any kind. His writings are lost, but later writers have credited him with a number of important inventions, ranging from musical instruments worked by compressed air to the metal spring. His most notable invention, the double-piston force-pump, came to be universally known as the 'Ctesibian device' (*Ctesibica machina*—see above p. 17 and Appendix 3).

Our principal source of information, Vitruvius, states (X.8.2 ff.) that he has drawn his accounts of the discoveries and inventions of this remarkably talented man from a book written by him, which gives some credence to his reporting. According to Vitruvius, the basic invention, that of the cylinder and piston, which led on to that of the force-pump, the water-organ and the compressed air catapult, arose out of a chance discovery: while attempting to set up a mirror adjustable for height in his father's barber's shop, he noticed a distinct sound, caused by the counterweight, a leaden ball, dropping through a narrow pipe, and causing the air to be compressed; as the air was released into the open, it escaped with a loud noise. There is nothing inherently improbable in the story; accidents leading to important inventions have been commonplace in the history of technology. The force-pump is described both by Vitruvius (X.7) and Hero (*Pneumatike* I.28). That its use was widespread is evident from the number of specimens that have come to light; the latest list (Battersby, 1981) contains twenty examples from widely separated sites the best preserved example is in the Museo Arqueologico in Madrid (Landels 1978, fig. 24, 81—part of a detailed account of the basic design and the extant varieties).

The water-organ *(hydraulis)*, also described by both writers (Vitruvius X.8; Hero, *Pneumatike,* I.42), employed both air and water under compression to produce the necessary flow of air, a set of valves and a keyboard to operate them, together with the supply tubes, and a set of pipes of the heights required to produce the range of notes—in fact, all the essentials of the organ as we know it except for the modern power source.

Philon of Byzantium is reported as flourishing soon after Ktesibios. He visited Alexandria and Rhodes, and his extant writings show familiarity with the work of Ktesibios. Of his comprehensive treatise on mechanics, the *Mechanike Syntaxis (Basic Elements of Mechanics)* only two books survive complete, together with portions of two others, the main subjects represented being compressed air machines and catapults. The wars of the third century provided a powerful stimulus to the design of more efficient forms of offensive weaponry (see above, pp. 46ff.), and Philon reports on some new designs of his own, as well as on those of other engineers. By far the most interesting item in the catapult section is his description of a rapid-fire repeater type invented by Dionysios of Alexandria. In this model a constant supply of arrows was provided by fitting a hopper above the groove in the slide mechanism from which the bolts were released one by one 'as in a nineteenth-century Gatling gun' (Sprague de Camp 1960, 145). The repeater system consisted of a windlass connected to the trigger mechanism by means of a chain-and-sprocket drive, the chains passing over pentagonal nuts (details, including diagrams, in Landels 1978, 183 ff.)—the first recorded attempt at this very important drive mechanism. Such a machine, with its interconnected movements which bring it into the category of an automatic wea-

pon (its Greek name was *polybolos*—'multishooter'), would be likely to have had teething troubles, requiring adjustements and modifications of the original design. But Philon's criticism of it give no support to the view that it was 'mechanically impractical', as Sprague de Camp (1960, 145) suggested.

Philon also exploited the siphon as the basis of a variety of ingenious toys, intended, no doubt, to amuse his leisured patrons and their guests. Among these are several mechanized toilet basins, one of which was cleverly designed to go through a cycle, which begins when the user reaches for a piece of pumice stone being held out by a bronze hand, which disappears when the pumice is removed, being activated to set up an automatic flow of water out of the faucet, sufficient to enable the user to wash his hands. The cycle thus completed, the flow stops and the hand reappears with another piece of pumice to recommence the cycle. Contemporary with Philon was Archimedes of Syracuse (287–212 BC), a commanding figure who made outstanding contributions to the development of both mathematics and mechanics. It happens that his surviving works are theoretical in content, a fact which has caused a serious imbalance in the assessment of his work. Modern writers have tended to underplay his achievements in technology, and have even sought, against a strong tradition, to deny him the invention of the compound pulley, and of the water-screw which bears his name! Archimedes was undoubtedly in the mainstream of Greek science, but we know from the lucky discovery in 1906 of his book *Peri Methodou (On Method)* that he actually used mechanical models in order to obtain mathematical results, discarding them later at the proof stage.

Of more concern to the student of technology is the other recently discovered treatise in two volumes, *Peri Ochoumenon (On Floating Bodies)*, which provided a firm theoretical foundation for the science of hydrostatics, which deals with the behaviour of liquids under varying pressures. This point is particularly interesting since, as we have seen, the third century

178

witnessed the appearance of numerous mechanical devices based on hydrostatics, notably the force-pump and the water-organ (above, p. 17). No incident in the entire history of science is better known, and none illustrates more clearly the personality of this great man than the filled bath affair, which ended with the scientist rushing naked through the streets of Syracuse, shouting 'Eureka' (I've found it!), the find, of course, being the principle enunciated in Archimedes' Law, that a body wholly or partly immersed in a fluid loses weight in proportion to the fluid displaced.

Archimedes' wide-ranging interests led him into many different areas of technology. He supervised the construction, by a Corinthian naval architect named Archias, of a three-masted 'twentier' (see Appendix 11) called *The Lady of Syracuse,* which was probably the first three-masted sea-going vessel ever built. That Archimedes' interest in astronomy was not exclusively theoretical is evident from his book *On Sphere-Construction,* now unfortunately lost, which contained directions for making a planetarium, in which the heavenly bodies were set in their various orbits, and put in motion through an elaborate system of gears. The ingenious 'clockwork universe' as it has been called, was evidently no isolated invention which left no progeny; we now have solid proof of this fact in the form of an astronomical computer, operated on the same principle as Archimedes' planetarium, but with a different and practical objective, namely that of providing a ship's navigator with accurate information on demand on the phases of the moon and the position of the planets at any given point in time. The Antikythera machine, named after the place of discovery, was found in 1900 among the scattered contents of a Greek ship that had been wrecked off the coast of the Peloponnese, but the exact nature of the find (it came up from the sea-bed in the form of a number of heavily corroded lumps of bronze) was only revealed, after patient research by D. de Solla Price (1958). It has been dated to the early part of the first century BC. This particular model (we may reasonably assume that they were in common use) had a large main wheel, which drove some twenty or more smaller gear wheels. These in turn moved pointers on separate dials which indicated the phases of the moon and the position of the planets. Evidence that it actually worked is provided in the form of signs of parts having been repaired. The estimated date of the wreck is around 65 BC. This ancestor of the Arab astrolabe must have had many successors, but no mention of it has survived earlier than about AD 1000, when a similar instrument was seen and described by a Persian who visited India at that time.

With the tragic death of Archimedes at the hand of a Roman soldier this highly productive phase in the history of technology comes to an end. Hellenistic engineering was merged in that of the Romans, as the Hellenistic kingdoms in which its exponents flourished were swallowed up by the might of Rome. Only one writer, the mysterious Biton, survives from the two centuries that separate Archimedes from the Roman writer, Vitruvius. His work, which is no more than a compilation, contains six chapters, dedicated to an unspecified King Attalus (there were three kings of that name, covering between them more than a century of rule), and the six machines described (four catapults (see Appendix 15), a scaling ladder, and a siege tower), are ascribed by the compiler to five different inventors, whose dates range from around 350 BC to 190 BC. Successive commentators have given up in despair the attempt to interpret the writings of this author, since the difficulties of the text are in no way illuminated by the accompanying illustrations, some of which are palpably absurd: 'The scaling ladder', wrote Drachmann (1963, 11), 'is certainly an armchair invention, for the container for the counterpoise will take 44 tons of lead where 4 tons were plenty.'

In spite of the loss of so much of the original writings, it is evident that the technical achievements of the Hellenistic period were impressive. In the judgment of J. D. Bernal, 'they were in themselves quite sufficient to have produced the major mechanisms that gave rise to the Industrial Revolution—multiple-drive textile machinery and the steam-engine' (Bernal 1969). There were technical difficulties standing in the way of both these developments (for a detailed discussion of the failure to harness steam as a source of power, Landels 1978, 8 ff). Our concern here is the assessment of what the Hellenistic engineers did achieve, and of the reasons for their success. One plausible hypothesis to account for this remarkable burst of technical improvements is that it came about through cross-fertilization between Greece, Egypt and Egypt's neighbours, particularly Syria and Iran. Bernal (1969, 102) draws a comparison between the stimulus to inventiveness produced by the ready market for money-making devices, in the shape of temple illusions, ingenious toys and clocks, with that of palace entertainment in the Renaissance on the scientists and engineers of the later period. In both cases it was royal patronage that gave tangible support to men who were now eager to develop the theoretical side of mechanics, and at the same time to think up practical applications. More important still, the writers of this period 'occasionally provide evidence of deliberate research undertaken to find the optimum solution to a practical problem' (Lloyd 1973, 91 ff., esp. 97–98).

Hero of Alexandria

This versatile engineer, who wrote in Greek books on such widely differing subjects as pressure mechanics, automatic theatres, catapults, and on a prototype of our theodolite, the *Dioptra,* was formerly included among the technologists of the Hellenistic period, on no better evidence than that he was born in Alexandria, or lived a good part of his life there. He was given a variety of dates from the second century BC onwards. The hypothesis of a Hellenistic connection was exploded in 1938, when the German scholar Otto Neugebauer drew attention to the important fact that the data given by Hero in his *Dioptra* for an eclipse of the moon fits a known eclipse of AD 62, and no other. His

Dioptra must, therefore, have been written later than that year; observation of the eclipse was used to calculate the Great Circle distance from Rome to Alexandria, and since Hero's methods are less advanced than those used by the geographer Ptolemy, who flourished around AD 150, we may draw the reasonable conclusion that Hero's active life fell between about AD 50 and AD 120. The question is fully discussed by Landels (1978, 200 f.). Although forty years have passed, popular contributors to the history of ancient technology (e.g. Hodges 1970), continue to repeat the erroneous dating. The range of inventions and improvements associated with Hero's name can be gauged from the frequency with which his name is mentioned in these pages. The five volumes of the Teubner edition of his works, from which two extant works, the *Belopoeika (On the Construction of Catapults)* and the *Cheiroballistra (Hand Catapult),* are excluded, cover a total of eight topics. Vol. I contains *Pneumatika (Pressure Mechanisms)* and *Automatopoietike (On the Making of Automata,* i.e. puppet theatres) in Greek; vol. II *Mechanica* (Mechanics) and *Katoptike (On Mirrors)* in Arabic only (a late translation from the Greek text, now lost); vol. III contains *Metrika (On Mensuration)* and *Dioptra (The Theodolite);* vol. IV has two works, the *Definitions,* part of a collection of technical terms, mainly geometrical, and *Geometrika,* an introduction to geometry (both in Greek); vol. V contains *Stereometrika (On Solid Geometry)* and *Peri Metron (On Mensuration)* also in Greek. It will thus be seen that only the first three items and the sixth *(Dioptra)* are relevant to our subject.

The content of the *Pneumatika* shows that its author was both a good theoretician and a competent practitioner. The *Introduction,* on the properties of air, on the scattered nature of vacua, and the impossibility of a continuous vacuum, and on the way liquids behave under the force of gravity, shows his evident dependence on previous writers. But the section is no stock scientific commonplace, having scant relevance to the body of the text: the seventy-five or so items described consist of devices which depend on either the behaviour of air under compression or in partial vacuum or on that of liquids under the conditions set out in the *Introduction*. Furthermore, he is not content to repeat the findings of earlier scientists, but shows his originality in drawing together the views of two contrasting schools of thought: that of Strato of Lampsacus and that of Archimedes (on this see Lloyd 1973, 15 ff.; Landels 1978, 201 f.). The body of the text, consisting of a remarkable array of devices based on air or water pressure systems, has given rise to the widely held impression that Hero was no more than a cunning deviser of toys and other gadgetry designed either to amuse the idle rich or to bemuse ordinary folk with 'magical' transformations, like the temple doors that opened of their own accord to admit the worshipper after he had placed a burnt offering on the altar, or profit-making temple slot-machines dispensing water for ablutions. But the making of toys and working models has been a spare-time occupation of scientists in many periods of history, and the picture sometimes presented of Hero in the pay of the crafty priesthood of Alexandria, 'with their purifying bronze wheels, their ever-burning lamps with asbestos wicks, and their spirit-reflecting mirrors' is not very convincing. (See e.g. Farrington 1980, 199f.)

The miniature puppet theatres of the *Automatopoietike* were worked on a very simple power-source, that of the falling piston which had inspired Ktesibios long before (above, p. 177). The piston was kept at the top of the cylinder by filling the latter with millet or mustard seeds. When the seeds begin to run out, the piston falls down, pulling with it a cord which turns a shaft, which sets the apparatus in motion: 'In the first device, the whole apparatus moves forward on wheels and stops in the right position; figures then revolve or move about; doors open and close, fires burn up on tiny altars, and so on. At the end it moves back out of sight. The second device does not move as a whole but has more figures, performing a greater variety of movements, some of them (rather oddly, in view of the likely audience) representing artisans at work. As it requires less power, dry sand is used instead of millet seeds in the cylinder, since it runs out more slowly and enables the programme to last longer' (Landels 1978, 204).

The *Mechanike* is a comprehensive treatment of the subject, and consequently of central importance to the student of classical technology; it is, therefore, most unfortunate that the only surviving text, apart from a few fragments of the lost Greek original, is a translation into Arabic of the ninth century AD. There are also a number of text figures; whether they are the work of the translator or of Hero himself cannot be determined: some of the more puzzling features may be the result of errors made by later copyists. The student is fortunate in having available the results of the painstaking researches of the doyen of the subject, Dr A. G. Drachmann, whose book, *The Mechanical Technology of Greek and Roman Antiquity* (Copenhagen 1963), contains a full discussion of the most important illustrations as well as translations with commentaries of the key passages. After working hard to produce a feasible interpretation of the figure which illustrates the single-screw olive press, the Danish scholar writes: 'When I consider the difficulty of getting a complicated figure like that of the composite *galeagra* (the cage of planking) right, I find it quite a wonder how good the MS figures are after all' (1932, 135).

The *Mechanike* is divided into three books. Book I is *Theory*—gear-ratios, the parallelogram of forces, how to reduce or enlarge plane figures by means of the pantograph, and how to do the same for three-dimensional figures, using a different machine; next the nature of pulleys, and how to gear down using block-and-tackle. The section concludes with chapters on the centres of gravity and on the distribution of loads. Book II is *Mechanical Advantage,* and the various devices used to obtain it—the windlass, the lever, the compound pulley, the wedge and the screw, and how they can be used in combination. This book also contains a section on

181

the mathematics of worm-and-pinion gearing, including the pitch and spacing of the screw-threads, the most effective profile for the worm-groove and the engaging teeth. Book III is *Practical Applications* of these devices, illustrated mainly from cranes and other lifting devices, and various types of press. 'This section takes us right into the practical mechanics of the ancient world, and is one of the most valuable for the historian of technology' (Landels 1978, 206). The most interesting single item gives detailed instructions for making a female screw, that is, an inside screw-thread needed to make a nut-and-bolt:

21 As for the female screw, it is made in this way: we take a piece of hard wood, whose length is more than twice the length of the female screw, and its thickness like that of the female screw, and we make on one part on half the length of the piece of wood a screw in the way we have already described, and the depth of the screw-turns on it should be like the depth of the screw-turns which we want to turn in this female screw. And we turn from the other part as much as the thickness of the screw-turns, so that it becomes like a peg of equal thickness. And we draw two diameters on the base of the piece of wood, and we divide each of them into three equal parts. And we draw from one of the two points a line at right angles to the diameter. Then we draw from the two ends of this line at right angles to this diameter, on the whole length of the peg, two lines at right angles; and this is easy for us to do, if we place this peg along a straight board and scratch it with the scribe (?) until we reach the screw-furrow. Then we use with great care a fine saw, till we have sawed down to the screw-furrow; then we cut off this third that was marked on the peg. And we cut out, in the remaining two-thirds, in their middle, a groove like a canal on the whole lenght, and its size should be half the thickness of the remaining part. Then we take an iron rod and sharpen it according to the screw-turnings; then we fit it into the peg with the canal in it. Then we make its end come out into the screw-turns, after we have fastened the two pieces together with a strong fastening, so that the two are fixed together and cannot come apart at all. Then we take a small wedge, and insert it into the canal-like groove and knock it until the iron rod comes out and lies between the two parts. When we have done this we fix the screw into a piece of wood

into which there has been bored a hole that corresponds exactly to the thickness of the screw. Then we bore in the sides of this wide hole small holes side by side, and then fit into them small, oblique, round pegs and drive them in until they engage the screw-furrow. Then we take the plank in which we want to make the female screw, and we bore in it a hole the size of the screw-peg; and we make a joint between this plank and the plank into which we have fitted the screw by strong cross-pieces, which we fasten very solidly. Then we insert the peg that carries the wedge into the hole that is in the plank in which we want to cut the female screw, and we bore on the upper end of the screw holes in which we place handles. And we turn it till it comes into the plank, and we keep on turning it up and down, and we serve the wedge with blows again and again, until we have cut out the female screw with the furrow we wanted. And so we have made the female screw. And this is the figure, and with it ends the book.

Anyone seeking to interpret the above description will receive no help whatever from either of the accompanying figures; neither the Leiden illustration, nor that from the British Museum MS (printed by Drachmann, 1963, 136 and 137), as he drily observes, 'gives us the slightest idea of the details of the screw-cutter as described in the text'. To prove, against some critics, that Hero's text could be satisfactorily interpreted, Drachmann constructed a model, based on his own interpretation, which involved a few corrections of the text, which worked quite well. The reconstruction appears in cross-section at p. 138 and a photograph of the model at p. 139. Hero's account is immediately followed by a description of the direct screw press which required a female screw-thread in the beam. Drachmann concluded that the screw-cutter was a recent invention, and thinks it probable that the inventor was Hero himself (see, however, pp. 31 ff.).

The extended treatment given to a single passage in a technical treatise is to be justified on several grounds: first, we have here a full description of the making of an important machine-tool, a very rare phenomenon in the history of ancient technology; secondly, the tool is evidently a recent development, and may

182

be linked with an important piece of machinery, the screw-press, and thirdly, there are problems of interpretation which cannot be resolved by textual study alone, but require the construction of a working model, made in accordance with the specifications and indications in the text. This is indeed the only way by which real progress can be made in our still very imperfect understanding of ancient machines and techniques.

The Roman writers

By far the most important source that has come down to us is the Elder Pliny's great encyclopaedia, the *Historia Naturalis,* published in AD 77. Its industrious and indefatigable author, who died, characteristically, while attempting to follow the course of the great volcanic eruption of AD 79 at close range, is a controversial figure in the history of science and technology. He has been variously attacked, and indeed by some dismissed as an 'untrustworthy compiler, who never handled Aristotle in Greek' (Rose 1875), 'a copyist who plundered the encyclopaedia of Verrius Flaccus' (Rabenhorst 1907), and more generally as an undiscriminating if tireless collector with an overpowering interes in the untypical and the bizarre. But these defects, which are obvious at a glance, should not blind the discriminating reader to the author's merits. In a balanced appraisal W. H. Stahl (1962, 105 f.) has shown that the charge of fraudulent invention of authorities cannot be sustained, and suggests that on the evidence Pliny probably used Aristotle and Theophrastus both directly and through intermediaries; there is nothing remarkable about such inconsistencies: how many scholars today can testify that they always consult their ancient sources at first hand, and never through translations? As regards his use of earlier Latin authors, we have proof that he used Cato both ways, that is, directly, as well as through citations in Varro (*HN* 14.47; cf.11.273–76) where a comparison between Aristotle and Trogus on the same topic makes it likely that both authors were in front of him as he

wrote (see my discussion of the problem of Varro's use of earlier writers in *Aufstieg und Niedergang der Römischen Welt,* I, 1973, 4.477 ff.). In any case, in assessing the value of Pliny's contribution to any particular subject, the relevant question is not the extent to which he copies the work of others with or without acknowledgement, but the quality of the product. Of course, Pliny's defects as a writer in technical matters are obvious enough! His immense industry is not matched by a developed critical faculty. But the work is astonishingly uneven: systematic sifting of the text relating to any topic will commonly reveal some gold among the dross, and sometimes even a nugget or two! In particular, his eye for the unusual has often caused him to record some new implement or technical innovation, which can be validated from archaeological evidence. The list of these items includes, in agricultural technology, the long-handled 'Gallic' scythe (*HN* 18.261) and the Gallo-Roman harvesting machine for grain (*HN* 18.296); in food technology the mention of brewer's yeast as a leavening agent (*HN* 18.68). The first two of these items are confirmed by monumental evidence. A surprisingly high proportion of these innovations is reported by Pliny as originating outside the Mediterranean area, and particularly in the north-western provinces, where high standards of craftsmanship had already been reached before the Roman conquest, notably in the design of road vehicles, in textiles and in metallurgy (White 1975). The reader who consults Pliny in search of information on technical processes is doubly handicapped; first by the author's style of writing, secondly by corruption in the text. On the former count, the style is extremely inconsistent: longer passages of 'fine writing' with rhetorical flourishes and intolerably pompous perorations are found alongside blunt, curt, but by no means lucid, sentences, which are often badly constructed, as if the author was in too much of a hurry to attend to his grammar! As for corruption, it is exasperating to find that the text is found to be defective precisely where important technical matters are under discussion.

But the fact that Pliny is slipshod and uncritical should not lead us to underrate his importance; not only does he provide valuable information on a great variety of topics in the technology of mining, civil engineering, hydraulics and food; his pages reflect a compelling interest in recent inventions and in their development—an aspect of our subject which has been neglected by commentators and historians. The following examples will serve to illustrate the author's concern (or Pliny's) for this side of the subject:

Pliny on ploughshares (*HN* 18.171–2)

The author begins his account of agricultural operations by stating that there are several different types of ploughshare; the list that follows begins with a simple ground-opening device, and ends with the first identifiable reference to a plough with wheels. The central portion is taken up with three varieties of ploughshare, representing adaptations of basic designs to different soil-texture and ground-cover conditions. The passage gains added importance from the paucity of other evidence: the agricultural writers have practically nothing to say about the design of shares; surviving shares, though common on Romano-British sites, are very rare for Italy, and representations of ploughs on monuments are far from easy to interpret. Pliny's account of ploughshares is important from the historical point of view; he shows some awareness of the evolution of the implement, and the order in which he represents his material here suggests that he was aware of a connection between the double-bladed share (the fourth of his five types) and the plough with a wheeled forecarriage. This last was a development of great importance, since it enabled farmers to exploit the heavier clay soils of the north-western provinces, and expand wheat production (e.g. in eastern Britain). This expansion, it would appear, not only satisfied the demands of the garrisons of this most heavily defended outpost of Roman power but made it possible for British wheat to be exported to those northern frontier areas which were subjected in late imperial times to increasing barbarian pressure.

Pliny on improvements in the technique of oil-making (*HN* 18.317)

The making of olive oil requires two successive operations, the first, that of removing the kernels from the berries without crushing them, and so affecting the flavour of the oil; the second, that of removing the oil and separating it from the watery scum or *amurca*. The oil had to be 'expressed', that is, squeezed out of the mush after the milling process, under pressure, either by means of a lever and ropes, a lever and screw, a direct screw, or a system of wedges. In the lever-press the material had to be placed as close as possible to the fulcrum, the free end of the long, massive lever being brought down either by ropes wound over a capstan, or by a screw which passed through the free end of the lever. In its simplest form, that of a tree-trunk anchored at one end, and weighed down with stones at the other, the lever-press was known in pre-classical times. The lever-press described in detail by the Elder Cato (*De Agri Cultura*, 18) represents two improvements on the primitive type: first, the construction of massive timber standards anchored into the floor to take the thrust; secondly, the substitution of ropes, pulleys and a capstan turned by handspakes for the clumsy stone weights. The next stage was the discovery of better methods of raising and lowering the press-beam or lever. This was achieved by using a screw instead of ropes and a capstan. The earlier version, mentioned by Pliny as a Greek invention of the first century BC, consisted of a wooden spar, which passed through the free end of the lever; the spar was fitted with a screw thread fixed to the floor, and a nut, which was attached to the lever. The lever could be brought down by turning the spar. The next improvement was to make the spar fast to a heavy stone weight, which could be brought down so as to descend into a socket in the floor. This arrangement also eliminated a defect in the earlier version: since the lever did not descend in a straight line but in an arc, the lowering process tended to force the screw out of the vertical, causing jamming. For effective working the lever had to be very long (40 ft, according to Vitr. *De Arch.* 6.7.3). The next improvement consisted of a complete change of design; if the screw could be adequately anchored to withstand the severe reverse thrust at the stage of maximum pressure, the clumsy space-wasting lever could be eliminated, and replaced by a direct-screw press. The problem had already been tackled by Greek designers, and two types of direct-screw press are mentioned, along with two of the older types of press, in the third book of the *Mechanike* of Hero, published around the middle of the first century AD, and therefore contemporaneous with Pliny. The new type of press had a horizontal beam, which descended as the screws were turned overhead (much in the manner of a letter-press). An indispensable technical requirement for this horizontal beam-press is that the beam should be fitted with a female screwthread. Drachmann (1932, 83 f.) argued that this development could well have been linked with, and dependent upon, the invention of a screw-cutting machine, described by Hero in the last chapter of his *Mechanike* it was this coincidence in time that led Drachmann to put forward the hypothesis that the screw-cutter was not invented before about AD 50. The 'Archimedean' screw, which is based on the use of a continuous external screw-thread, is at least three centuries older. Attractive though it is, the hypothesis cannot stand; where the thread is to be of metal, a screw-cutter is needed, but female threads can easily be cut in wood with normal carpentry skills and tools.

The arrangement of Pliny's encyclopaedia is as follows: Book I consists of an exhaustive table of contents, book by book, which is specially designed, as the author explains (*Preface* 33) to save the time of busy readers, together with impressive lists of the authorities used in the preparation of each book; II–VI are concerned with cosmology, geography and ethnology. Zoology, covered by four books, is pre-faced by a book on man (VII); next comes a vast series of books on botany (XII–XIX), a comprehensive heading which includes both agriculture (XVIII) and horticulture (XIX). This is followed by a run of thirteen books grouped together under the heading of *materia medica* (XX–XXVII deal with botanical remedies, and XXVIII–XXXII with zoological). The last four books deal with mineralogy and metallurgy, but include information about painters and sculptors, and the materials they use. In his *Preface* Pliny describes his *Historia Naturalis* as a compendium of twenty thousand facts, compiled from some two thousand volumes and one hundred selected authorities.

We have already noticed (above, p. 183) how Pliny's critics have sought to discredit him as a fraudulent collector and compiler. He has not fared any too well at the hands of his translators: the Bohn Library version, by Bostock and Riley, is usually more reliable than the Loeb, and has the additional benefit of useful footnotes on many of the numerous technical problems. To sum up: Pliny's encyclopaedia deserves the serious attention of all students of classical science and technology; in spite of its obvious shortcomings, it is much more than an assemblage of curious information to amuse the idle antiquarian or dilettante.

Vitruvius

The importance of his book *De Architectura,* published early in the reign of Augustus, and dedicated to that emperor, is two-fold: there is first, its rarity, or rather the fact that it is a rare survivor from a category both numerous and important, that of the technical monograph (for other survivors see below, s.v. Frontinus); secondly, since the qualifications and aptitudes of an architect, according to Vitruvius, included those required of a town planner and civil engineer, so that the ten books include sections on water supply and the various machines used for raising, storing and transmitting it to the user, civic defence and the design of the necessary siege-engines, catapults and ballistae, as well as valuable discussions of the various building materials available to Roman architects (see Appendix 9). Vitruvius declares in his *Introduction* that he intends to write like an architect, not like a rhetor or grammarian. If we ignore some of the more pretentious introductions and excursuses, in which the author, somewhat against the grain of his own nature, appears to be making rhetorical gestures appropriate to the rules of the contemporary literary game, and concentrate on the body of the text, we find that the monograph is in no way unworthy of its important position. The introduction to Book V (on the elements of town planning) is cited here as giving a good indication of the author's practical and unpretentious attitude:

Those who have filled books of unusually large size, Emperor, in setting forth their intellectual ideas and doctrines, have thus made a very great and striking addition to the authority of their works. I could wish that circumstances made this equally permissible in the case of my subject, so that the authority of this treatise might be enhanced by increasing its size; but this is not so easy as might be supposed. Writing on architecture is not like writing history or poetry. History enthralls the reader from its very nature; for it holds out the prospect of various novelties' [here one is reminded of a strikingly similar passage in Livy's *Preface*]. 'poetry, with its rhythms and metres, its refinement in the arrangement of words, and the delivery in verse of the sentiments expressed by the several characters to one another, delights the reader's senses, and leads him smoothly on to the very end of the work. But this cannot apply to architectural treatises, because those terms which originate in the special needs of the art give rise to obscurity of ideas from the unusual character of the language. Hence, while the things themselves are unfamiliar, and their names not in common use, if in addition, the principles are set forth in a very diffuse manner without any attempt at concision and exposition in a few clear sentences, such fullness and amplitude of treatment will be only a hindrance, and will give the reader nothing but hazy notions. Therefore, when I mention obscure terms, and the symmetrical proportions of parts of buildings, I shall provide brief explanations, so that they may be committed to memory; for when thus expressed, the mind will be enabled to grasp them more readily. Furthermore,

since I have observed that our fellow-citizens are preoccupied with public affairs and private business, I have thought it best to write briefly, so that my readers, who have only small intervals of leisure, may be able to comprehend the subjects in a short time.

Readers who have to rely on translations of classical texts will find themselves in difficulties with Vitruvius. The heavy incidence of technical terms and descriptions has opened the way for manuscript corruption; and more work is needed on the text before we can say with any authority how much of the obscurity is to be ascribed to the author's technical or stylistic shortcomings. What is clear is that Vitruvius, while rather more than a mere compiler, is often imprecise in presenting to his readers the Greek theories by which he sets so much store. His editors and commentators have been unfair to the author in several ways: thus he has been taken to task for inappropriate application of important scientific discoveries to practical problems; the illustration he gives of the practical value of the Theorem of Pythagoras in designing staircases for private houses (IX, *Introd.* 7) has been dismissed as 'fatuous'. The complaint seems unjustified; a proportion of 3:4 between risers and treads gives an inclination of 36°; a staircase built on this formula would certainly not be dangerous; and the method proposed would make it easy for the builder to get his steps vertical. The reader must be thankful that this eminently practical author has confined his conventional philosophical genuflections to the *Prefaces,* and conveyed in the body of the text so much of practical value to builders and planners. We must also be thankful that his broad view of the knowledge and training required of the architect has resulted in the inclusion in his work of valuable chapters on hydraulic engineering and military technology.

The contents of the work are as follows: Book I introduces the subject by treating of the scope of architecture, and the training required of those who intend to practise it, and goes on to discuss the principles of town planning and civic design, including a section on the art of fortification. Book II deals with building mater-

186

ials, and is prefaced by some remarks on the life of primitive man, and on the progress that has been made in providing him with shelter and other facilities for community life. Books III and IV are concerned with the design of temples, with details of the four classical orders of architecture used in their construction, Doric, Ionic, Corinthian and Tuscan. Book V relates to secular buildings, Book VI to private houses, and Book VII to interior decoration. Book VIII is concerned with water in all its aspects: how to find it, how to assess its quality for various purposes, how to use it in levelling devices, and how to convey it to the urban consumer. Book IX deals with sundials and other methods of measuring time, and Book X with machines used in the building industry, followed by a section on catapults and other military devices. Whether Vitruvius was, like other writers of text-books before and since his day, an unsuccessful practitioner of his art, is debatable (we know of no building designed by him apart from a basilica in a modest country town). Great teachers are not necessarily outstanding practitioners of the craft they teach. In any case this question is not immediately relevant to our purposes. For the student of technology the importance of Vitruvius' contribution is twofold: first, he is an essentially practical man: as one critic has put it 'he writes execrable Latin, but he knows his work'; secondly, his curriculum includes both engineering and architecture, with the result that Books VIII and X have provided valuable information on topics which a more conventional 'classical' approach to the subject would have excluded.

Frontinus

In Sextus Julius Frontinus we encounter a politician, a public figure of some importance in the late first century AD, who, at the peak of his career accepted a senior administrative post in the service of the City of Rome, which prompted him to write a book on the water supply. Born around AD 35 Frontinus passed through the stages of an official career appropriate to a man of the senatorial class, including a spell of four years as governor of Britain, and no less than three consulships, in two of which he was honoured by having the emperor Trajan as his colleague. In his early sixties he was invited by the emperor Nerva to take on the post of *curator aquarum*, or Administrator of the Water Supply. In this capacity he had the ultimate responsibility for the efficient operation of a public undertaking of great importance to the health of the community, with a staff of seven hundred slaves (including the technical managers on the job). Thus far, a perfectly normal official career for a man of his rank. The unusual feature is that he ended his tour of duty by writing a book—not a memoir enlivened by personal anecdote or made tedious by reminiscence, but a solid two-volume survey of the water-supply system, which has earned him a place of some importance in the history of this branch of technology. In this work, the author not only provides a detailed account of the organization of a fairly complicated supply system, but tells us a good deal about the difficulties he encountered on the job. He also explains very clearly how he came to write the book. The pertinent observations he makes at the outset concerning the relations between himself as administrator and his subordinates make fascinating reading, and would be appreciated by any contemporary Minister of State taking over the portfolio of a department of which he had no previous experience. He writes:

I therefore considered that my first and most important task was to learn thoroughly what it is that I have undertaken ... there is nothing so debasing for an honourable man than to perform the duties of the office entrusted to him according to the directions of his subordinates. This however is a course that must be followed whenever an inexperienced official relies on the practical knowledge of his assistants; these services, though necessary ... should nevertheless be regarded as no more than a sort of hand and tool of the responsible official.

The first and most important thing to notice about this treatise is that it is not a standard compilation or text-book of the subject, but a highly personal document, embodying much shrewd comment based on his own experience of the system as he found it. In several passages the author deplores the carelessness of some of his predecessors, in others he alludes to the positive dishonesty of some members of the staff who have unlawfully diverted public supplies for private use, the net result of negligence and dishonesty being a serious loss of water at the distribution end. In his detailed commentary on the work Clement Herschel, the hydraulic engineer, inferred from internal evidence that Frontinus' main purpose was to reform the administration of the water supply, and suggested that his appointment by the emperor Nerva was in line with imperial policy which aimed at removing this sector of the administration from the control of the freedmen who had previously been in charge. In support of this opinion we may notice that the work contains numerous recommendations designed to check abuses and to secure a better supply. Among the most important of these is a proposal to employ the water from several aqueducts to serve individual districts, so as to avoid a total breakdown of supply to any one area when essential repairs were undertaken; an increase in the public supply by the provision of additional reservoirs and fountains (Ch. 88); and measures to obtain effective control of the workforce in the interests of efficiency, providing both work schedules and records of work actually done (Ch. 117).

The military career of Frontinus had also borne literary fruit in the shape of a four-volume compilation entitled *Strategemata*. This earlier work, published about AD 84, soon after his return from Britain, is a treatise on the art of war, consisting of a collection of the sayings and actions of the most famous generals of the past. Tediously anecdotal and thoroughly second-rate, this work is vastly inferior to its successor. Frontinus also made important contributions to the collection of manuals for instruction in the art of land surveying, known as the *Corpus Agrimensorum*, dealing with types of land, centuriation, and land disputes; they are the earliest accounts we have of these important topics; none of them is complete, but all are illustrated (see Dilke 1971, 105–8).

In using Frontinus as a source, it is important that we should bear in mind that he was not an engineer, but an administrator (see the excellent picture of this many-sided man in action drawn by Landels 1978, 213–15). When he discusses technical points, we must not expect him to be faultless. There are examples of carelessness in calculation, and on the important topic of rates of flow in open channels he is notably defective (see Appendix 12), but we are not to infer that Roman water engineers were unfamiliar with the criteria needed to estimate them. With these qualifications, and in an area where the products of technical skill are so much in evidence, and the literary remains so scanty, we must be grateful for the accident that has preserved his most important work.

Appendices

APPENDIX 1

Industry: distribution and location in the Roman Empire

Workers: occupational distribution in Roman and Modern Egypt (1927)

Distribution and location of industry

The coherent political unity which we call the Roman Empire provided those who lived within its Mediterranean borders with the tangible benefits of peace, security by sea, and a vast network of roads, which opened new trading opportunities, along with easier access to old, and fresh access to new sources of raw materials. It scarcely needs to be emphasized that only a privileged minority were able to enjoy these benefits; to extend them down the social scale would have required an economic revolution, fuelled by technical innovation and a fuller exploitation of resources. That this change did not come about is a fact of history; why it did not happen is a difficult question to answer, but one or two focal points may be mentioned here.

The first limiting factor was the size of the 'market' for manufactured goods. We have earlier drawn attention to the importance in the history of the Greek city-states of the doctrine of economic self-sufficiency, to which the vast majority of those states, with rare exceptions such as Corinth and Athens, clearly conformed. The economic needs of the most of their citizens could readily be met from local resources, through the labours of small, independent craftsmen, or by means of workshops operating on a modest scale.

The conquest of the Mediterranean by the Romans provided opportunities in the new provinces, both east and west, of which merchants of many different nationalities were not slow to take advantage. Italian businessmen established themselves in Asia, Jews and Syrians in Gaul; nor were the old boundaries of status observed in this expanding area of economic activity, as Martin Frederiksen reminded us, 'the organization of trade and manufacturing in the Roman world moved easily across the status boundaries, involving men at all social levels' (*JRS* 65 (1975), 167).

What were the main commodities that were being traded? To what degree did the range extend beyond the traditional export of corn, oil and wine, together with the more important metals? The fact that pottery containers and tiles bear marks indicating both the centre of manufacture and the maker's name has made it possible to trace specific trade links, and the easily recognizable design of certain types (e.g. 'Arretine' red ware) has thrown valuable light on the rise and decline of manufacturing centres in response to changing consumer demand. What commodities then were being traded? We noticed earlier that in the seaborne trade of classical Greece, the basic agricultural products, corn, wine and oil dominated exports, and the traffic in animal skins and those 'mute instruments' the slaves provided the return cargoes. Exports of manufactured goods did play a part in the economy of the Empire, notably pottery and textiles, though the extent of the traffic in these commodities has been much debated. Those who hold that the export trade in virtually all manufactured products consisted exclusively of luxury articles can point with satisfaction to Diocletian's *Edictum de pretiis* where the entries under manufactured clothing show that both cheap and luxury garments were produced in many different centres (Jones 1964, vol 2, 848 ff.). It should, however, be noted that 'it was only high quality garments whose production was concentrated in a few towns, whence they were exported to all parts of the empire' (Jones, *loc. cit.*). As for ordinary clothing, the evidence is that little of the vast quantity needed was made at home, even the poor buying their clothes ready made (see Jones, *Econ. Hist. Rev.* 13 (1960), 183–92). In the light of this fact, it is not at all surprising to find evidence for the view 'that there probably was a market for cheap textiles in the largest cities, and a long-distance trade to supply them' (Jones, 1964, vol. 2, 850).

The Industrial Revolution of the eighteenth and nineteenth centuries was in its earlier phase bound up with two technical inventions in the process of spinning textile fibres, Arkwright's spinning jenny and Crompton's mule. These developments took place in a situation of increasing demand by consumers. No such increases in the level of demand occurred in the Roman world. The vast majority of the population, living on the products of subsistence agriculture, had few wants beyond those that could be easily satisfied either by home production or by that of small workshops. The demands of the military, which had certain distinct effects in stimulating invention and innovation in the armaments industry (see Appendix 15), would *prima facie* be likely to promote centralized production of clothing and other standard equipment. 'Clothes', it has been pointed out, 'are easily portable objects, even more portable than the raw materials out of which they are made, and here if anywhere large-scale manufacture and trade might seem to have been practical and profitable.' (Jones, 1964, 848.) In fact, the manufacture of both civilian and military clothing was widely dispersed, and the famous centres, it seems, produced only high-grade garments. However, a single but significant passage in the *Life of Melania (Vit. Mel* (G) 8; (L) 8) implies regular importation of cheap workmen's garments from Antioch to Rome in the fifth century AD. While we have plenty of evidence both from Greek and Roman sources of centres that were well known for the manufacture of particular items of clothing (see Michell 1957, 187 ff.; Jones 1964, 849 f.), and numerous references to items of armour and weaponry coming from a variety of places, we are not entitled to infer factory production of such items, still less a flourishing trade in these products. The exceptions were, of course, the state factories for the production of woollens *(gynaecia),* and for linen garments *(linyphia),* set up by Diocletian along with armament factories *(fabricae)* and dyeworks *(baphia)* (see below p. 191).

Occupational distribution of workers

The only statistical information we have on this important topic comes from Egypt, which may be said to occupy a middle position, 'more industrialised than the Danubian Provinces, less than Syria or the

valleys of Meuse or Moselle' (R. Remondon, in *Histoire générale du travail,* (1959), 334, who also provides the tables on p. 191).

Some observations on manufacturing industry

Attention has been drawn in an early chapter of this book to the vigorous and continuing controversy concerning the nature of the economies of the classical world and the role and scale of industry within them. The now conventional view, which goes back to J. Hasebroek (1933), and indeed to Max Weber before him, is that the ancient economies were essentially household economies, and that factory production, with the notable exception of the state-run factories of the Roman Empire (see below) was precluded by the prevailing conditions of the economy: 'widespread prevalence of self-sufficiency in necessities was enough to put a brake on extensive production for export' (Finley 1973, 138). Reference has been made in several chapters of this book to the various factors which worked against the development of export markets, and to other factors working in their favour. In an appendix concerned with various aspects of industry, it seems appropriate that comments on the location and distribution of industry and those employed in different occupations should be complemented by some observations on the ways in which manufacturing industry was conducted, drawing together such indicators as we have of movement away from the generally prevailing pattern of household economy. Information is hard to come by, and is unevenly distributed as between different areas of manufacture. Under such conditions, to speak of industrial organization is highly misleading, to write of it impertinent, the paucity of information on all the major heads, viz. structure, role of the entrepreneur, the manager, the supervisor, and the work-people giving rise to fruitless debate over issues which are incapable of exposition because they can scarcely be formulated.

In his fundamental study, undertaken many years ago, A. W. Persson wrote:

in order to understand the fragmentary information we have concerning the relations of the State with manufacturing industry under the Roman Empire, we must first look at Ptolemaic Egypt, where similar (but better documented) conditions existed. Here the royal monopolies controlled food, luxuries and manufactured goods, while the temples had the royal patent to manufacture textiles. There are no signs of royal factories. Before the Roman take-over the monopolies had been replaced by a taxation system. In Roman Egypt

190

taxes were levied on manufacture and sales. The State got its supplies of textiles (a) from state factories, (b) from taxes in kind. (1923, 55).

Italian industries were concentrated in·a limited number of regions: in Campania, and round Tarentum in the south; in the north, in Etruria, around Bologna and Mutina, and at Aquileia. Two fundamental trends appear, which have characterized industrial development throughout its history; first the exploitation of local raw materials (the local clay at Arretium, and the wool of Samnium and Apulia processed at Pompeii and Tarentum respectively); secondly, the importation of raw materials from territories opened up by Rome's expansion; an outstanding example is that of Aquileia, where the mines of Noricum, Rhaetia and Dalmatia provided her craftsmen with the means of making the varied array of arms, agricultural implements, of bronze and silver ware that make up the rich contents of her museum.

The study of industrial organization is beset with problems. It is easy enough to set forth the topics that need to be discussed: the role of slaves and free citizens, the division of labour and the supervisory function; the scale of enterprises and the role of the state as manufacturer. But broad generalizations are no substitute for detailed analysis of evidence, and studies of individual industries are badly needed. The textile industry, which has received welcome attention in recent years, provides interesting evidence of the operation of forces that do not fit into the 'orthodox' pattern of the ancient economy.

The State factories of the later Roman Empire consistute a classic instance of an all-powerful government wielding its authority to ensure a regular supply of items considered to be of vital importance to the armed forces, viz. armour, weapons and uniforms (see e.g. *Cod. Theod.* 10.22.2—imperial order dated AD 388 issued to arms factories; *Not. Dig. Occ.* 11.45.73; 12.26–7). The latter document contains a full list of woollen mills in the western empire in the early fifth century; there were four in Italy, three in Illyricum, five in Gaul, one in Africa and one in Britain. Little is known about the management of these establishments. The workers, who were state slaves bound by hereditary requirements to their occupation, were probably supplied with rations; each factory had to reach a fixed production quota of garments per year (Jones 1964, vol. 2, 836–37, and p. 31). The raw wool was obtained either by compulsory levy or by purchase.

The Pompeian wool trade has been the subject of a detailed study by W. O. Moeller (1976). The relevant findings of his enquiry are as follows, (1) division of labour, male weavers, 'and therefore factory production' are found in the industry at Pompeii; (2) all classes, including the municipal aristocracy, were involved in the industry; (3) the local aristocrats did not have a negative attitude towards trade (*contra* Finley 1973); (4) nor did they suffer socially or politically as a result of being in industry; (5) Italy was a major textile centre (*contra* Wild 1970), important both for output and technical innovation. Not all the contentions of this provocative study will survive the onslaught of the critics; but they provide an opportunity for fresh debate and more detailed analysis of the problems they raise.

Percentages of total employed

Occupations	Roman Egypt	Modern Egypt (1927)
agriculture	*68%	62%
mines and quarries	0.15	0.50
industry	**9.50	10
transport	1.10	2
commerce and finance	4.55	6
public service	5	3
private services	2	2
domestics	3.20	3.50
various	6.50	11

Percentages employed in different industries

food	22.50	22
textiles	25	39
wood	17	4
leather	3	4
glass, pottery	2.50	0.75
building materials	6.50	3
metal	9.50	5.25
various	12.50	20

	USA 1820	France 1827
* cf. agric., forestry, fishing	72.3%	63%
** cf. industry, mines, building	6.3%	

APPENDIX 2

Roman techniques:
inheritance, adoption, invention

Many of the claims made in the following lists of inventions are highly speculative, resting either on a literary reference to a supposed inventor (see Ch. 4) or on an alleged first appearance of a device on a bas-relief, painting or the like; conversely, claims that the classical world was ignorant of this or that principle or technique are frequently advanced on the flimsiest of evidence, and the plea of *non liquet* seldom occurs. In an area of enquiry dominated by claims and counter-claims, by dogmatic assertions and equally dogmatic denials, far too much attention has been given to the matter of inventions, and far too little to that of innovation and development. Numerous references have been made in the body of the text to these questions in relation to the specific area of technology concerned (see Index, s.v. 'crank', 'pivoted axle'), but an attempt might be usefully made to examine the question as a whole.

The most comprehensive list is that furnished by the editor of the *Histoire générale des techniques* (Paris, 1961), Henri Daumas, in the first volume of that encyclopaedic work, on pp. 248–49. The entries are set out in three categories:

Inherited from the Hellenistic period and improved
(a) cement (including hydraulic cement), enabling architectural innovation and development (e.g. vault, dome, staircase, bridge, aqueduct, sewers).
(b) arboriculture (including viticulture), cadastration, earthenware (moulded), glass-blowing, harbours, hypocausts, manuring, mosaic, oyster-breeding, pins [the reference is presumably to the *fibula,* which is in fact a safety-pin], road-building, rotation of crops (partial), steering oars, tiling, timber framing, war engines, water-mill.

Adopted from the barbarians
acclimatization of plants (especially chestnut, peach, apricot), barrel, chairs (basketry), enamel (polychrome), garments (sewn), plough (asymmetrical), reaping machine, semi-rigid nail (?), soap, steel (for swords), most wheeled vehicles. [I have treated the topic of road vehicle design, along with some other aspects of the material culture of the Celtic peoples in 'Technology in North-West Europe before the Romans', *Museum Africum* 4 (1975), 43 ff. Of outstanding importance are the 'one-piece' wheels for passenger vehicles, which will have required a very high degree of skill on the part of the wheelwright.]

Invented by the Romans
aqueducts, awnings (for arena), carpenter's brace, codex book, candles (wax), weaver's comb, curtains (for theatre), vaulted dome, drainage wheel (for mines), chemical fertilizer, force-pump, frame-saw, gimlet, window glass, martress, milestone, plan, punch, ratchet-bow, pivoted scissors, screw-press (direct action), shorthand, steel yard, turnkey.

Professor Finley (1973, 146–47) gives a remarkably brief list covering both Greek and Roman inventions, 'gear screw, rotary mill, water-mill, glass-blowing, concrete, hollow bronze-casting, the lateen sail and a few more'. This mere handful stands in sharp contrast with the French historian's list of twenty-three inventions credited to the Romans alone! Finley offers no evidence, nor does he indicate that much larger claims have been made, but goes on to discuss the question of blocks to technical progress, that is, technical advances or improvements which ought, in his opinion, to have been made, since they 'affected essential and profitable activities' (*op. cit.* 147). His first example, taken from mining, is quite specific: removal of water from mine-workings was vital, yet 'no one found a way to improve on hand bailing, the water-wheel operated by foot-treadle and perhaps the Archimedean screw for drainage devices; so technically simple a device as the chain-pump with animal-power is unattested' *(loc. cit.)*.

The first comment to be made on the above passage is that there is no 'perhaps' about the use of the Archimedean screw for mine drainage; widely used for raising irrigation water, especially on the Nile, where it can still be seen, it made its first appearance in the mines in Spain (early first century AD) Diodorus (V.37.4, cited by Healy 1978, 95) refers to it as 'an exceptionally ingenious machine by which an enormous amount of water is thrown out, to one's astonishment, by means of a trifling amount of labour'. As for the chain-pump, this device, powered by an undershot waterwheel, is described by the Hellenistic engineer Philon of Byzantium, and was well known when Vitruvius wrote his treatise *De Architectura*. Landels (1978, 75 f.) estimates the output of such a pumping device as 540 hl or 12,000 galls per day, running 24 hours a day with low-cost supervision. But an animal-powered pump of this type will have involved the excavation of an underground hall big enough for an ox to move around in. The bucket-chain 'must almost certainly have been more expensive to construct than the bucket-wheel' (Landels 1978, 73). His discussion of the bucket-hoist and indeed his entire chapter on water-pumps (Ch. 3) will repay careful study. The extensive claims made by Daumas are, for the most part, set out without the supporting evidence, of which by far the most important is that furnished by archaeological sources, whether in the form of artefacts or of representations. What is needed here is less of dogmatic assertion, and a more rigorous sifting of evidence, less emphasis on invention and more on developments of inherited techniques, a point well made by Professor Finley in an important paper written half a generation ago:

Paradoxically, there was both more and less technical progress in the ancient world than the standard picture reveals. There was more, provided we avoid the mistake of hunting solely for great radical inventions and we also look at developments within the limits of the traditional techniques. There was less—far less—if we avoid the reverse mistake and look not merely for the appearance of an invention, but also for the extent of its employment. (Finley 1965, 29.)

Technical development under the Romans: the barbarian contribution

The technical accomplishments of the Celtic peoples of north-western Europe are well attested, especially in metallurgy (from the La Tène culture of the Iron Age onwards), in the art of the wheelwright and the design of vehicles for road transport, and in that of agricultural implements and machines. E. A. Thompson, reviewing F. W. Walbank's *Decline of the Roman empire in the West* in *CR* 63 (1949), in a discussion of the effect of slavery on technical progress writes: 'It is no coincidence that the major inventions and discoveries of the period of Roman greatness were not made by Romans at all, but by barbarians.' In support of this thesis he cites Lynn White, jr., *Speculum* 15 (1940), 144 (with extensive bibliography) and notes from White's list the following: trousers, fur coats, cloisonné jewellery, felt-making, skis, soap, butter, barrel- and tub-making, the cultivation of rye, oats, spelt and hops, the heavy wheeled plough, stirrups, horseshoes and falconry. A glance at Daumas' list (above, p. 192) shows that only four items are common to both. There is no mention of wheeled vehicles! Returning to the topic in the introduction to his edition of the anonymous *De Rebus Bellici* (Oxford 1952), Thompson quotes the author of that treatise on the remarkable fact that the barbarian

peoples make up for their lack of eloquence and other civilized accomplishments by their inventive capacities, and refers to an incident during the siege of Perra in Colchis, when Huns showed Romans how to construct a new type of battering-ram (Procopius, *Gothic War* 8.2.27 ff).

Some tentative conclusions

Such lists of inventions and innovations are of very little use, unless the claims are supported by evidence; in justice to Daumas, it must be said that he has attempted to cite the evidence: nine items in his list of alleged Roman inventions are carpenters' tools, and all are provided with references, archaeological or literary or both. Several of them are known from that rich source of information on crafts and working conditions, the corpus of Gallo-Roman funerary reliefs. But how valid is this category as evidence of first appearance, and therefore of invention? The fact that Gallo-Romans of the second and third centuries AD were unusual among contemporary peoples in depicting on their tombstones the implements and the occupational environment of the deceased should lead us to exercise caution in the use of their evidence. Lists unaccompanied by evidence should be ignored.

APPENDIX 3

On the power output and efficiency of various machines

The assertions made by modern contributors on this subject are not always soundly based. To take a key example, that of grain-mills, the 'Pompeian' animal-operated mills are of two sizes, the smaller being driven by donkeys, the larger by horses or mules (both types are noticed in the *Edictum de pretiis* of AD 301). The larger size will evidently have produced more meal per hour, yet Forbes (*Studies* II, 88) gives only a single figure (see table below) and offers no calculations to support it. Such estimates, if they are to serve any useful purpose, must be based on experiments with mills of different dimensions and power sources. In the case of water-mills, the variables are more numerous, and estimates will inevitably be tentative: output will depend on (a) rate of flow, (b) whether the wheel is undershot or overshot, (c) revolutions per minute (dependent not merely on (a) and (b), but on size (d), and number (e) of vanes), (f) diameter of wheel, and (g) gear ratio. A glance at the data presented will show how very inadequate is the available information concerning even the best preserved mills on the list. The tables on mills and water-raising machines should, therefore, be treated with reserve.

Mills

Type	Horse-power	Output	Source of report
1 Saddle-quern			
2 2-man hand-mill		kg p.h.	
3 Donkey-mill	0.5 hp	kg p.h.	
4 Horse mill	0.6 hp (in theory 3 hp)	?	
5 'Vitruvian' undershot water-mill fed by aqueduct at 45 rpm wheel diam. 2.2 m			
6 Barbegal multiple system (2×8 wheels) width 70 cm, ht 2.2 m (overshot)	$2-2\frac{1}{2}$ hp per unit	15–20 kg per set (= 240–320 kg p.u.) (Forbes, *Studies* II, 91)	Landels (1978, 22)

Water-raising machines

Type	Horse-power	Head	Output	Source of report
1 Water-screw (1 man treading)	0.1	1.2 m (4 ft)	*c.* 2,000 galls p.h.	
2 Tympanum (compartment wheel) diam. 3 m (2 men treading)	0.2	1 m (3.3 ft)	400 l (88 g) per rev. (at 2 rev. per min.) *c.* 10,000 g.p.h.	Landels (1978, 65)
3 Bucket-chain (? 1 man treading)	0.1	15 m (52 ft)	180 g.p.h.	Landels (1978, 74)
4 Same, operated by A(5)	3		*c.* 12,000 g.p.d. (= 24 hrs non-stop)	*ibid.*
5 Ctesibian force-pump[i]		4.9 m (16 ft)	140 g.p.h. (at 1 str. per sec. & 95% efficiency)	Landels (1978, 79)
6 'Hollow-rim' waterwheel diam. 4.4 m		3.65 m	116 g.p.h.[ii]	Landels (1978, 69)

Based on the 'Dramont D' pumps (Landels 1978, 81). The Silchester pump was estimated to have produced 300 g.p.h. at 20 strokes per minute and 80% efficiency. Much larger pumps are now known, the largest of the type with oak cylinders producing more than 1,100 g.p.h. The eight-pair multiple system used for mine-draining on a sector of the Rio Tinto complex could lift about 2,400 g.p.h. at a head of almost 100 ft. (Landels, 1968, 69). By way of comparison the earliest steam-operated beam-engine (Newcomen and Savery, 1712), ran at 12 strokes per minute, lifting 10 gallons of water through tiers of pumps to a height of 153 ft, equivalent to about $5\frac{1}{2}$ h.p.

APPENDIX 4

Why the ancient world did not produce a steam-engine

John Landels (1978) includes among the factors which prevented industrialization in the Graeco-Roman world the failure of the Greeks and Romans to harness steam as a power source. In an interesting attempt to discover how near they came to a workable steam-engine, he enumerates the minimum essential elements: these are (1) metal cylinders with (2) pistons that fit, without being so loose as to lose too much steam, or so tight as to cause too much friction, (3) a workable valve-mechanism, and (4) a boiler for raising the head of steam. A crank, he points out, which is required for converting the rectilinear movement of the pistons into a rotary movement, is not essential, adding that it was almost certainly a mediaeval invention.

All four items listed above were known to engineers of the first century AD, when Hero of Alexandria was working; each of them appears in one or another of the devices he constructed: the piston and cylinders in his fire-fighting pump, the required valve-mechanism in the device known as 'Hero's Fountain', and the boiler in his steam-machine, which, as he himself points out, was never more than a high-speed toy. All the required elements were available, so why no steam-engine? Landels mentions three reasons that have so far been given by modern writers on the problem: (1) Hero never thought of reversing the action of his pump, so as to admit liquid or air under pressure, and so forcing the piston up the cylinder, and setting a beam in motion, as in the earliest beam engines (Newcomen and Savery, 1712); (2) perhaps he made the attempt at (1), but either blew himself up, or took fright; (3) he lacked the necessary high-grade fuel. Landels rejects all three of the above reasons, indicating in a curiously oblique reference that the real causes were 'probably economic and social'.

APPENDIX 5

Technical improvements and innovations in agriculture and the processing of food, wine and oil

Agriculture

Implements

1 Winged ploughshare, with vertical cutter to make the furrow, and sharp lateral edges to cut off roots of weeds (Pliny, *HN* 18.172).

2 Plough with forecarriage, 'recently invented in Rhaetia (E. Switzerland), fitted with spade-shaped share' for heavy soils (Pliny, *HN* 18. 172).

3 Improved harrow fitted with iron teeth (Pliny, *HN* 18.186).

4 Improved whetstone for sharpening scythes (Pliny, *HN* 18.261).

5 Long-handled Gallic scythe: 'used on the latifundia; economizes by cutting the stalks at middle height and missing the shorter ones' (Pliny, *HN* 18.261)

6 The 'Great Chesterford' scythes. Fitted with very long blades and heavy flanged backstrips; probably fitted with bar handles for wide sweep. Known from three sites in Britain. No literary record. (White 1967, 208 f.).

7 'Heading' machines for grain: used on Gallic latifundia; two types *(vallus, carpentum),* and several variations of the *vallus* type (Pliny, *HN* 18.296 *(vallus)*; Palladius 7.2.2–4 *(carpentum)*.

8 Improved threshing-sledge *(plostellum punicum)* with driver mounted on sledge. From eastern Spain: Carthaginian origin? (Varro, *De Re Rustica* 1.52.1).

Techniques

9 Ploughing-in of panic and millet for bumper crops; adopted by Roman farmers from Alpine tribe of Salassi (Pliny, *HN* 18.182).

10 Fertilizing with marls: discovered by farmers in Britain and Gaul (Pliny, *HN* 17.42).

11 New method of grafting fruit-trees, combining layering and grafting: invented by Columella, *(De Re Rustica* 5.11.13); reported by Pliny (*HN* 17.137).

Food processing

Grain-milling

1 The 'hopper-rubber': 4th-century Greek. Improvement on the ancient 'up-and-down' quern.

2 The rotary hand-mill: ? 3rd-century Greek. Improvement on (1).

3 The animal powered 'hour-glass' rotary mill.

4 The undershot water-mill: lst century BC. Most important invention to date; cheap to install—no mill-race; no aqueduct; slower than (5).

5 The overshot mill. Date of invention unrecorded. Expensive to install; ? slow to develop; hence no. 3 persists in competition.

Wine and olive-pressing

1 Improved lever-press using capstan and leather rope.

2 Lever-press substituting drum + pulleys for capstan.

3 As No. (2) but with suspended weight to increase pull.

4 Substituting vertical screw for drum; screw firmly fixed in floor, so that it can revolve and draw down the lever by means of the nut above it.

5 Direct screw-press using pressure from above as in letterpress.

APPENDIX 6

Technical development of the water-mill

The view presented earlier on this controversial topic (Part I, pp. 65 ff.) does not conform to the thesis originally put forward by Marc Bloch in 1935, and generally accepted by writers on this subject. According to Bloch, the water-mill, an early application of the undershot water-wheel, which was already known to Vitruvius, remained largely unexploited during the next three centuries, to be fully developed several centuries later by the more resourceful technologists of the mediaeval period. This appendix includes two annotated tables, the first containing literary references (15 in all), the second archaeological (18 in all). They owe much to the labours of two scholars, Dr E. Maroti (1975) for the first, and Dr O. Wikander (1979) for the second. Most of those who have written on this topic have contented themselves with a handful of references. They make no claim to completeness, but in view of the emphasis laid by so many scholars on the alleged 'time-lag' between invention and application, it seemed appropriate that we should have the available evidence set out in a

convenient tabulated form, with some observations on the points that appear to be particularly significant, leaving readers to decide for themselves which of the two accounts of the matter is the more acceptable.

Archaeological evidence

Most writers on the subject have confined their archaeological references to no more than five well-known examples, namely the excavated mill-sites of Venafro (Campania), Barbegal (S. France), Athens (the Agora) and Rome (the Janiculum site), together with a piece of mosaic from the Palace of the Emperors at Istanbul, which displays, in very poor perspective, a wheel and its mill-house. The actual number of remains is, however, considerably more than four; two certain specimens and a possible third have been identified from a single area, Hadrian's Wall, which is situated in a remote sector of a distant province of the Roman Empire, and therefore of special importance in relation to the much debated question of the extent of the development of waterwheels and mills in the Roman period. The most recent contributor to the subject, Dr Wikander (1979, 14) reports that ten are known from Britain alone, and that the scarcity of finds elsewhere need cause no surprise, since the best sites were chosen first and seldom abandoned, so that all traces of the earlier structures, at least in populated areas, will have been obliterated by later developments on the same site. The blank spaces in the table on p. 200 f. point to the obvious gaps in the record; more than half the entries contain mere lists of finds. Now that the old view that the Roman period saw virtually no development cannot be sustained any longer (see e.g. Maroti 1975), renewed interest may produce some of the missing evidence.

Types of water-mill
and areas of development

Of the two types of mill known in classical times, the undershot variety, described by Vitruvius (10.5.2), though much cheaper to install than the overshot type with its wall and mill-race, requires a steady flow of water; but the seasonal variations in the level of the major rivers of Italy will have immobilized the mills for considerable stretches of the year (Forbes, *Studies* II, 89 and 95). Smaller streams, however, could have been provided with the necessary dams and mill-races to ensure a constant, if modest supply,

196

to meet local requirements in the rural areas. Wikander (1979, 14) suggests that it is these rural mills that Pliny has in mind in the admittedly corrupt passage (see *Table of Literary Sources*, No. 5. on p. 198) which nevertheless implies that water-mills were by no means rare in Italy. Unfortunately there is as yet no archaeological evidence to support his theory. If Wikander's reports are authenticated, the orthodox view of a long time-lag, and a very slow take-up of the invention will be difficult to maintain, since the literary evidence can hardly be said to support it. Nor does the theory fit the evidence for the early mediaeval period, as is shown by the following comment: 'I know of only four water-mills from the next centuries, three of them in the Athenian Agora, and a large and sophisticated mill with three vertical waterwheels working in parallel at Old Windsor from the VIIIth or IXth century' (Simpson 1976, 184).

The reported sites

1 *The Venafro mill* The material remains include (i) the mill-race, (ii) part of an aqueduct, (iii) prints on the tufa wall of the housing of the millstones, which were covered by 3 m of sediments.

Dimensions: undershot wheel: diam 25 Roman digits (= 1 m 86). Nave diam. 10 digits (=74 cm). The wheel had 18 spokes.

Output: Forbes, *Studies* II, 93 gives 3 h.p. At 46 revs per minute the mill could grind 150 kg per hour. Compare the donkey-mill with 21–25 kg per hour and the 2-man hand-mill with 7 kg per hour.

References: Mostra Augustea della Romanità. Catalogue. Rome, 1938, 3rd ed., 591, no. 27; P. Jacono, *Ann. Lavori Pubb.* no. 2, 217; C. Rendl 'Ein römisches Wasserad'. *Wasserkraft und Wasserwirtschaft* 34 (1939), 142 ff.; Forbes, *Studies* II, 90.

2 *The Barbegal multiple system* The substantial visible remains of this remarkable complex have been known since the seventeenth century, and variously interpreted. They include an aqueduct (fourth-century in design) running parallel to its first-century neighbour which supplied the nearby colony of Arelate (Arles) with water, and leading into a large reservoir (122 m² in area); a series of stone-built enclosures ran down a slope of *c.* 30°, housing the eight pairs of wheels and the mill-lades which connected each of the two sets. All the elements except the wheels are represented by remains *in situ* and the last-named can be identified, and their size determined, by mineral deposits from the water falling on the walls.

The reservoir, which resembles that of the Janiculum series in Rome. (No. 3 below), and the Viroconium series in Britain (Ashby 1935, 46), performed the essential task of regulating the flow of the water the speed of revolution of the wheels.

Statistics: diam. of wheels 2.20 m (7 1/3 ft); cf. Agora Mill (No. 4 below) 3 m 24. Width of wheels: 0.70 m (2½ ft); cf. Agora 0.54 m (1⅔ ft); The overshot wheels were turned in an anti-clockwise direction by means of attached scoops (contrast the vanes *(pinnae)* of the undershot 'Vitruvian' wheel).

References: F. Benoit (1940), 19–80; R. J. Forbes, *Studies* II, 91–92.

3 *The Janiculum series* These mills were set up on the slope of the Janiculum Hill, and were fed by water from the Aqua Traiana. Their existence is attested by several literary references (see *Table of Literary Sources*, No. 14, p. 199), and by an important edict issued at the end of the fifth century AD by the City Prefect (see same table, No. 12). The installations suffered both from the Gothic invaders, and from later works aimed at restoring the Aqua Traiana supply, but some portions were uncovered when new streets were being laid out near the site on which the American Academy was later erected, and the finds were recorded by Lanciani in 1880. Further discoveries were made when the foundations of that institution were being laid in 1912–13 (part of the aqueduct passes under the Academy). The physical remains are insufficient to furnish any information as to size or output, but their importance is stressed by Procopius, since their loss provoked a food crisis which was met by the floating mills set up by Belisarius (*Gothic Wars* 5.19).

References: *CIL* VI.1711 (The Edict of 480); Lanciani in *Monum. Ant. dei Lincei* I. 480 ff.; A. W. Van Buren and G. P. Stevens in *MAAR* I, 59 ff.; pl. 15.

4 *The Athenian Agora mill* This fine example of an overshot mill was discovered in 1933, south of the Stoa of Attalos, and fully described by A. W. Parsons in *Hesperia* 5 (1936), 70–122, pl. 1. The mill-race, like the mill-house, was sunk into the bedrock. The brick courses in the latter provided support for the cross-beams, which was and still is in Greece the common method of scaffolding support as we can see it on the tomb painting of Trebius Iustus (Burford 1972, 176, pl. 69). As with the Venafro mill, the heavy deposits of lime made by the water splashing continuously on the wall betrays the size and type of wheel used. The wheel was overshot, its outline beautifully preserved

'as a series of concentric grooves formed by the rim not running true, giving the exact diameter of 3.24 m (10½ ft)' (p. 80). The diameter of the vertical gear wheel can also measured from scorings into the stone (its diameter was 1.11 m (3⅔ ft)), and the diameter of the horizontal gear can be approximately calculated from the marble bearing block; it was ± 1.36 m (4 ft 11 ins), a striking confirmation of Vitruvius' statement that the horizontal gear was the larger, with a ratio of less than 1:1 (see Landels 1978, 23–24 and fig. 4).

5–7 *The mills on Hadrian's Wall* 'In Britain examples are few but significant. The army was driving mills by water at three points on Hadrian's Wall, using for the purpose respectively the North Tyne[5] and Irthing[6] rivers, and the Haltwhistle Burn.[7] The stone core of the waterwheel's hub still exists at the North Tyne, and two hub-cores of the same kind have recently been found at Lincoln.[8-9] At the villa of Wollaston Hill[10] power-driven stones were reused as flooring in a building close to an artifical water-leat; one occurs at the villa of Chedworth,[11] while an iron spindle for similar stones from Great Chesterford[12] is in the Cambridge University Museum of Archaeology and Ethnology. The wide distribution of these relics is impressive.' Thus writes Sir Ian Richmond in a round-up of the evidence available in 1955. Wikander's tally presumably includes some finds reported since that date. The information obtained is set out in the table below.

Most of the remaining references come from British sites. A systematic review of all known sites would undoubtedly produce a long list, making the coventional notion of a long time-lag between invention and widespread use even more difficult to maintain. The slow development of the invention may have been due to two factors: first, the low output of grain from the slower undershot mill; secondly, the high cost of installing the much more efficient overshot mill, which also required both a strong and a permanent supply of water.

Table of Literary Sources

No.	Date	Source	Text	Technical data
1	2nd half of 1st cent. BC	Antipater of Sidon (*Anth. Pal.* IX. 418)	Verses celebrating the ending of the drudgery of hand-milling by invention of the water-mill.	Implies overshot mill in common use by now?
2	*c.* 50 BC	Lucretius, *Nat.* V. 516.	Water-wheels used as a source of comparison with unfamiliar rotary movement.	Waterwheels obviously common by now; no certain reference here to water-mills.
3	Internal date *c.* 80 BC	Strabo, *Geogr. XII.* 3.30	Mithridates VI, K. of Pontus (*c.* 120–63 BC) had a water-mill on his country estate.	Proves (?) that the water-mill appeared first in Hellenized Anatolia.
4	*c.* 25 BC	Vitruvius, *De Arch.,* X. 5. 2.	First precise description of a water-mill. The 'Vitruvian' wheel is certainly of undershot type, used to drive a bucket-chain hoist, but 'water-mills are turned on the same principle' *(loc. cit.)*.	Much cheaper to install than the overshot type. Gear ratio of less than 1 : 1 means slow rate of grinding, but no lack of power.
5	before AD 79	Pliny, *HN* 18.97.	Water-mills cited with the simple pestle-and-mortar as commonly in use over most of Italy (?) (the text is uncertain).	Brooks and streams of rural Italy could be easily used; cf. the archaeological evidence from rural Britain (Wikander 1979, 14 & refs.).
6	AD 301	Diocletian, *Edictum de pretiis* 15.54.	Water-mills listed along with animal- and hand-mills, with a price-tag of 2000 *denarii*. The donkey-mill was priced at 1250.	Water-mills must have become very common by this time: note small difference between price of horse-mill (1500) and water-mill.

No.	Date	Source	Text	Technical data
7	4th cent. AD	Palladius, *Op. Agr.* I.41.	Recommends water-mills (when bath-waste water readily available) as dispensing with labour of men and animals.	Use of water-mills increases with increasing manpower shortage.
8	AD 398	Edict of Arcadius and Honorius: *Cod. Theod.* XIV. 15. 4.	Heavy fines against those demanding water destined for the water-mills on which Rome relies for provisions.	Water-mills now more important than animal mills in supplying Rome (Rickman, 1980, 206 & refs).
9	c. AD 420	Cassianus of Marseilles (c. 360–435), *Conlatio* I. 18. 1–2.	The swift movement of the mind compared with that of the water-mill.	Both are driven by powerful forces, which neither can arrest. Water-mills obviously common in Gaul.
10	c. AD 480	Edict of Dynamius. *CIL* VI. 1711.	Against dishonest millers. (1) official weighing-machines to be set up on the Janiculum; (2) fixed payment for millers; (3) heavy penalties for abuse.	Water-mills on Janiculum vital to food-supply of Rome (cf. No. 8).
11.		*CIL* III. 14 969/2.	On the right to use water for water-mills near Promona (Dalmatia).	Not noticed by writers on water-mills, but discussed by Wilkes, *Dalmatia,* London, 1969, 239.
12	c. AD 450	*Book of Senchas Mor*	MS (contemporary with St Patrick) refers to horizontal mills in Ireland.	Sprague de Camp (1960), 232.
13	c. AD 534–538 C. *praef. praetorio*	Cassiodorus *Varia* XI. 39. 2 .	Large number of water-mills in Rome cited as evidence of size of population still living in the city.	Not noticed by most writers on water-mills, but cited by Wikander (1979).
14	c. AD 538 (*Vivarium* founded)	Cassiodorus, *De Inst.* I. 29.	In planning the *Vivarium,* C. mentions a stream suitable for water-mills as an important aspect of his community project.	
15	AD 537	Procopius, *Gothic Wars* V. 19. 8–19.	The Janiculum water-mills, fed by an aqueduct, were located below the summit and fortified by a perimeter wall, which was destroyed by the Goths, who then broke up the conduits.	This crisis was met by the construction of a series of floating mills on the Tiber (below, No. 16).
16	AD 537	Procopius, *Gothic Wars* V. 19 ff.	'Belisarius... fastened ropes from the two banks of the river, stretching them as tight as possible, then attached to them two boats side by side and two feet apart' (at points of maximum flow) 'and, placing two mills on each boat, he suspended between them the usual mechanism for turning mills. Below this point he fixed other boats, each attached to the one behind in order...'	
17	c. AD 575	Gregory of Tours, *Hist. Franc.* III. 19.	Description of the water-mills at Dijon (Côte d'Or): 'Before the gates it' (the smaller stream) 'turns mill-wheels with marvellous speed'.	

Table of Archaeological Sources

No.	*Date*	*Place*	*Finds*	*Technical data*	*References*
1		Venafro, Italy	mill-race; part of aqueduct prints of millstones on wall of housing.	undershot wheel of 18 spokes; diam. of wheel 1.86 m (6 ½ ft), of hub 74 cm (3 ft).	Reindl, in *Wasserkraft...* vol. 34 (1939), 142.
2	late 3rd cent. or early 4th cent	Barbegal, near Arles, Provence	8 pairs overshot wheels on a sloping site, fed by aqueduct.	wheels: diam. 2.20 m 7 ½ ft) width 0.70 m (2 ½ ft) overshot type, with anti-clockwise movement.	Benoit (1940), 19–80; Forbes, *Studies* II. 90–91.
3	after AD	Janiculum, Rome	numerous millstones; remains of sluice		Van Buren and Stevens, in *MAAR* I (1915–16), 59 ff.; XI. (1933), 69 ff.
4	mid or late 5th cent.	Athens, the Agora	mill-race, mill-house, lime deposits defining wheel; scoring on wall defining vertical gear wheel.	wheel: diam 3.24 m (10 ½ ft) gear wheels: vertical: 1.11 m (3 ⅔ ft); horizontal: *c*. 1.36 m (4 ft 11); ratio: less than 1 : 1.	Parsons, in *Hesperia* 5 (1936), 70 ff.
5		Chollerford, N. Tyne	stone hub of mill-wheel; mill-race *in situ*.		Simpson (1976).
6		Willingford R. Irthing	paved mill-race; stone spindle		Simpson (1976).
7	3rd cent. AD	Haltwhistle Burn	part of wooden mill-wheel millstones.		
8		Lincoln I	part of stone hub of mill-wheel.		Richmond, in Wacher (ed.) (1966) 1. 83, pl. IX.
9		Lincoln II	part of stone hub of mill-wheel.		Richmond, in Wacher (ed.) (1966) 1 pl. IX, p. 83.
10		Chedworth Villa, Glos.	large millstone (pierced) of millstone grit; pieces of others, of old red sandstone.	diameter of stone.	Mus. Cat. No. 139.
11		Wollaston Villa, Glos.	part of millstone reused as flooring; artifical mill-lade.	diameter of stones: 75 cm (3 ½ in).	*Arch. Camb.* 93 (1938), 122.
12		Great Chesterford, Cambs.	iron spindle (the vertical shaft of the water-mill).		*Arch. J.* vol. 13 (1856), pl. 3, n. 28. Liversidge (1968) 104 in *Mus. Arch.* & *Eth.* Cambridge.

No.	Date	Place	Finds	Technical data	References
13	(1) 2nd cent AD (2) 4th cent AD	Ickham (Kent)	2 water-mills on river (1) 2nd cent. AD (eastern); (2) 4th cent. western); the latter technically more advanced.	No. 1 has planked revetments on both sides of the river; midstream structure set on piles; No. 2 has mill-race 1.4 m wide × 13 m long; pieces of upper and lower millstones.	*Britannia* 6 (1975), 283 ff.
14		Saalburg (Zugmantel)	2 iron spindles; 'lantern' pinion of oak; upper and lower millstones. 760 mm (31 ½ in) (lower), depth 15 cm	spindles: length: 80 cm (2 ft) thickness: 3 cm (1 ¼ in), stones: diam. 755 mm (32 in) (upper); (6 ¼ in).	*Saalburgjahrb.* 3 (1912), 88–95; fig. 38, p. 82; pl. XVII. on lantern pinion see Moritz (1958) (1975), 283 ff.
15		Witton (Glos.)			Jarrett, *JRS* 52 (1962), 176.
16		St Albans (Herts) Verulamium			Wacher (1966)

Appendix 7

Sea trade

There were three categories in this area of trade: first, essential minerals, iron, copper, lead and tin which were needed everywhere, and only mined in specific and limited regions; second, wheat, oil and wine (exportable surpluses also confined to certain areas); lastly, a range of luxury and semi-luxury articles, from high-grade textiles to perfumes, unguents and other exotic products, as well as incense, which was certainly not a luxury item! To this third category, which was greatly extended under the Roman Empire, belong also the valuable marbles carried regardless of cost, as Jones reminds us (1964, 846), by wealthy senators as well as by the imperial government. There were, in addition, two items which do not fit into any of the above categories: first, the long timber beams for the building trade, and secondly, ship-timber.

For ancient Greece we have abundant evidence of a widespread trade in timber (Meiggs, 1982, 350 ff., in a chapter well stocked with detailed references and discussion of all aspects of the timber trade). We also have Theophrastus' valuable classification of the varieties of timber used in the building trade. In the fourth century BC the principal external supplies, all described as superior to the local products, came from Macedonia, Pontus and the region south of the Sea of Marmara (*HP* 5.2.1). Meiggs *(loc. cit.)* notes that Syria, Cilicia, Cyprus and Phoenicia, all well-known suppliers in the Roman period, are absent from the list. Among later documents is a delightful exchange of correspondence from sixth-century Egypt (a treeless area), between the patriarch of Alexandria, Eulogius, who needed timber, and Pope Gregory the Great, who supplied it. When the consignment arrived, the patriarch complained that the beams were too short. Gregory replied that the ship he had sent to fetch them was not long enough! (Greg. *Ep.* 8.28; 9. 175, quoted by Jones, 1964, 846, n. 54). While there is 'no (*scil.* literary)' evidence or any regular trade in building stone or timber' (Jones, itals.) there is at any rate some important evidence for the first of these in shipwrecked merchantmen (see p. 155). 'In spite of the great gaps in our evidence', writes Meiggs (1982, 354), 'it is clear timber played a significant part in the sea-borne trade of the Mediterranean'. But what was the scale of the commercial activity of which it formed a sizeable part? Revision of the now established and fashionable 'minimal' view of commercial activity has recently been called for by Hopkins, who reminds us that farming was a hazardous occupation, so that famines and gluts occurred in situations which would inevitably have stimulated sizeable flows of trade when transport conditions permitted. Hopkins contends that the main difference between marketing in the ancient world and ours was the absence in antiquity of regular trade routes for the exchange of commodities between areas of surplus agricultural production and areas of manufacture (cited by Peacock, 1982, 153). The map on p. 28. plots some of the main centres of supply of foodstuffs and raw materials, and of manufacture. Our knowledge of the organization of long-distance trade is derived mainly, though by no means exclusively, from literary sources. The topic has been trenchantly reviewed in a recent summary by Peacock (1982, 158 ff.), who also stresses the importance of the inland waterways of the north-west, and the evidence of accumulated wealth in the region furnished by funerary inscriptions, adding: 'it does not follow that because these are scarce elsewhere, civilian markets failed to generate significant long-distance trade' (*op. cit.* 159). See also Table 13 on p. 238.

APPENDIX 8

Architects and engineers

The great public buildings discussed in Chapter 7, and the fine combinations of engineering and architecture of Chapter 8, were not the work of specialists in our sense of the term, but of workers in many different crafts and trades assembled for the purpose.

Their diverse efforts were co-ordinated, as Dr Busford explains (1972, 102 ff.), by men who were neither architects nor engineers, but whose knowledge and experience covered both disciplines. An *architekton* was literally a controller and organizer of the labour force. For confirmation we need only turn to the contents table of Vitruvius' great treatise *De Architectura,* and to the (probably embellished) list of accomplishments required in such a director given in the Introduction (Burford has translated the major part of the disquisition at pp. 103 ff.). What does clearly emerge is the conviction that a good general education is a prime requirement, as is a range of practical experience. Recent reports on the need for more practical training in the curricula of prospective architects and engineers prove that Vitruvius' view of the matter was well founded: 'An architect should not be as good a master of language as Aristarchus, but he should not be illiterate either ... nor a painter of the calibre of Apelles, though competent in drawing; nor a sculptor such as were Myron and Polycleitus, though possessed of some knowledge of the plastic art ...' (*De Arch.* I 1.13).

Training
Architects

Where did an architect like Iktinos (who built the Parthenon) get his training? Did he and the other famous architects of the Periclean Age cut their teeth on the temple of Zeus at Olympia, 'the only significant building to have been put up on the Greek mainland between 480 and 450'? (Burford 1972, 103, who speaks of 'an international pool of temple-building experience' on which the various specialists could draw *(ibid.)*).

In Rome we know of certain families in which the profession was handed on from father to son. From these establishments came the men who filled posts in the Public Works Departments, the *praefecti fabrum,* etc. This pattern was upset by the arrival of Greeks in Rome after the conquest in 146 BC, and the profession became mixed. Some of the Greek *maestros* were slaves, and many of these could boast of a variety of skills as is testified by inscriptions, e.g. *CIL* XIV.472—the epitaph of a house-born slave who became a professor of military and civil engineering. Note the frequency with which troops were used on large construction works (*CIL* VII mentions no less than 128 works carried out by the legions of Roman Britain). In the surviving inscriptions referring to military architects only one is an officer, the rest are referred to *milites;* many were *veterani:*

CIL XI.20 T. Flavius Rufus *centurio. CIL* VI.3182 T. Aelius Martialis *architectus equitum singularium. CIL* VI.2725 C. Vedennius Moderatus legionary; he was *architectus armamentarii imperatoris. CIL* VIII.2850 M. Cornelius Festus legionary and architect. *CIL* XII.623 Delius *architectus navalis.*

Engineers

Here again, although literary evidence is lacking, we may assume that, as in many other crafts, the training of civilian engineers was via apprenticeship. In the Roman period, however, army services provided a wide range of facilities for the training of engineers, from the regular laying out of the camps (see Stuart Jones, 1912, 22–43) to the construction of roads, bridges, canals, siege works, including mines (on the civil engineering aspects see R. W. Davies, *Peace-Time Occupations of the Roman Army,* diss. Durham University, 1959). Julius Caesar's engineers built a pile bridge across the Saône in a single day: his famous trestle bridge over the Rhine, with its 56 tents of piles and a total length of 1,400 feet took only four days to erect (BG 4.17). Civil engineers are referred to in the following inscriptions:

Ingenui (free) *CIL* II.2259; III.6588; V.3464; VI.9153; X.1443; 6126; 8093. *CIL* XII.186. See Iul. Caecilianus Architect. *Liberti* (freedmen) V.1886; 2095; VI.872–45; 9151; 9154; IX.1052; X.841; X.1614; XI.3945 (of these four were *liberti Augustorum). Servi* (slaves) VI.8726; VII.1082; IX.2886; X.8146; 4587; XII.2993 (of these only the first was an imperial slave).

Organization of work

As Dr Coulton has pointed out (1977, 51), the fact that construction begins at the bottom of a building means that initial mistakes, whether in the actual design or in the execution of that design by the builder and his team will be ruinously expensive: 'an architect more than any other artist needs a technique of design', to enable him to visualize the finished building; he also needs a satisfactory method of communicating his intentions to the builders; for these purposes modern architects use a variety of techniques, including scale drawing, ground-plans, elevations and sections, maquettes and models. What techniques did Greek and Roman architects employ to secure these results? The problem is somewhat complicated: we know from Vitruvius (I.1.4 and L.2.2) that ground plans, elevations and perspective drawings were standard practice in his day. But there has been disagreement about Greek practice, and this cannot be easily removed, since the meaning of the Greek technical terms is itself a matter of dispute (for the details see Coulton 1977, Ch. 3 'The problem of design', at pp. 54–57). The power of tradition was so great that the early Greek temple builders had no need of these technical aids. Even as late as the fourth century BC, when Philon designed his arsenal for the Piraeus (see Ch. 7, p. 75), a building which 'did not follow a closely-defined convention' (Coulton, 1977, 55), the specifications contain no reference whatever to drawings, plans or elevations. With regard to models and maquettes, full-sized specimens and templates were certainly used for tiles and other decorative features (see Coulton, pl. 15, for the tile standard found at Assos in Asia Minor, one of three which have turned up on different sites). Scale models are attested for catapult design (Appendix 15), an area where radical experiments were being carried on in the late third century, but the evidence for their use in architecture is ambiguous. Under the Roman Empire, on the other hand, when architects were competing with one another in selling their schemes to prospective patrons, the use of scale models is appropriate and well attested (Coulton 1977, n. 74, p. 72). As for the supply of stone and other building material, the inscriptions make it clear that quarrying was done according to specification, sufficient margins being allowed to give room for manœuvre to the stonemason working on site. Professor Heyman (*JSAH* 31.1 (1972), 3–9) shows in fascinating detail how this actually worked out on the Parthenon architraves. Precise measurements of the blocks reveal that the end block is significantly shorter than the remainder; this means that the mason fitted the blocks as they came (with minor trimming where needed), and was left with major cutting on one block only.

Working with varieties of stone requires varying degrees of skill (marbles are notoriously difficult). There is evidence that the workers usually knew the local materials—thus at Epidaurus Argive and Corinthian masons worked with tufa and limestone with which they were familiar. Foremen were, then, as now, important for successful operations, where much depends on good labour relations and reasonable working conditions (on these aspects see Martin 1965, 46–54): some extant contracts make provision for food and clothing allowances.

APPENDIX 9

Building materials and methods

Materials

Brick and tile: Greek

Sun-dried brick as a building material has a very long history, stretching back into early Near-Eastern practice, where it is still carried on, using age-old techniques. Mud-brick walls with vertical timbers embedded in them to form a system closely resembling the later Elizabethan 'half-timbering' make an appearance in Greece in the first temple of Artemis Orthia at Sparta, dated to as early as the ninth century (Dawkins 1929), and were still being used in the Heraeum at Olympia (*c.* 600 BC). Here the mud-brick walls rested on a masonry base: to this primitive material we owe the preservation of the superb statue of Hermes by Praxiteles, which was securely buried in the remains of the disintegrated brickwork.

The mud-brick walls of these early temples, which comprised only a front portico (the *pronaos*), giving access to the rectangular *cella* 'were very inadequately protected by the slightly overhanging eaves' (Dinsmoor 1950, 47). The subsequent extension of the architecture of the portico to cover the flanks and rear, which established the characteristic peripheral colonnade was probably due as much to the need to protect the walls (unprotected mud-brick surfaces would have a very limited life even in the drier parts of Greece) as to the desire to make the building more impressive. With the rapid change-over to stone construction brick virtually disappears as a building material, except for defence, where its capacity to absorb the blows of battering-rams made it a first choice for city walls (see Pausanias 8.8.8). Baked tiles, on the other hand, played an important role, being moulded into a variety of shapes and sizes for water conduits and drains, for pavements and roofs, and as revetments for cornices. The fine terracotta acroterion that crowned the gable of the Heraion at Olympia, which was 2.2 m or 7 ft 7 ins in diameter, has a

central hole, presumably to allow for contraction during baking; a further sign of technical innovation is the use of rougher clay for the core, the finer material being applied, before baking, to form the decorative mouldings. On surviving roofing tiles this exterior coating resembles a hard vitreous glaze.

Brick and tile: Roman

Roman bricks were characteristically slim (varying between 3.5 cm and 4.5 cm (1½ ins—2 ins) in thickness), and were lightly baked, and so more porous than Roman tiles. This enabled them to absorb mortar and so make a firmer bond. Tiles, on the other hand, were baked longer to make them waterproof. Both bricks and tiles were cut for a variety of purposes; typical of late Republican and early imperial building are wall facings made from roof tiles cut into triangles *(opus testaceum)*. Now sun-dried brick, if it is to last, needs a coating of plaster; but Roman builders, having graduated to baked brick, continued to plaster them. A notable exception to this hidebound conservatism is to be found, along with other architecturally exciting developments, at Ostia, where brickwork was employed with consummate skill and freedom as a firm surface to a highly developed system of construction. There brick was used for external and partition walls in the ordinary coursed form, but strongly bonded brick arches were used over shop windows and for carrying staircases; and many floors are of laminated brick and concrete. The three standard sizes of bricks comprised the 20 cm square *bessales,* the 44 cm square *sesquipedales* (= '1½-footers'), and the giant 60 cm square *bipedales* (= '2-footers') which are cut (or sawn) into triangles for wall facings, and into rectangles for arches. It is worth noting that some of the most distinguished brick buildings of the present century have been carried out in the slim and elegant Roman brickwork, 'a structural unit of this kind capable of being put to so diverse a use would cheapen and facilitate modern construction' (Atkinson and Bagenal 1926, 162).

Roman concrete

The development and application of concrete mixes of varying composition and appropriate strengths rank among the most outstanding contributions of the Romans to structure. Its use in foundation courses and in walls may well have developed from the long established method of filling the space between solid stone facings with a mixture of rubble and clay. Rome's original contributions consist of two techniques: first, the replacement of the clay by mortar, and especially the immensely strong and waterproof variety made from pozzolana; secondly, the use of wooden shuttering (for all-concrete walls) and of elaborate types of formed timber-work for the construction of all-concrete vaults and domes. In wall construction the method was not to mix the aggregates with the liquid cement before pouring, but to lay courses of rubble (*caementa*—the meaning is quite different from our word 'cement') or of brick-bats or broken tile in courses, then to pour the liquid cement over it so that each course was penetrated and solidifid. The resultant mixture was so strong that it could be left with no facing at all or with a surface studding of stone or brick. Close examination shows the amount of care given to the arrangement of the aggregates, layers of larger stones alternating with a mix consisting of broken lumps of tufa, peperino and broken brick—a cheap and economical system of construction, in which the materials were ready to hand, and the waste products could be incorporated! The marks of the shuttering and lateral struts are still clearly visible on numerous buildings. In the three well-known systems of faced concrete construction (*opus incertum, opus reticulatum, opus testaceum*) the concrete core was strengthened by bonding courses of bricks passing right through the wall from face to face; a wall of this type, was a series of superimposed cells filled with concrete. Roman fondness for veneer facings involved them in an extremely laborious process: inserting nails or marble plugs into the facings was, it seems, their only method of keying them into the plaster beneath!

Construction methods: Greek

The 'measured drawing' myth

'The Greek method of building can best be described as a three-dimensional geometric construction, executed in the actual marble blocks in the course of erection with the help of very simple instruments, the cord, the plumb-line, the levelling instrument, the square, etc., and based on the fewest possible specifications' (Bundgaard 1957, 139). What is the evidence on the information supplied by the architects to the masons?

Columns

The *Arsenal Inscription,* which describes the proposals of two architects for a building to be erected at the Piraeus (it was, in fact, built about 330 BC under the direction of one of them, Philon), supplied, at the planning stage, only two items relating to the columns: (1) the full height, and (2) the lower diameter. Information on the tapering, the degree of curvature of the entasis, and the inclination of the upper surface of the first column drum, in relation to the horizontal plane 'can be established between architect and mason in course of erection' (Bundgaard 1957, 138).

Ashlar

Ashlar masonry, according to Bundgaard (1968, 232, n. 288), shows, on close examination, three stages of working, corresponding to the three stages implied in the Prostoon inscription from Eleusis: (1) rough hewing in the quarry, (2) 'finishing' on the building site, (3) adaptation during the actual building of the wall.

'*Refinements*'

Discussing conventional views on 'refinements' Bundgaard (1957, 233, n. 293) thinks Goodyear (*Greek Refinements,* Yale 1912), 'was probably right in saying that the Greek buildings are improved by this quality of "free-hand work", but as he did not explain how this could be present in buildings created with such precision, and on the contrary confirmed the reader's belief that they were built according to extremely exact calculations, he could not fail to give the impression that he visualized the Greek architects and craftsmen as *fin-de-siècle* romantics of the Ruskin type, who with great labour erected accurately planned buildings inaccurately. Anyone who knew anything about Athens in the fifth century was forced to revolt against this.'

How then did these 'free-hand' craftsmen set about their work? As Coulton has pointed out, exact calculation of dimensions (and especially of proportions) could not be worked out for two reasons: first, lack of an accepted system of numeral notation; secondly, the imprecise system of measurement, based on a foot divided into four palms, each consisting of four fingers, which made it extremely difficult to calculate accurately the various dimensions of a building beforehand. On the basis of Vitruvius' rules: the architect calculating the height of an architrave to match a 24 ft high Ionic column would have to use a measurement of 1 foot, 3 palms, 2 dactyls (fingers) plus an amount which could not be written in numerals or measured with a normal ruler.' (Coulton 1977, 67). Even in that model of precision, the Parthenon, some quite striking variations in measurement have been discovered: the height of column drums, for example, varies from 0.73 m to 0.99 m.

'Approximate calculations would presumably be made for the orders to be placed with the quarry, but where real precision was required it could more easily be achieved by working directly on the appropriate part of the building ...' (Coulton, *loc.cit.*).

Construction methods: Roman

An instructive example from Roman building practice is the water-supply system introduced at Aspendos in Pamphylia during the late second century AD, when that city, along with many others in southern Anatolia, enjoyed a period of increased prosperity Notable here were two specific features, the remarkable pressure-system incorporated into the supply-line (Ch. 12, pp. 161 f.), and the variety of materials and methods employed in the construction of the aqueduct and the tall pressure-towers which formed an essential part of the system. This variety in materials and methods has been taken as evidence of repairs on an original structure, but, as J. B. Ward Perkins has demonstrated in an important discussion ('The aqueduct of Aspendos', *PBSR* 23 (1955), 115–23), the visible remains all belong to the original scheme. The main aqueduct-bridge, which took the pipeline across the River Eurymedon, had a core of mortared rubble, and was faced with squared masonry of the local conglomerate, the massive blocks being carefully finished and pointed. The width of the structure (5.50 m) shows that it must have been used both as aqueduct and road-bridge. It has been generally assumed that pressure-pipes were made of lead—indeed, the capacity of lead pipes to withstand high pressures has been much debated.

The pipes at Aspendos were made of limestone, and consisted of cubes with circular channels 26 cm in diameter, the blocks being fitted together by flanges and sockets. The bases of the pressure-towers of ashlar masonry, and the fine proportions of the south tower closely resemble that of the main bridge. The upper portions of the towers, and the voussoirs of all the arches, except the two pairs either side of the towers, are in brick. The ramps which raise the water to a height of some 30 m from the bridge are of hard, mortared rubble, resembling concrete, interspersed with brick courses for added strength, and faced with small stone blocks to provide a smooth surface. The use of through courses of brick to strengthen a structure, noticed at Ostia, is paralleled elsewhere (see e.g. Biers, 'Water for Stymphalos', *Hesperia* 47.2 (1978), 171–84).

Roman vaults and domes

Apart from a single passage in Vitruvius (7.3.1–3) the meaning of which is not entirely clear, we have no texts describing the methods used in building these structures, and the surviving examples have been interpreted in differing ways (see Ward Perkins 1977, 98 ff.).

There is some evidence for the following sequence of developments in technique: *1,* a shift from full timber centering to methods that were more economical in the use of timber; *2,* reduction of weight by the insertion of rows of empty amphorae into the concrete; *3,* complex combinations of brick courses (used as ribbing) linked by connecting rows of tiles, the purpose of which is as yet not understood.

In the construction of vaults and domes two main problems had to be solved: *1,* how to get a suitable concrete mix (we do not know how many structures collapsed before the right mixture was achieved); *2,* how to cope with the weight problem (done either by buttressing or by lightening the concrete).

To be of any use against overturning a buttress must project more than half the thickness of the wall at its base, but some surviving examples are mere pilasters, and quite useless (see Atkinson and Bagenal 1926, 96). The best way to get at the problems involved is by discussing a number of buildings in which these features occur. Drawing reasonable inferences as to purpose, however, is notoriously difficult—as difficult in fact as reconstructing a mechanism from the parts without ever having seen the finished product.

The physical appearance of the earliest surviving vaults and domes indicates that they were supported on full timber centering; the observed changes seem to be concerned either with strengthening the structure and reducing the thrust or with economizing in the use of timber. Thus tiled vaults in which rows of tiles were laid on the timber scaffolding, and remained in the concrete after it dried, appear as early as the reign of Nero (in the Golden House), and had become quite common by that of Trajan. They were not used in the dome of the Pantheon, however; here the dome was lightened by reducing progressively the aggregate in the concrete mix. By the third century we find rows of amphorae inserted into the concrete head to tail as a way of reducing the load. Much more difficult to interpret is the practice, which begins with the Colosseum, of breaking up the concrete fabric with brick ribbing, forming a series of small compartments. The most likely explanation is that the division into compartments was economical both of time and material, enabling a skeleton to be constructed quickly, and 'fleshed out' by gangs working simultaneously to fill the compartments with concrete, and with considerably less timber. In some vaults the ribbing seems to have been designed to take the thrust and convey it to the supporting members in apparent ignorance of the fact that a concrete vault is a static structure (see the following paragraph), but the unnecessary buttressing (it occurs in the temple of Minerva Medica, *c.* AD 250), and that of Portunus at Porto suggests a refusal to take risks which is not without parallels (see Robertson 1945, 232 f., part of a chapter which is still a good introduction to the structural problems involved). In a much earlier study of the subject, Middleton (1892, 172–73) made an important comment on Roman vaulting methods:

The construction of the various arches and vaults affords many interesting examples of the Roman method of using the arch form without the principle of the arch. They were in all cases cast in one solid mass and had no lateral thrust. Some of the great vaults could not have stood for a moment if they had been built with true arches, as the thrust of such wide spans would have inevitably pushed out the lofty walls on which they rest.

On foundations

Greek

While private houses had virtually no footings, much more substantial provision was made in the case of temples and other public buildings. While the foundation courses might be laid in a variety of materials—from sand or pebbles to polygonal masonry, a clear distinction can always be seen between those portions of the foundation courses which were set beneath columns or walls and those with no superstructure. The former might be either megalithic, with one block supporting several columns, or monolithic, as in Temple 'C', Selinus, where the blocks are equal in length to the interaxial spacing of the columns, or, in what became the standard classical pattern of one block beneath each column and one beneath each intercolumniation. The advantage of the second method is that the columns 'act as cover joints and thus protect in a measure the sub-structure' (Marquand 1909, 40). The non-load-bearing portions were filled with earth or chippings left by the stone-cutters. The most interesting feature of the

visible portion (the *krepidoma*) was undoubtedly the convex profile, which probably originated as a rain shedder, but had important aesthetic effects in giving life to the severely geometric form of the colonnaded temple. For this and the other (vertical and rectilinear) 'refinements', see the brief but lucid account in Coulton 1977, 108–10, and for the parabolic method of construction refer to Dinsmoor 1950, 167 and fig. 59.

Roman

Vitruvius' important chapter on the subject (Book 3, Ch. 4 'The Foundations and Substructures of Temples') begins in a very general way, but it very quickly proceeds to discuss what steps are to be taken to provide good foundations where there is no solid ground:

If however no solid ground can be found, but the place turns out to be nothing but a heap of loose earth down to the very bottom, or a marsh, it must be dug up and cleared out and set with piles made of charred alder or olive-wood or oak, and these must be driven down by machinery, close together like bridge-piles, and the intervals between them filled in with charcoal; finally the foundations are to be laid on them in the most solid form of construction.

According to Jüngst and Thielscher (*Comm.* on Vitruv. 3.4.2) the two terms used, *pali* and *sublica,* are quite distinct, the former being used to consolidate waterlogged ground, while the latter were used as load-bearers.

Notes on the roof-spans of Greek temples
from Hodge 1960, Table I, p. 39.

Temple	Span	Between columns
1 Paestum: Temple of Athena ('Demeter'/'Ceres')	5.70 m	same
2 Paestum: Temple of Hera ('Basilica')	11.44 m	5.6 m
3 Selinus: Temple 'G'	17.93	7.15
4 Agrigento: Olympieum	12.85	same
5 Athens: Parthenon	19.20	11.05
6 Bassae: Temple of Apollo	6.81	5.23
7 Aegina: Temple of Aphaia	6.28	3.85
8 Olympia: Treasury of Gela	9.68	—

The evidence from temples in mainland Greece indicates that when the architects had to cover a span wider than 6.5 m, they broke it by the insertion of an interior colonnade. Hodge (1960, p. 38) draws attention to the difference, as shown in the above table, between mainland practice and that of Magna Graecia and Sicily, where buildings as large as those on the mainland nevertheless dispensed with internal supports; he suggests that the Sicilians either used a tie-beam truss or had access to much better timber for their beams. See further Coulton (1977, 157 f.). On the important Hellenistic modification of the traditional bearer-beam roof, as found in fourth-century stoas, see Coulton (1977, 155 and pl. 54).

APPENDIX 10

On road transport in Roman times

Range of vehicles used

The severely restricted use of animal-drawn vehicles in Greece is reflected in the small number of vehicle types recorded. The Romans, by contrast, disposed of a great variety of types, designed for different purposes, and to suit varying road conditions; Isidore's list of vehicles (*Origines* 20.12) contains nine names, many of them of Celtic origin (see below).

Evidence on vehicle design

The high standard of craftsmanship displayed by the Celts of north-western Europe is well known; the archaeological evidence includes a number of wheels, some of them remarkably well preserved. The recently discovered fragments of Diocletian's *Edictum de pretiis* include a section on parts of vehicles, providing fresh, or confirmatory evidence on road vehicle design, particularly in relation to the wheels, and on the vexed question of a swivelling front axle (see *JRS* 63 (1973), 104–7). The most important class of evidence is provided by representations, mainly on tombstones, and sculptured relief. Until recently, they have received scant attention, but the position has been remedied to some extent by the photographs, with accompanying notes, published in Vigneron (1968, vol. II). Details of some of the vehicles

depicted are discussed in an excellent study not immediately related by Albert C. Leighton (1972). The notes on eight commercial vehicles that follow are drawn largely from Leighton's work. He points out an important distinction between representations of ox-drawn wagons, and those drawn by mules or horses. The ox wagons look like boxes with wheels at each corner, while the horse-drawn wagon usually has an undercarriage or chassis, which suggests that it may be derived from putting two two-wheeled chariots together. There is some archaeological evidence to support this suggestion: see Haudricourt-Delamarre (1955, 163, fig. 9).

1 The Pompeian wine-wagon

The authoritative representation is that of Roux, *Herculaneum et Pompeii*, Paris 1840, vol. III, pl. 126. Others are 'purely fanciful'. The capacity of the skin might be of the order of 8–10 cubic feet depending on the size of the animal.

2 The Langres wine-wagon

Contents of cask 106 galls (or 400 litres (sic!)—should be 525 litres KDW). Has undercarriage—a pivoted front axle cannot be proved. Other notable features are (1) the massive wheels; (2) the strong, projecting naves; (3) the effective working posture of the mules (see 4 below).

3 Igel monument

West side (lower); freight wagon with undercarriage, heavily loaded with large bundles, carefully secured with ropes; pulled by 3 mules, 2 of which are connected by a yoke (the 3rd may be a trace-animal used for heavy climbing—KDW).
Ref: Dragendorff–Kruger, 'Das Grabmal von Igel', *Röm. Grabmäler des Mosellandes*, Berlin–Leipzig 1924, 79–80.

4 Neumagen (a)

Monument of Lucius Securius. Unfortunately only the front portion has survived, so we have just the animals and no sign of the wagon. 3 mules are shown, 2 pulling effectively with their heads down! They are exerting pressure against the yoke, not the collar, which acts merely as support for the ring through which the guiding reins pass.
Ref: W. von Massow, 'Die Grabmäler von Neumagen', *Röm. Grabmäler des Mosellandes,* Berlin–Leipzig 1932, 2.1.141–42.

5 Neumagen (b)
Special-purpose vehicle, like 2 above with wine barrel laid lengthways. Animals and wheels are both missing. Undercarriage: some portions survive, including 2 axles connected by long pole, but front axle almost completely effaced. Leighton thinks it may have duplicated the rear axle, and therefore have had no swivel. The size of the front wheels is irrelevant to the question of whether the vehicle had a swivel.

6 Vaihingen (now in museum at Stuttgart)
4-wheeler with light 8-spoked wheels, loaded with an indeterminate large object; there are 3 horses.
Ref: Espérandieu, *Germanie,* no. 258.

7 Baden-Baden (now in Karlsruhe Museum)
4-wheeler with pole joining front and rear axles; at least 12 spokes per wheel. Heavy load pulled by 2 horses. Wagon body has staked sides.
Ref: Espérandieu, *Germanie,* nos 297–98.

8 Langres
4-horse wagon with tandem harnessing. Precursor of the 'four-in-hand' harness of coaching days.
Ref: Espérandieu, *Gaule,* vol. IV, no. 3245.
Note that it is often asserted (e.g. by Lynn White, Jr. 1962) that the ancients could not harness animals in file. This wagon is sufficient to disprove the theory!

Road conditions

Chevallier (1976, 180) mentions, among possible reasons for the severe loading restrictions imposed by the Theodosian Code (200–600 lb for light carts; 1000–1500 lb for heavy carts used by the Imperial Post), unsuitable teams, vehicle-bodies not strong enough for the conditions, slow speeds and poor road surfaces. The most important factors would seem to have been, in the case of the vehicles, lack of any suspension, inadequate lubrication, and lack of an effective braking mechanism; and in that of the roads, the unsuitability of the road surface for fast wheeled traffic; the wickerwork bodies which were in common use (see Varro, *De Re Rustica* I.23.5) were light and flexible, but Horace's upstart freedman who 'wears away the Appian Way with his cobs' (*Epod.* 4.14) will surely have had a rough ride on that famous highway!

Construction methods
Chevallier, R. (1976, 82–82).
Duval, P.-M. 'La construction d'une voie romaine d'après les textes antiques', *BSAF* (1959), 176 86.
Forbes (1964)
Forbes, *Studies* II, 2, and 1964 volume.

Order of construction, as given by Statius (*Silv.* IV.3.40–55): *1,* Tracing of furrows to define the road; *2,* Trenching to get a firm foundation; *3,* Reinforcing foundation by ramming, pile-driving or layering with brushwood (marshy ground required timber, stone or even criss-cross logs (e.g. Caesar, *Gallic War,* VII, 57–58); *4,* Packing with layers of stone, gravel and sand; *5,* Cambering for drainage; *6,* Surfacing with paving stones held in place by kerb-stones; *7,* Binding with lime mortar (unusual elsewhere).

Texts
Regulations controlling access of vehicles to the city of Rome. Ref: L. Friedländer, *Roman Life and Manners,* 1913, vol. IV, 28 ff (Appendix VI, vol. I), trans. J. H. Freese.
The *Lex Iulia Municipalis* forbade the use of vehicles in the streets for ten hours after sunrise; exceptions: *1,* vehicles used for public buildings, temples and demolition works; *2,* vestals, flamines and triumphing generals; *3,* circus processions; *4,* empty vehicles which had arrived overnight, and *5,* rubbish carts. No change in 'these regulations during the first two' centuries AD.

The poet complains of too many distractions and interruptions to work in Rome:

'Poor men's blouses, not long mended, are torn afresh as a lorry comes by, carrying a tall fir swaying to and fro, and waggons besides loaded with pines, which nod on high and threaten the public. Suppose a dray, loaded with Ligurian marble, has turned over and, upsetting its contents, has showered the trooping crowds with a mountain of rocks, what will be left of their bodies?' Juvenal, *Satires* III.254–9

'Here in hot haste drives the builder with his mules and porters, there a huge crane uprears a massive stone or beam, here mournful hearses jostle with powerful waggons.' Horace, *Letters* II.2.72–4

'Two laden waggons were being hauled by hinnies up the Clivus Capitolinus. The muleteers of the front waggon tried to support it with their shoulders as it swung round, to make it easy for the hinnies to pull it. At this point the front waggon began to run backwards, and when the muleteers who were in between the two waggons abandoned their position, the second waggon was struck by the first one, ran backwards and crushed someone's slave; the slave's owner took advice to find out against whom he should bring an action.' Alfenus in *Digest* IX.2.52.2

APPENDIX 11

Notes on Roman ships

Design and construction of Roman merchant ships

The lucid and comprehensive description by G. S. Laird Clowes (1932, 36–38), is here reproduced verbatim, corrected and updated, from evidence that has subsequently come to light.

'When, however, we leave the difficult question of actual measurements, and come to the details of construction of these Graeco-Roman merchant ships, we reach much surer ground. They were built with a very delicate *stem,* and with a less pronounced *sternpost,* which were obviously connected by a strong external keel. There is no indication of the exact arrangement of the ribs or timbers, but some of the beams of the upper deck are shown with their ends projecting through the outer planking, in the style characteristic of Mediterranean ships of the 14th and 15th centuries, and which even now persist in the cargo-junks of Japan.

'The stern-post curves gradually upward and is prolonged into the head of a swan, looking aft—the use of this bird as a stern finish seems to have been practically universal in merchantmen—while the stem, which is also curved, though to a lesser degree, projects above the bulwarks and is sometimes prolonged forward so as to carry some sort of *look-out post* or small fighting-castle. Above the water-line, additional longitudinal strength is provided by two or more *wales,* heavy external timbers which run continuously from stem to stern. A curious feature was that on each quarter, at about the level of the upper deck, the run of the ship's was carried aft so as to form a trough-like projection on each side, open at the after end. This truncated *quarter-galley* was obviously derived from the *parodos* or *outrigger* of the galley, the after end of which served to support the *steering-oars* and into which these oars were drawn up and housed when not in use. These projections on each quarter are surprisingly heavy, but are extremely chararcteristic of Roman sailing ships. Their heaviness seems to be due to the fact that the two beams, one a foot or two above the other, which formed their after end, served as bearing surfaces to keep the *loom* of the steering-oar in position. It must be remembered that steering was effected by rotating the oar about its longer axis, by means of a short *tiller* set at right angles to the loom, and not by sweeping the blade of the oat to port or starboard—in short, by the method still used in whaleboats in normal conditions, but not that by which they achieve their sharpest turns, and which is employed in heavy surf. Insecure as this method of fitting a quarter-rudder may seem, it is similar to that still employed by the Japancse in their fishing craft, where the rudder, although fitted in the stern *transom,* and on the centre line of the boat, is similarly arranged and inclined for lifting. The habit of shipping one or both of the steering oars persisted in the Mediterranean as long as did the use of these very inconvenient attachments, and even in the late fifteenth century examples of sailing ships with their quarter-rudders lifted may be found. Above decks there was a considerable *cabin* aft, together with a curiously *inclined gangway* or platform which projected *abaft* the swan-shaped stern, and whose use must still remain in doubt, although its position is similar to that in which *gang-planks* were carried in some of the Greek vessels of 500 BC.

'The rigging of Roman merchantmen is of great interest, and attained a degree of perfection which has only recently been understood. The *main-mast* was stepped centrally and reasonably vertically, while projecting over the bows was a subsidiary mast or *bow-sprit,* which may be conveniently referred to as the *artemon.* The main-mast was well stayed by means of a very heavy stay which led forward to, or near, the stem, and also by some four pairs of *shrouds* which, together with the stay, were set up by means of *dead-eyes.* A very heavy *yard* was employed, made of two spars lashed together at the centre, as is still usual in the yards of lateen sails, and this set a large square mainsail, made with the cloths horizontal, and fitted with some nine sets of *buntlines* or *brails.* The sail was controlled by means of sheets, with the addition of *braces* to the *yardarm,* while *lifts* led from each yardarm to the *masthead.* Above the yard in fine weather some vessels at least were set two triangular *topsails,* like the 'skyscrapers' of 1794 or the more modern 'raffee' sails. The *halyard* led aft in the fashion of ancient Egypt, and was arranged in the same way as was common in English ships of the sixteenth century, a method which still persists in Arab dhows. Just as in these instances, the Roman halyard consisted of a double tie, with the *falls* leading through a *multiple block* or *dead-eye,* with another block or *knighthead* on deck. Whether these blocks were or were not fitted with *sheaves* is doubtful. The artemon already mentioned, which projected over the bows and played a part intermediate between those usually associated with a *foremast* and a *bowsprit,* served to set a small *foresail* or *spritsail.* The Portus relief shows the sail furled and the yard lowered, the Sidon and Ostia reliefs show it set and *drawing,* while in certain other contemporary representations the sail is nearly as large as the mainsail, and the artemon appears like a normal foremast. Like the mainsail, the spritsail was set on a yard *athwartships* and was formed of horizontal cloths. It was fitted with some five buntlines and controlled by *sheets* which led in or near the stem. For its length the artemon was a very stout spar and, like the bowsprit in the fifteenth and sixteenth centuries, there is no indication that it was provided with any *standing rigging.* Small as this spritsail seems normally to have been, its introduction must have exercised a most important effect on the development of navigation, for it enabled a vessel to sail *with a beam wind,* or even perhaps with the wind a little *forward of the beam.* From the way in which it would facilitate manoeuvres under sail, it made it possible for large ships to rely on sail alone—with perhaps a few subsidiary oars for occasional use—and so enabled merchantmen to dispense with large numbers of oarsmen which practically prohibited long merchant expeditions, and entirely prevented voyages at any considerable distance from land ... Considering the value of the artemon and spritsail and their importance in making possible the regular voyages of Roman cornships between Egypt and Italy, it is one of the mysteries of the history of ships how these valuable adjuncts came to disappear, with the destruction of the Roman Empire, so completely that the spritsail was not seen again in square-rigged vessels until the end of the fifteenth century, and even then a bowsprit had been in use for other purposes for more than two hundred years before the idea of setting a spritsail on it was once again evolved.'

Shipbuilding techniques and methods*

The following indications of the various stages in the construction of a merchant ship of classical times have been gathered almost entirely from the study of excavated wrecks. Greek evidence is unfortunately rare, but representations of Greek merchant ships give no ground for supposing that any fundamental changes in technique took place during the Roman period.

Homer's account (*Odyssey* V.246–51) of how Odysseus built his ship refers clearly to the edge-jointing method of constructing the shell (Stage II(b) below). *Stage I* Setting-up of *keel, sternpost* and *stern*. *Stage II* (a) Curvature of the *hull* laid out by the master carpenter (see the funerary relief of the shipwright Longidienus, *ill. 149*, p. 147) (b) Hull built up by setting the horizontal *strakes* edge-to-edge with close-set mortises. *Stage III* The hull being now complete, the *frames (ribs)* and *strakes* (longitudinal timbers) were shaped and fitted to the hull. *Stage IV* To strengthen the hull against heavy seas, and to protect it against rubbing, strakes known as *wales* were attached so as to encircle the entire hull; in some examples they were secured to the frames by long bolts. *Stage V Caulking of the seams,* sometimes of the entire hull, with pitch, or with a mixture of pitch and beeswax. *Sheathing with lead* as a protection against marine borers ('part of a major development in shipbuilding technique of the 5th–4th centuries BC') (Casson 1971, 185. n.9; 209 f.)

Note The Kyrenia wreck had the entire outer hull surface sheathed with lead oo3 m thick, but she was 'riddled with holes by the time she sank' (Blackman, in *IJNA* I (1972), 117).

Attention is drawn to the following important techniques: (1) use of mortise and tenon joints, often so closely set that there is hardly any space between them; (2) the use of pegs with bronze spikes to secure frames to planking; (3) attachment of reinforcing frames by 'darning' with cords known as 'sennits'.

Materials used

For planking, frames and keel shipbuilders of the Hellenistic period preferred red fir, cedar and pine (Theophrastus *HP* 5.7.1). The Byzantine ship wrecked off Yassi Ada in Turkey (see Bass, 1972), had her sternpost and wales made of cypress, while her frames were of elm. Frames in the Roman period seem to have been commonly made of oak.

Sizes, tonnages and cargoes

Note The evidence available from wrecks is variable; some well-known wrecks have never been properly excavated; in many cases evidence has been destroyed by pillaging skin-divers before proper excavation has taken place; as for cargoes, all identified wrecks were carrying either building stone or finished works of art in stone or bronze or amphorae, the contents of which (wine, oil, garum, etc.) are casily identified; estimates of tonnage vary widely, depending on the basis of calculation used (see Casson 1971, 183–90). Casson's list of 32 wrecks (1971, 416) has been greatly extended during the last decade, as the pace of development of underwater archaeology has quickened. The notes that follow will serve to indicate the range of tonnage and content.

Lists of wrecks

1 Casson (1971, 189–90): a list of 12 ships whose dimensions are known, 'at least approximately'. They range widely in date, from Kyrenia (end of 4th BC) to Yassi Ada (7th century AD). Their cargoes were limited to stone or amphorae (filled with wine, oil or other liquids), 'for only such cargoes have been able to survive and mark the spot of an ancient wreck' (Casson 1971, 189).

2 Throckmorton, P. in Bass (1972, 66 ff.): a select list of larger cargo vessels, divided into two categories: (A) wrecks yielding structural wood that could be studied; (B) wrecks whose cargoes were sufficiently intact to enable the tonnage to be estimated, though nothing remains of the hulls. They are defined as big, single-keeled vessels, of 150–200 tons, too heavy to be beached easily, often sheathed with lead, and so large that they would have been unloaded in Ostia. *A1 Antikythera* ship (Throckmorton in *TAPS* 55.3 (1965), 40–47): May have exceeded 200 tons, but never properly excavated. Strakes 7.9 cm thick; lead-sheathed: 1st century BC.

A2 Mabdia ship (Taylor 1965, 48 ff.): Carried at least 230 tonnes of cargo; keel 30 m (close to the length of keel of a 300 ton *perama* recently built in Syros with keel 80 ft, length overall 100 ft, planks 2½ in thick'): 1st BC.

A3 Albenga ship (Lamboglia, N. in *Riv. Stud. Lig.* 27 (1961), 213–20): A large Roman amphora carrier found in 1949 off the Italian Riviera. Still unexcavated, but visible portions measure 100 ft × 26 ft.

A4 Torre Sgarrata ship (Throckmorton 'Torre Sgarrata' (preliminary report 1969): Sunk off Taranto en route for Asia Minor with a cargo of *c.* 170 tons of marble.

B(1–6) Ships carrying cargoes of blocks or drums or monolithic columns, almost all of marble. Lightest cargo 120 tons, heaviest *c.* 350 tons. T. notes that it is now the practice not to load ships with stone beyond 2/3 of gross tonnage.

3 Manacorda, D., *Relitti sottomarini di età repubblicana: stato della ricerca con appendice bibliografie* (1979). This comprehensive study of a single category, viz. Roman wrecks of the Republican period, illustrates graphically the vast increase that has taken place in the past decade in the number of wrecks identified and/or excavated; the bibliography runs to 91 entries, and the ten periodicals used in compiling it include two devoted exclusively to nautical archaeology, *Cahiers d' Archéologie subaquatique* (1971–74), and *Nautical Archaeology (IJNA),* which first appeared in 1972 and *Archaeonautica* (1977–).

Casson (1971, 171 f.) identifies four categories of freighter:

1 Small galleys used for short-haul transport. P. Cair. Zen. 59015. Dated 259/8 BC. This vessel was a keles (fast, single-banked merchant ship powered by 10–12 oarsmen; she carried 145 jars of 18 choes capacity) (= 58 litres or 13 gallons) and 34 half-jars of oil from Asia Minor to Alexandria. Weight around 14 tons of 31–35 cu m of liquid capacity.

2 Medium-sized freighters: grain-carriers powered by sails (c. 80–130 tons): (a) IG II² 360 (*c.* 325/4 BC), carrying 3000 medimni (120 tons) of grain from Cyprus to Athens (the records include several others of similar capacity); (b) IG II² 903 (126/5), carrying 1500 metretes of oil (m = 39 litres) cargo will have weighed 100+ tons.

3 Large merchantmen (up tp 340 tons (the 'standard' size of grain-carrier under the Empire (see *Digest* 50.5.3—on exemption from public compulsory service of shipowners prepared to build and furnish carriers of not less than this capacity: the date is late 2nd century). Even larger vessels (up to 500 tons) are reported. The Albenga wine-carrier (A3 in Throckmorton's list) seems to have belonged to the established category known as '10,000-ers', identified by Wallinga as referring not to weight in talents, but to capacity in medimni or amphorae. The Albenga's cargo was distributed over an area of 25 m (82 ft) in five super-imposed racks of amphorae, which will have totalled 10,000; these, when full, will have weighed over 400 tons. The evidence suggests that

freighters of this size were not at all exceptional (Casson 1971, 172, n. 23 for references).

4 Very large merchantmen ('superfreighters'): (up to *c.* 1900 tons) Three of these are mentioned. The first was built by Hiero, king of Syracuse, under the supervision of the leading scientist of the day, Archimedes, and an account of her was transcribed four centuries later by Athenaeus (*Doctors at Dinner,* 5.206d–209b). The report does not give her dimensions, but the cargo she carried, which comprised grain, pickled fish and wool, ran to nearly 2,000 tons! Next on the list is Caligula's superfreighter which brought the obelisk from Egypt which now stands in front of St Peter's, Rome (see Pliny, *HN* 16.201, who mentions also the four blocks that served as its base, and the ballast of lentils—the latter needed because the obelisk was carried on deck). The weights work out as follows:

obelisk	— 322 tons
base	— 174 tons
ballast	— 800+ tons (130,000 *modii,* according to Pliny)
total	— 1,300+ tons

The third member of this select group is the *Isis,* the big grain carrier that was blown off course, and put in at Piraeus, where the writer and publicist Lucian had the opportunity of going over her (see Ch. 11, pp. 155 f.) He provides three dimensions (length 55 m (180 ft), beam 13.7 m+ (45 ft+), and depth (deck to lowest point in the bilge) 13.25 m (43½ ft)—enough to satisfy an interested visitor (Lucian says he got them from the ship's carpenter), but not sufficient to give the capacity. To get that you need the length of the keel, for this, as Casson points out (1971, 187), determines the stowage capacity. Using a comparable vessel (a 16th-century Venetian man-of-war) Casson arrives at a figure of 1,228 tons, which would make her bigger than any ship afloat between the fall of Rome and the arrival of the big East Indiamen in the 18th century. Roman shipbuilders had the technology to build ships of this size; and the imperial grain service, which was responsible for shipping huge quantities of grain every year from Alexandria to Rome, had the motivation. Ptolemy's monster may well have been an expensive white elephant, and Caligula's superfreighter, which was probably bigger than Isis, only made the one voyage; her shape can be identified from the mass of solidified concrete (now high and dry, see Testaguzza 1970, pp. 90–91) with which she was filled by order of Claudius to form part of his new harbour. If few vessels like Isis were built (and so far no wrecks with estimated capacities above 450 tons have been identified) the limiting factor may have been the manpower needed to handle the enormous square sails (Throckmorton in Bass 1972, 77), who points out that a present-day schooner-rigged vessel can be handled by a crew of three against ten required for a Roman square-rigger of the same capacity.

On the design and development of anchors

Problems of dating and sequence

'Anchors come next to amphoras in the category of objects that are most subject to arbitrary salvage and disposal' (Gianfrotta 1977, 285). Datable finds are notably scarce for the period 1st century BC to 1st century AD.

Materials and types

Earliest anchors were rough stones with holes through which cables were passed and attached to the prow: cf. Homer, *Iliad* 1.436: Odysseus brings Chryseis back home to her father with peace-offerings; when they reach harbour, they furl and stow away the sail, then drop the lowered mast into its crutch, row the vessel into her berth, cast the anchor-stones *(eunai)*, make fast the hawsers, and jump ashore. This was the earliest type of anchor known to the Mediterranean area. The two major types may be subdivided as follows:

Type I Stone. (a) roughly rounded natural stone with a hole for the cable; (b) ovoid; (c) trapezoid (1) plain, (2) with lead shaft.

Evidence: many examples of I(b) and (c) were found at the Punic port of Motya in Sicily (heyday from 8th to 6th centuries BC, Motya fell in 398). Three examples of I(c) (2) were raised by Throckmorton from the Taranto wreck of early 3rd century BC (de Vries and Katzev in Bass 1972, 48 f.).

Type II 'Hook' or true anchors (*agkura* = 'hook'). (a) with wooden shanks and arms and a variety of stocks: (1) stone stock; (2) lead stock (i) fixed stock, (ii) movable stock; (b) of iron.

Evidence: de Vries and Katzev (1972, 49, fig. 15) (carpenter on a Sardinian gem shown shaping a wooden anchor). 5th-century anchors of type (1) from Aegina (de Vries and Katzev (1972, pl. 20); of type (2) from Syracuse harbour (*ibid.* pl. 19).

The true anchor (*agkura* = 'hook') with flukes for gripping the sea-floor appears first in the 6th century BC. Invented according to Strabo (7.3.9) by Anacharsis; according to Pausanias (1.4–5) by Midas; Pliny (*HN* 7.57) supports Strabo. Materials used: shank and flukes of wood; stock and cross-brace of lead. Confirmed by archaeological evidence (G. Kapitan, in *Sicilia Archaeologica* 4 (1971, 13–22). This type persists with modifications (e.g. lead casing of stock over wooden core) for more than eight centuries.

Examples: 1 Antibes wreck (probably early 6th century BC. Earliest known anchor with lead stock (Benoit 1961 170; Casson (1971), 254 n. 119). Fits the Anacharsis story. 2 Brindisi I. Fixed type (IIa(2) (i)), length 44 cm; weight *c.* 5 kg. 3 Brindisi II. Movable type (IIa(2) (ii), length 103 cm; weight *c.* 93 kg (204 lb). 4 Messina wreck. Lead bars from centre of shaft: lead collar to hold arms to shank; weight probably 1½ tonnes.

Development

From stone to lead

Gianfrotta (1977), agrees that lead stocks are technically more evolved than stone ones, but points out that the latter were fixed in exactly the same way as the former. The new form is probably signalized by the appearance of the term *agkura* (the true anchor) alongside the earlier *eunaion* or *eune* (= anchor-stone) in the 7th century BC. The new term denotes the new form, with its stock, shank, and flukes.

Iron anchors

Mercanti (1979) reports no datable finds of lead or wooden parts later than the 2nd century AD, and takes this to be evidence of the dominance of the iron anchor. The 'Nemi-type' anchor with semi-circular flukes and almond-shaped points persists up to the end of the 3rd century AD. By the 4th century the technique of manufacture shows a vast improvement, leading to the fine anchors of the Byzantine period (see e.g. the Yassi Ada wreck) with fine-grade iron and movable stocks.

APPENDIX 12

Water supply problems

Pressure problems in closed-pipe systems

Landels' discussion (1978, 35 ff.) demonstrates very clearly the difficulties associated with closed-pipe systems of supply, which made them exceptional in the Greek and Roman world; his discussion of the Pergamum system, as originally planned by Eumenes II, and of the Roman modification, states that we have no clear evidence concerning the material of which the pipes were made. All that has been found are the stone blocks with holes (appropriate either to earthenware or lead pipes), but the fact that the pipes themselves have not been found weighs in favour of lead or bronze, since earthenware pipes 'could not have been re-used elsewhere' (1978, 47). 3 km of bronze pipes would have taxed the resources of the kingdom. Therefore the pipes are likely to have been made of lead. Landels estimates the pressure at the bottom of the first bend at about 260 lb per sq. in. or 18.5 kg/cm² (Forbes, *Studies* I,160, gives a pressure of up to 20 atm (= 292 lb per 59.m) in the two siphons). On the 'elbows' (Vitruvius VIII. 6.8, calls them 'knees' *(genicula)* see Landels (1978, 45).

Sediment disposal in closed-pipe systems

See Landels' discussion of the problem *(ibid.* 46–47) where he accepts the emendation of Vitruvius, *De Architectura* VIII, 6. *colliviaria* to *colluviaria* which he interprets as = sludgecocks for draining the system while repairs are being effected, not air-valves for releasing air-locks. Note the use of settling tanks in pairs for alternate use.

Problems connected with measurement of rate of flow

Landels notes that Frontinus, while recognizing that if the gradient of the aqueduct supplying water into a regulated system of distribution to consumers via nozzles of varying diameters is steeper, the rate of flow will be faster, does not attempt to find out how much faster (1978, 49). It appears that any increase in supply was regarded as a bonus to the consumer, while there is some evidence of adjustment (? reduction of the charge) if the rate was slower than usual. Some

light may be thrown on the problem raised by Frontinus' apparent neglect of the rate of flow aspect in relation to water charges by the following passage in Hero's *Dioptra* (Ch. 4) where a practical problem in the use of the instrument is set out:

Given a spring, to evaluate the amount of water it could provide. First we must determine with the *dioptra* a point low enough to receive the whole content of the spring, and then construct a lead conduit of rectangular shape, of greater diameter than that of the current, which will take the flow. Let us then take at the end of the conduit the water which enters, let us suppose that it rises to a height of two digits and that the width of the opening is 6 digits. Multiply 6 by 2, making 12; we see then that the cross-section of the stream is 12 digits. We must always bear in mind that in order to estimate the amount of water provided by the stream, it is not sufficient to determine the cross-section of the stream, which we say is 12 digits; we must in addition have its rate of flow; the faster the rate of flow, the more water the spring will provide; the slower the flow, the less it will give.* For this reason, after digging a reservoir beneath the stream, we must discover, by means of a sundial, how much water enters it in one hour, and to deduce from that figure the amount of water provided per day; in that case one need not measure the cross-section of the stream; the time calculation is alone sufficient to show how much water the spring provides.

Siphons or bridges?

Reference was made in Chapter 12 (n. 254) to the 'standard' view that siphons were much less common than bridges (e.g. Forbes *Studies* I, 161). What considerations induced Roman engineers to choose one or the other? In a recent paper, Norman Smith (*HT* (1976), 45–71) gives the following comparative statistical tables for the heights of a number of valley-crossing, (A) by means of a bridge, (B) by means of a siphon; they are reproduced here with the author's kind permission:

A Bridges	*feet*		
Pont du Gard (France)	160	Euilly (Lyons)	210
Alcantara (Spain)	165	St Gènes (Lyons)	270
Narni (Italy)	120	Grange-Blanche	290
Cherchel (Algeria)	115	(Lyons)	
Segovia (Spain)	100	Soucien (Lyons)	304
Mérida (Spain)	85	Tourille (Lyons)	375
B Siphons		Beaunant (Lyons)	405
		Rodez (France)	300
St Irenée (Lyons)	155	Alatri (Italy)	340

Comparison of the tables shows that the deciding consideration must have been that of the depth of the valley, stability of arcaded bridging not being guar-

---

* perhaps a lacuna here. Hero has, in fact, given no instructions for measuring the rate of flow. The common assertion that Roman engineers did not know that quantity of flow

was effected by velocity of flow rests on the dubious authority of Frontinus. Hero was an engineer, Frontinus an administrator.

anteed above 150+ ft. By contrast siphons could evidently be safe at more than twice the depth; Smith's calculations (*HT* (1976), 61, n. 51) of the atmospheric pressure at the bottom of the Soucien siphon show a figure well below the point at which lead piping would collapse, viz. 2000–2100 lb sq in (= 18 atmospheres), at a depth of 600+ ft of water.

Some important texts

1 Herodotus, *History* III. 60 (on the water-tunnel at Samos):

I have dwelt rather long on the history of the Samians because they possess the three greatest works of the Greeks. One is a tunnel through the base of a 900 ft high mountain. The tunnel's length is seven stades (= 1400 yds or 1.5 km); its height and width are both 8 ft or 2.4 m. Throughout its length another tunnel has been dug 3 ft wide and 3 ft deep, through which the water flowing in pipes is led into the city from a copious spring. The builder of the tunnel was the Megarian Eupalinos, son of Naustrophos.

2 *Inscription* from Saldae (Bougie), Morocco (*CIL* VIII.2728 = *ILS* 5795. AD 152). Letter from the Procurator Varius Clemens, from Lambaesis. To M. Valerius Etruscus:

The renowned city of the Salditani and myself in association with the Salditani entreat you, my Lord, to urge Nonius Datus, retired civil engineer, of the 3rd Augustan Legion, to come to Saldae and complete the unfinished portion of his work: 'I set out for Saldae, and suffered at the hands of brigands on my journey; I and my party escaped, but I was stripped and wounded. I arrived at Saldae, where I met the Procurator Clemens. He escorted me to the mountain, where they were complaining about the badly constructed tunnel. It appeared that the project would have to be abandoned, since the length of the tunnel already exceeded the breadth of the mountain. What had evidently happened is that both tunnels had deviated from the straight line to such an extent that the upper tunnel was veering to the right in a southerly direction, while the lower one was also veering right in a northerly direction. Both sections were thus out of the true alignment, and the line had wandered off across the mountain from East to West. In case any reader may be in doubt about the tunnels, the use of the terms 'upper' and 'lower' should be understood as follows: the upper part is where the tunnel receives the water, the lower where it discharges it. When I assigned the work to give each gang its own tunnelling area, I set up a competition between the marines and the local mercenaries, and they assembled to complete the tunnelling through the mountain.

3 Frontinus, *De aquis urbis Romae,* 1.35 (on maintaining flow),

Let us not forget in this connection that every stream of water, whenever it comes from a higher point, and flows into a delivery tank through a short run, not only comes up to its measure, but moreover yields a surplus; but whenever

it comes from a lower point, that is, under less pressure, and is carried for a comparatively long distance, it will actually lose some of its volume owing to the resistance of the conduit (?); that is why, following this principle, it needs either a check or a help in its discharge.*

* Forbes *Studies* I, 170–1, comments as follows: 'The influence of the height of pressure, the grade, the resistance of the channel or pipe and the velocity of the water was not properly known, as will be seen from this passage from Frontinus.' But all that may be legitimately inferred from the passage is that Frontinus' knowledge of these matters was defective! (See *Sources,* pp. 187 ff.).

APPENDIX 13

Communications and commerce in Roman Gaul and Spain

Roads were built by the Romans for military purposes; other requirements were entirely of secondary concern. This is correct; but highways originally planned and built for strategic purposes came to be used in increasing numbers by traders and travellers after the end of the Civil Wars and the establishment of the Roman Peace. If military needs were paramount in the choice of routes, access to important minerals was hardly less important, and the need for those engaged in profitable trading activities was also recognized, as the following examples demonstrate.

Spain

Access to minerals by road

1 Republican road from Barcino to Ilerda, and thence to the mining district of Osca; 2 Augustan road from Bracara via Lucus Augusti to Asturica Augusta (Astorga); 3 Emerita to Hispalis (serving the mining district of Montes Mariani); 4 Carthago Nova to Castulo; 5 Tiberian road from Caesaraugusta via Turiassio, Clunia, Ocelodurum to Asturica; 6 New road connecting Asturica and Bracara; 7 Short link road from Cordoba to Castulo and the Montes Mariani (in AD 33, following the confiscation of these mines); 8 Transverse road from Emerita up the Tagus to Augustobriga, Toletum, Complutum, Segontia, Caesaraugusta, providing a second land access to the Montes Mariani and western Baetica.

Evidence of much relaying and repairing in these areas indicates the volume of traffic from the mines: 'These roads cannot have been built for any other purpose than that of encouraging trade' (Charlesworth 1970, 154).

Routes for transport of agricultural produce

1 Via Augusta; *2* Caesaraugusta—Juliobriga (Ebro Valley); *3* Ilerda—Caesaraugusta.

Gaul

In the Province, river communications were much more important than roads, and this has remained true right down to the present day.

Access to minerals by road

1 In the north-west, in central Brittany a network of roads was developed to link the Veneti With the Loire valley, and serve the ironworks; *2* From the main road Lugdunum—Burdigala a branch turned off at Augustodunum leading to Avaricum, serving the ironworks and the silver mines around Bourges; *3* The coast road communicating with Spain via Nice—Narbo—Arles—Tolosa shows evidence of extensive repairs.

Trade with Britannia—four routes

1 Narbo—Burdigala—thence by sea; *2* Up the Rhone Valley, then across to the Loire via Auvirca and thence by sea; *3* Up the Rhone to Arar—Doubs—Seine—Normandy ports; *4* As *3,* but by more direct route through Lutetia to Portus Itius.

Access to minerals by canal

1 Marius' Rhone canal (see Table 6, 12); *2* Drusus' canal from Rhine to North Sea (see Table 6, 9); *3* Dyke to control lower Rhine begun by Drusus and completed by Nero (see Table 6, 10); *4* Corbulo's canal from Meuse to Rhine (see Table 6, 11); *5* Ambitious plan of Vetus to link North Sea with Mediterranean by means of a 60-mile 'Grand Junction' canal from Moselle to Arar abandoned (see Table 6, 15).

APPENDIX 14

On the consumption and output of iron in the Roman Empire

The first serious attempt to study this problem in depth was made in 1975 by R. Aiano (*The Roman Iron and Steel Industry at the time of the Empire,* M. A. diss., Aberystwyth). His estimates and tentative conclusions are summarized as follows:

Amount of iron required

There are no statistics, but a possible basis may be taken from the known requirements of the UK in the 17th century (well before the changes in demand wrought by the Industrial Revolution), which were 25,000 tons per annum for a population of $5\frac{1}{2}$ millions, with an average consumption of around 10 lb per head; if we scale this down to say 5 lb per head, we get an annual requirement for the Roman Empire (estimated population, *c.* 50,000,000) of 125,000 tons.

Ore requirements

About 7 tons of ore are needed to produce 1 ton of iron; to fire 1 ton you need approximately 1 ton of charcoal. To meet the annual requirement the industry has to produce 875,000 tons of ore, and the same quantity of charcoal.

Timber requirement

To produce 1 ton of charcoal you need 4 tons of timber; to produce the 875,000 tons of charcoal requires $3\frac{1}{2}$ million tons. According to Oliver Davies (1935), the Romans extracted from the island of Elba 11 million tons of ore, from which was produced some $1\frac{1}{2}$ million tons of iron, representing no more than 12 years' supply.

Loss of timber reserves

Aiano notes only that the timber requirements are enormous. Healy (1978, 152) points out that a proper estimate of timber consumed must include not only the charcoal used in the smelting, but what was needed for the preliminary roasting and subsequent forging; he does not attempt an estimate. The extent of consequent deforestation has been the subject of strong controversy; it can easily be exaggerated. Healy's estimate for the area deforested annually by smelting alone is 5,420 hectares (nearly 14,000 acres).

Meiggs (1982, 380) takes a more cautious view.

In a recent paper Henry Cleere, who had already produced results from an experimental furnace designed on the basis of 2nd-3rd-century furnaces found in the Kentish Weald (*Britannia* (1971), 203 ff.), has taken the whole subject of iron-making a stage further by examining the entire milieu of a specific site in the same area, the Bardown site, which has been intensively studied over a period of 15 years. On the basis of the amount of slag found, Cleere *(art. cit.)* gives the following estimate of iron produced:

The working life of the site was almost 80 years AD 120—AD 200). The total quantity of slag deposited was $10\frac{1}{2}$ thousand tonnes, which gives an annual rate of 125 tonnes, and a production rate (*c.* 50 percent of this figure) of 60–65 t.p.a. The three furnaces on the Bardown site will have produced 20 tonnes apiece at the rate of 90–100 kg per day over an actual smelting period of *c.* 7 months p.a. These results have been extrapolated from the only one of the three furnaces to be closely studied. On the question of timber consumption, Cleere notes that the modern retort method of producing charcoal demands a much higher ratio, 5–6 tonnes per tonne, and poses the question whether deforestation of the area was a part cause of the abandonment as early as AD 200. This important study points the way for future research, which must not be confined to the strictly technical aspects of iron production; in particular, the organization of the major industries of the classical world deserves to be more closely studied for the light it can throw on the economic development as a whole.

APPENDIX 15

Development of the catapult

'A number of ancient treatises, and a few surviving illustrations, together present a coherent picture of the design of various weapons of this type, and of the ways in which these designs were gradually developed and improved.' This statement, made only a short time ago (Landels 1978, 99) has been overtaken by recent discoveries, and the studies based on them, some of which are still unpublished, have both clarified and complicated our problems, attention being now focused on developments in the Roman period which promise to be very significant. The variety of types of missile-shooters mentioned in the various sources, and the variations in design among the different types, together with the lack of secure dating, make the task of assessing innovation and development extremely hazardous, the more so as much of the recently discovered material has yet to be fully investigated. The subject is, therefore, introduced by reference to the earlier discoveries at the frontier fort of Saalburg, West Germany, and at Ampurias in north-east Spain, goes on to review the latest finds, and concludes with a Table of Development, which should be described as strictly provisional.

Historical

The discovery of parts of several different types of catapult during the excavation of the fort of Saalburg on the German limes aroused the interest of a German officer, Major Schramm, who was able to make accurate working replicas, and to produce a definitive publication (*Die Antike Geschütze der Saalburg,* Bad Homburg, 1918, 2nd edition, with new introduction by Dietwulf Baatz, 1978). After some years of neglect, the subject was put on a firm foundation by Eric Marsden (*Greek and Roman Artillery,* vol. I *(Historical Development)* Oxford 1969, vol. II *(The Technical Treatises),* Oxford 1971), who had only the Saalburg reconstructions and the Ampurias finds to help him.

The archaeological evidence

Prior to 1968, the only archaeological evidence, apart from representations of catapults on Trajan's Column and other monuments, was that from the Haltern and Ampurias finds. The former site, a legionary camp on the German frontier, yielded a large number of arrow-heads of a type that could only have been fired from a catapult, and which prompted the fine series of models made at Saalburg. At Ampurias in north-east Spain, which was first colonized by Greeks from Phocaea, the remains of an arsenal, discovered in 1912 yielded a remarkably well preserved catapult, complete with iron frame and cylinders, and an assortment of washers and underplates, with holes so punched as to make possible very fine tensioning or 'tuning' of the sinew-ropes of the torsion-type weapons.

The new evidence

The series of new discoveries began with the excavation of two forts on the Danube frontier, one at Gornea, the other at Orsova, both in what is now Romania. The Gornea excavation produced no less than three examples of a man-operated cross-bow (*cheiroballistra, manuballista*—see table below). These were of an advanced design, with short torsion-skeins, and other features not mentioned in the texts. The main find at Orsova consisted of a light arrow-shooter, with the same calibre (79 mm) as that at Ampurias; the type is well known from the literary record (e.g. Vegetius 4.22; Ammianus 23.4.1–3). Of the other recent discoveries (eight are listed in Dr Baatz's new edition of Schramm (Baatz 1980, vi), the most important are those from Cremona in North Italy, and Hatra in Iraq. As we shall see, these finds provide valuable testimony to the Roman advances in catapult design which have been inferred from the monumental sources. Further evidence of important technical developments has emerged from recent work on a collection of catapult material in the museum at Cremona, North Italy, parts of the debris left behind on a neighbouring battlefield, where savage fighting took place during the Civil Wars of AD 68–69, as reported in detail by Tacitus (*Histories* III.22–25), who includes the story of a monster stone-thrower *(ballista)* put out of action by cutting the torsion cable!

Two improvements in design are evidenced by this material: first, in the design of the bronze boxes *(modioli)* which held the torsion cables; secondly, a distinct forward curvature of the catapult frame, allowing for an increase in the stretch of the arms, and therefore of the range of the weapon (see below). This latter evidence confirms that had been suggested by the shape of a catapult depicted on the funerary monument of C. Vedennius Moderatus. All this recent evidence, as Baatz points out, reflects the new design of catapult on Trajan's Column, with its low wide frame, and more widely-spaced springs protected by metal cases both against weather and enemy attack. These modifications will have effected improvements in the field of vision, as well as in range and penetrative power. More recent, and as yet unpublished, work by M. J. T. Lewis suggests that the typically wide frames were designed to make it possible for the arms of the bow to be set on the inside, and pulled back much further giving an increase of perhaps as much as 70 per cent in the range—a genuine revolution in military technology!

Roman improvements

In catapult design
When the Roman armies first began to use these weapons early in the 3rd century BC Greek engineers dominated the field—a situation which continued up to late Republican times. By the end of the 1st century AD a distinctively Roman design of catapult had appeared: the types on Trajan's Column show a wide, shallow frame replacing the tall, narrow Greek type. Both the Romanian and the Hatra models are of the same design.

In stone-throwers
Another improvement in design which is definitely of Roman origin is a stone-throwing torsion-spring catapult, the recoil 'kick' of which gave it the appropriate name of *onager* (wild ass). The horizontal tor-

sion-spring was housed in a heavy timber frame, and it had a long throwing arm carrying the sling, which was winched down to the frame (the historian Ammianus says it needed eight men on the windlass to lower the arm). No writer earlier than Ammianus mentions the machine, which seems to imply that it was not in general use before the 4th century AD. This may well be the case, as Landels suggests (1978, 132) of a less complicated weapon gaining ground at the expense of the more complicated one. The critical factor here was that of weight: a long-range onager, with only a single spring, would need a very powerful spring, and a very heavy frame to absorb the recoil, even when placed on a turf platform, as Ammianus advises (*History*, 23. 4.5). The trajectory could be varied by altering the length and positioning of the sling which carried the stone, but, as Landels points out (1978, 132), 'it is difficult to see how the *direction* of aim could be altered, in view of the fact that the whole machine would have to be shifted round.' The recent discovery of a large and very powerful stone-thrower at Hatra shows that as late as the middle of the 3rd century such weapons were still being developed and improved. The front and sides of this ballista, which was a two-armed version, were protected by sheets of bronze about 2 mm thick. The massive frame (2.40 m wide, 0.84 m high and 0.45 m thick), makes it easily the largest and most powerful ballista so far discovered, and the fact that is was actually in use during the siege of Hatra is prima facie evidence that the difficulties to which Landels refers (1978, 132), had been overcome. Perhaps Ammianus was guilty of some exaggeration when writing of the massive recoil.

Two important technical improvements

Automation
The efficiency of the torsion catapult was also increased very considerably by the automation of three of the six steps needed to fire it, and by the introduction of a chain-drive on the slider mechanism (for its application to water-lifting, the bucket-chain, see Ch. 10.).

Standardization
Greek military engineers discovered the formulae by which torsion-spring catapults and ballistas could be turned out on a systematized basis. The key dimension, they found, was the size of the skeins, which was determined by the diameter of the holes that

received them. All the other components were found to be proportionate to this diameter, the proportions in each case being established by altering the sizes and testing the results, as Philon explains (Belop. 3 = Cohen and Drabkin 1948, 319). The formula for the arrow-shooting type was found to be simple, the diameter being 1/9th of the length of the arrow, but that for the stone-thrower was more complicated, and it is no surprise to be told that 'some workshops were lucky enough to have lists of measurements for the various sizes of shot, worked out by a tame mathematician' (Landels 1978, 120–21), whose lucid account of the matter includes details of the two other methods employed (*ibid.* 121–23).

Development of the catapult

CLASS A. TENSIONED BOWS

Dating	*Name*	*Pros and cons*		
Not before 4th cent. BC (not mentioned by Thucydides)	*Gastraphetes* Mk I	*Pro*	(a)	uses body-push to draw a stronger bow than could be drawn by an archer's arms
			(b)	mechanical release mechanism improves on the archer's fingers
			(c)	pawl and ratchet for holding 'on the draw'
		Con	(a)	only single shots
			(b)	slow rate of firing
	Gastraphetes Mk II	*Pro*	(a)	capstan and handspakes instead of stomach-push for tensioning
			(b)	remains in position between shots: Mk I has to be lifted down after each shot
	Gastraphetes Mk III (or *katapeltos*)	*Pro*	(a)	same as Mk II but increased size, and
			(b)	increased stiffness of bow, giving greater power of penetration: (*katapeltos* means 'hide-piercing' i.e. shield-piercing); could also penetrate mantlets covering siege-engines
Charon of Magnesia Biton Sidorus of Abydos (? d)	*Stone-throwers* (*ballistae*)			same basic design, but modified to take stones thrown by a sling (Landels 1978, 105); larger models had handles at either end of windlass for 2-man operation with block and tackle

CLASS B. TORSION-SPRING CATAPULTS

Begin *c.* 350 BC	T.S.C. Mk I	timber frames with tensioning sinew-rope wound round them
	Mk II	torsion springs tensioned by rotating iron rods
	Mk III	improved washer with guide and underplate, making tensioning still easier
	Mk IV	providing a vernier-type fine tuner for the springs
	Mk V	redesigned frame (Landels 1978, 116) for greater strength, and shock-absorbers fitted to the arms: no increase in trajectory, but much heavier missiles can be discharged.

219

APPENDIX 16

Technical aspects of the siege-engines and counter-devices used at the siege of Syracuse (213–212 BC)
(Polyb. 8.4–6; Plutarch, *Marcellus* 17)

Complete command of the sea enabled the Roman commander, Marcellus, to attempt to breach the strong defences of Achradina from the sea, but the most powerful type of battering-ram, the *helepolis* or 'city-taker', could not be mounted on ships,* and his hopes of success rested on another innovation, a much-improved scaling-ladder, called the *sambuca* or 'lyre' from its resemblance to that musical instrument. To enable them to be used in an assault from the sea, they were carried on two quinqueremes lashed together, and probably hoisted up the mast by block and tackle. A horizontal platform at the top could then be swung out at the height of the wall. The defence, which was under the direction of Archimedes, the most distinguished scientist of the time, was based on the use of three main devices: first, *lithoboloi* (stone-throwing catapults), both long- and short-range types, range and angle being adjusted as the Romans approached; next, after the *sambucae* had been attached to the walls, the *organa* (presumably cranes) which were made to swing out from the wall and drop heavy stones or lead weights on the ships which were now stationary under the walls; these machines were fitted with a universal joint, and a trigger mechanism. Finally, having cleared the forward decks with stone shot, he let down *grabs* (lit. 'iron hands') which hoisted the prow, and caused the ship to capsize. Polybius' account (closely followed much later by Plutarch, in his Life of the Roman commander, Marcellus (Ch, 17)), contains the predictable rhetorical embellishments, but nothing beyond the capacity of the technology of the time. For details of the grabbing device see Landels (1978, 99 ff., and fig. 33), and for interesting comments on the use of the graded stone-throwers, Tarn (1930, 135).

---

* Alexander must surely have used them more than a century earlier, at the siege of Tyre.

TABLE I

FOOD PRODUCTION—CEREALS

Date	Class of grain	Operation	Instrument	Technological innovation/development	Comment
mid 1st cent. AD	wheat (hulled) emmer (*far*)	breaking fallow sowing cultivating weeding	plough; hoe hoe	'Rhaetian' wheeled plough (Pliny 18.172 calls it a 'recent invention' in Switzerland	first literary reference to the heavy wheeled plough which was later 'to dominate farming in NW Europe' (White, 1964, 134).
1st cent. AD (Pliny)	wheat (naked) a) breadwheat b) *triticum vulgare* (hard) c) *siligo* (soft)	harvesting	1) sickle	improved 'balanced' sickle claimed as Roman invention	backward set of handle makes cutting less tiring
			2) 'heading' machine (a) *vallus* (b) *carpentum*	two-wheeled cart strips off heads of grain / pushed by mule/donkey; improved model has a larger container pushed by ox: much larger grain capacity than (a)	'destroys straw by trampling: only useful in large fields or where straw has no economic value' (Pall. 7. 2. 2–4), brief description of (a) survives plus several representations on stone. Full description of (b); no representation.
	barley (2-rowed) (6-rowed)	threshing	1) animals trampling 2) threshing sledge 3) flailing	improved 'Punic cart' has operator mounted on sledge now fitted with wheels (Varro I. 52. 2)	used in E. Spain; has advantage of man's weight and control
	millet				
	beans	winnowing	1) winnowing shovel 2) winnowing fan		requires steady breeze much slower than (1)
	grass (hay) mixed forage (*farrago*)	cutting	scythe 1) 'Gallic' with long handle	this type still used in E. Europe straight blade and handle	
?4th cent AD			2) 'Italic' short handle scythe 3) long 'British' scythe	single-handed—half-way between sickle and scythe giant size with very long blades and very heavy; hafting probably curved important development in response to increased demand for fodder for army (cavalry arm much enlarged at the time)	14 found in one blacksmith's hoard in 1875; several more of identical design in 1971 and 1973

TABLE 2

FOOD PROCESSING I

Date	Class of Grain	Operation	Instrument	Product	Technological innovation/development	Comment
Prehistoric (Moritz 1958, 18 ff.)	wheat (hulled) emmer (*far*)	roasting pounding	pestle and mortar a) of stone b) of wood	porridge *puls* groats	'iron-clad' pestle (Pliny 18.97)	on distinction between mortars and querns, Moritz 1958, 18 ff.
	wheat (naked) *triticum* (hard)	milling	a) hand quern (*mola trusatilis*)	meal	the most ancient 'up-and-down' rubber rudimentary development of (a) to overcome the difficulty of fitting the mill-stones so that grist can be fed in while working	Near East in origin, early representation from Egypt the most important specifically Greek contribution to the history of corn-milling' Moritz (1958, 47).
Greece 6th cent.			b) 'hopper–rubber' ('Olynthian')	flour bread		Moritz thinks rotary mill discovered once; rapid spread helped by Roman armies
early 3rd cent.	*siligo* (soft)		c) rotary hand-mill (*mola versatilis*—Cato)		two types to be distinguished: (1) thick; small diameter; horizontal socket, (2) with vertical handle set in vertically slotted projection: pre-Roman, origin Spanish (?)	
3rd cent. but earlier than donkey mill			d) rotary, vertical, man-operated, but lacks the reversible 'hour-glass' upper stone of type (e)		eight out of a total of 26 mills of four distinct types found at Morgentina, Sicily	Fully discussed by K.D. White 1967 who regards them as prototypes of the 'Pompeian' mill.
late 3rd cent. (not certain)			e) donkey mill (*mola asinaria*) 'Pompeian' mill larger type (horse-mill)		Very important invention, speeding up the grinding process; very much faster than hand-mill; cheap motive power of poorest kinds of animal	from first appearance Italian references predominate; could be an Italian invention
? late 1st BC			f) water-mill (*mola aquaria*) (1) undershot ('Vitruvian') type		developed from the undershot waterwheel described as such by Vitr. (early AD 1?) and common by 301 (in *Edictum de pretiis* with (c) and (d)), or even earlier	gearing (by wooden crown and pinion) may not have exceeded 1:1; slow workrate but cheap with continuous flow
			(2) overshot type		this development much more efficient than undershot (at least $\times$ 8 with same rate of flow)	much more expensive to instal than type (1): needs 'head' of water and millrace, etc.

FOOD PROCESSING II

Class of Grain	Operation	Instrument	Product	Technological innovation/development	Comment
wheat (siligo)	sifting meal after grinding	flour sifter (cribrum)		'The Gallic provinces invented a horse-hair sifter' (Pliny 18. 108) 'Top quality bread depends on (a) the goodness of the wheat (siligo) (b) the fineness of the sieve' (Pliny 18. 105)	even after sifting of the meal, the best Roman bread was probably dark in colour; what was the quality of the panis secundarius, eaten by rank and file?
	baking	oven	bread	Romans had great variety of fancy breads (short list in Pliny 18. 105; longer in Athen. 3. 109 ff.)	
emmer (far)	boiling	pestle and mortar	groats (alica)	special chalk as additive from 'White Earth Hill' near Naples (Pliny 18. 114)	
barley (hordeum)	roasting pounding		groats (alica)		
2-rowed (distichum)	boiling		pearl-barley (polenta)		
6-rowed (hexastichum)	baking		barley cake (maza)		
millet (milio)	boiling		white porridge; leaven (fermentum)	'When Gallic and Spanish wheat is steeped in water to make beer, the froth forming on the surface... is used for leaven; that is why their bread is lighter than that of the others' (Pliny 18. 68)	
	wheat (vitis vinifera)				

WINE PRODUCTION

TABLE 3

Raw material	Operation	Instrument	Product	Technological innovation/development	Comment
cultivated grapes (*vitis vinifera*)	trenching	spade	wine		
	planting	– – –			
	staking	mallet			
	cultivating	hoe			
	pruning	pruning-hook		improved multi-purpose hook (*falx vinitoria*) Colum. IV. 25.	remains standard pruner for almost 1800 years; superseded by the secateur
	harvesting	(1) cutters (2) wooden troughs			
	treading	vats			
	pressing	(1) lever-press (2) Pompeian frame-press (3) screw-press (a) single (b) double (4) 'congeries' press		development of (1) from Greek (weighted by man + stone) to Roman (a) by use of capstan (b) of screw-down collar; screw-press proper developed (?) from the frame-press (2)	volume of grapes ripening simultaneously gives impetus to increasing speed of pressing operation
	fermenting	(1) large jars (*dolia*) (2) barrels (*cupae*)			
	'bottling'	(1) *amphorae* (2) large skins (*cullei*)			

TABLE 4

OIL PRODUCTION

Raw material	Operation	Instrument	Product	Technological innovation/development	Comment
cultivated olive berries	digging holes	spade	oil		full account of oil-making in White 1975, 225–233
	planting	(1) plough (2) hoe			
	cultivating				
	harvesting	ladders, mats canes, baskets			enough to break skins and facilitate milling
	storage (light treading)	bins in loft vats			
	milling	(1) 'Catonian' mill (trapetum) (2) 'Columellan' mill (mola olearia) (3) small sledge (4) 'clog and vat' (canalis et solea)	'mush' (sampsa)	Note: essential aim of improved design of mill was to avoid crushing the stones and spoiling the flavour (1) is defective: no adjusting mechanism for fruit of different sizes; (2) allowed for raising or lowering the flat millstones; no description given of (3) or (4). See White 1975, 226 f.	Colum.'s 'oil-mill' almost certainly fitted with round stones running on their rims, so easily adjustable. Colum. (12.52.6), mills (2) are more practicable for making oil than a trapetum than (1), and a trapetum than (3) or (4)
	pressing	lever-press (as for grapes) frails (fisci) of soft basketry to hold the mush	oil (various grades); scum (amurca)		
	separation of oil from amurca	(1) sets of pans and ladles (2) lead tank (3) double settling tank (structile gemellar)		(1) preferred by Colum. (12.52.10) improvements in oil-making process recomm. by Colum. a response to demand for higher grade oil (?)	straining through 30 pans ensures aeration and removal of amurca pairs of settling tanks (2) are reported from excavated farm sites

TABLE 5

CIVIL ENGINEERING

Item	Operation	Material used	Tool	Technological innovation development	Comment
1 high ways	road-making	sand, clay, rubble, flint, flagstones, gravel, timber	surveyor's *groma* pick and shovel rammers levelling instruments	development can be traced in surfacing; from gravel to flint paving, secured by kerbing Roman innovations include 'corduroy' roads in waterlogged country (NW Gaul, Germany)	construction of properly surfaced roads begins with the Romans variety of construction, in terms of available materials and conditions and, durability are the out-standing features, as revealed by extant remains
2 bridges and viaducts	bridge-building	cement, rubble (infills), dressed masonry	mason's kit carpenter's kit	Roman innovations: 'cofferdam' method for constructing bridge-piers (Vitr. V. 12. 3 on harbour works); downstream buttressing of piers to check erosion	
3 road tunnels	tunnelling	shuttering	carpenter's kit surveyor's equipment		Roman tunnels show a remarkable capacity to survive; some are still in use

TABLE 6

ROMAN CANALS

A canal is usually defined as either an artificial watercourse for inland navigation, or as a synonym for a dyke or drainage channel. The earliest canal on record was the first of many attempts to cut a channel through the narrow neck that divides the Nile from the Gulf of Suez, and so provide a link between the Indian Ocean and the Mediterranean. The earliest documented canal here was dug on the order of Darius, King of Persia (five extant inscriptions found along the route between the Bitter Lakes and the Gulf testify to the work), but for some unrecorded reason it remained incomplete at his death, and more than two centuries elapsed before Ptolemy Philadelphus completed it. The latter is said to have provided the canal with a lock.

Of the canals recorded in the Roman period, three (Nos 2–4 in the list) were dug for the purpose of draining and controlling the dangerous floodwaters of the Po, as was the Car Dyke system in Britain (No. 5). No. 9 (the Fossa Mariana) began life as a strategic supply line, developing later into an important commercial waterway.

	Date	Location	Type	Purpose	Builder and comment	Reference
ITALY	1 2nd cent. BC	S. of line Modena-Parma	4	drainage of lower Po: control of flooding from tributaries; land reclamation	M. Aemilius Scaurus censor 109 BC; builder of Via Aemilia Scauri	Strabo 5, 1. 5
	2 2nd cent. BC dates uncertain	Bologna, Piacenza and Cremona areas	4	flood-control	M. Aemilius Lepidus consul 187 BC; censor 179. BC built Via Aemilia from Placentia to Ariminum A. Postumius	Strabo 170
	3 1st cent. BC	Forum Appii-Terracina	4	drainage of Pomptine marshes: used for navig. when Via Appia unusable		Strabo 5, 3, 6. Hor. *Sat*, 1. 5.
	4 end of 1st cent. BC (date uncertain)	Ferrara-Padua	2	flood-control (?) navigation: direct connection between Ravenna and Po estuary	Augustus: the Fossa(e) Augusta	inscr. *Not. Scav.* 1915, 126 f. 119, 482. n. 57
		other drainage canals in the Po estuary: Fossa Flavia; Fossa Carbonaria; Fossa Philistina; Fossa Clodia (Pliny's text is corrupt)	?	some were diversionary into main branches; subseq. change of landscape by erosion and siltation makes identification impossible		Pliny 3, 120, Smith, 'Roman Canals', *Trans. Newcomen Soc.* 49 (1977–78), 77.

TABLE 6 (continued)

	Date	Location	Type	Purpose	Comment	Reference
BRITAIN 5	1st cent. AD ? AD 50–60	R. Cam—R. Ouse. The so-called Car Dyke depth 7ft; width (bottom) 30 ft; (at water level, 45 ft)	4	part of drainage scheme to release Fenland for farming; also navigation?	perhaps part of a dual-purpose canal system (navigation between Cambridge and York)	Clark, in *Ant. Journ.* 29 (1949), 145–63.
6		R. Ouse—R. Nene (Northants)	4	purpose of whole scheme strategic; note evidence of increasing cultivation of Fenland	strong evidence of increased population in Fens in early period of Roman occupation	
7		R. Nene—R. Witham (Lincs.)				
8		R. Whitman—R. Trent —York (so-called Foss Dyke)				
GERMANY 9	BC 12	Rhine—Yssel (14 km; 9 m) the Fossa Drusiana	2	strategic, connecting the Rhine near Arnhem with the old Yssel near Doesburg (Furneaux, *comm.* on Tac. *Ann.* 2.8)	to speed movement of troops moving N. to the Frisian coast and the R. Ems	Tac. *Ann.* 2.8; 11.1 Suet. *Claudius* (Tac. *Ann.* 13.53 refers to completion, by Pompeius and Paulinus of a long dyke begun by Drusus (Tac. *Ann.* 13.53)
10	? 9 BC	Rhine Dyke. Built by the Elder Drusus in connection with his canal.		to retain sufficient water in the northern arm of the Rhine/to navigate his canal.	demolished by rebel Civilis, AD 70 (Tac. *Hist.* 5.19)	
11	AD 47	Rhine—Meuse (23 Roman m. possibly the Vliet, which leaves the old Rhine at Leiden and reaches the Meuse via Delft (Furn.* *Comm. ad loc.*)	4	'to prevent the rivers flooding the land, by having their waters driven back by high tides' (Dio** 61.30)		
				* see Col. 6	** see Col. 5	* Tac. *Ann.* 11.20; built by Corbulo ** Dio gives 170 *stadia* as the length

	Date		Location	Type	Purpose	Comment	Reference
GAUL	BC 101	12	Fossa Mariana—from Rhone via the Crau (Pliny's 'stony plains') to Foz (NB the name) on the coast east of the Bouches du Rhone by-passing the Rhine	2	Marius had great difficulty getting supplies for campaign v. Teutones, the river 'being choked with mud, sand and clay' (Plut. *Marius* 15)	Foz (Fos-les-Martigues) certainly marks the seaward end! Smith, (*Dict. Geog.* s.v.) gives full discussion of its possible course	Strabo *Geogr.* 2.5.183 Pliny 3.34 ('the Marian canals') Plut., *Marius* 15.
	?	13	Narbonne—R. Aude giving N. access to river system same route as modern canal (13 km; 8 m)	2	to make Narbonne a river port		Pliny 3–4 Grennert II, 485 ff.
PROJECTED	late 1st cent. AD	14	by Nero? Ostia—Rome	1	there was a canal linking the new Ostian harbours with the Tiber, but details are not clear		Suet. *Nero* 16.
			(projected by Nero) Ostia—Puteoli 160 Roman m.	1	intended to run as far as Lake Avernus where it would connect with Agrippa's Fossa Augusta		Suet. *Nero* 31.3; Tac. *Ann.* 15.42. (traces remains near Avernus of Nero's impossible scheme
	AD 55	15	R. Saône—R. Moselle	1	to enable goods from the Med. to reach the Ocean via Rhone, Saône, Moselle, Rhine	a logical extension, since all four rivers were currently used for traffic, but presupposes capacity to construct locks*	planned by Lucius Vetus: Tac. *Ann.* 13.53.
	AD 111	16	Lake Sophon, near Nicomedia (Bithynia) to the sea-coast	1	to avoid cost of transferring goods to carts after crossing the lake	plenty of labour available; alleged fall of 60 ft not regarded as an obstacle	
			Isthmus of Corinth	3	to improve sea communication between E. and W. Med. avoiding the dangerous Cape Malea	projected earlier by Caesar (1) and by Gaius (2). Begun by Nero (3) but never finished	
					linked with Caesar's plans for E. Med. trade, based on revival of Corinth as trade entrepot	* see Col. 6.	* On this see Pliny *Letters* 10.41 and 62; interesting exchange of letters between Pliny and Emperor Trajan. Trajan's concern is the risk of emptying the lake (*Letter* 42) Pliny (*Letter* 41)

TABLE 7   TRANSPORT VEHICLES

Material	Tools	Product	Development/Innovation	Comment
timber	carpenter's kit	frames	Med.: animals yoked in pairs to central pole; from NW Europe came shafts with single animal; attempts to adapt ox-yoke to equines unsatisfactory innovation: four-in-hand with pole and trace harness (Gallo-Roman)	evidence of use of all-round trace-harness to overcome collar problem; inefficient harness continues to limit load-pulling capacity of equines
basketry (for bodies)	wheelwright's kit	bodies		
iron (for tyres)		wheels		
	blacksmith's tools	tyres		
leather (for harness)		harness		
timber (for yokes)		yokes	improved design of wheels, both passenger—and commercial vehicles known from Gallo-Roman monuments, surviving wheels and *Edictum de pretiis* specifications	
soft basketry (for muzzles, panniers)		muzzles panniers		

TABLE 8   WATER TRANSPORT—ROMAN

Operation	Material	Tools	Product	Technological innovation/development	Comment
barge-building	timber	axes	river-barges drawn by (1) men (2) oxen	Roman barges fitted with hauling-posts (normal gear) also had capstan aft for winching upstream	
	esparto grass (for cables)	saws			
		adzes			
		chisels			
boat-building	linen (for sails)		coastal and river cargo boats	invention of sprit-sail (first appears in Med. 2nd cent. BC) and lateen sail (perhaps late 1st cent. AD, certainly early 2nd AD) helps increase speed and manoeuvrability of small ships	Arab origin of lateen now exploded (Casson 1971, 243 f. and pl. 181)
			harbour tugboats		
ship-building			heavy freighters	sail development: increased size of *artemon* improves performance	
				improved keel and hull design makes possible big increase in tonnage and greater stability	alternation of hinged ribs with rigid ribs promotes flexibility and 'give'
				improved gear, e.g. anchor design including moveable stock	
				improved harbour facilities, including revolving cranes (Vitr. 10.2.10)	

TABLE 9

HYDRAULICS(1) Obtaining supply and lifting

Lifting machine	Power source	Output	Lift	Technological innovation/development	Use	Comment	Date
WELLS*							
(1) rope and bucket	man or animal				(1) domestic supply		
(2) windlass	man				(2) watering animals		
(3) 'Ctesibian pump	man working 2 pistons with rocker-arm	2-3 gall /minute (105 litres) — 140 galls/hour (650 litres)		(1) bronze cylinders worked on lathe (Hero). Final fitting of pistons done with oil as lubricant (Vitr.) (2) fitted with swivelling nozzle and lugs to prevent pipe being blown off (Hero). The example in Madrid has more efficient disc valves. (3) wrecked Roman cargo vessel of c. AD 1 had 4 bronze pumps, showing improvements in fit of cylinders and pistons—? use of grinding paste. An 8-cylinder model, use unknown, was found in Brittany in 1971	(1) as (1) and (2) (2) fire-fighting (3) organs (4) bilge-pumps	described by Vit. (10.7) and by Hero (Pneumatika 1.28) banks of these 'manual' engines were fitted to fire engines up to mid 19th cent. There is one in the London Museum	3rd cent. BC.
WHEELS							
(1) of pots	animal and gear wheels				irrigation		
(2) bucket-wheel	animal and gear wheels						
(3) hollow-rimmed wheel	men on outside of wheel	1 man 19 gall/minute (86 litres)	12 ft (3.65 m)		draining mine shafts. A Rio Tinto system has 8 lifting 97 ft (29.6 m)		
(4) bucket-chain	(1) men on treadwheel (2) waterwheel	3 gall/minute (13 litres)	(97 ft 29.6 m)				
(5) drum (tympanum)	men on outside of wheel	2 men 176 gall/minute (800 litres)					
SCREW							
'Archimedean' screw (cochlea)	1 man treading on cleats round the cylinder	35'/50 gll/minute (160/235 litres)			pumping water out of mine shafts in Spain		
SWIPE (shaduf)	man raising and lowering bucket with counterweight on swing beam				low-lift irrigation (Egypt)		

TABLE 9 (continued)

HYDRAULICS (2) Distributing purifying

Method	Building material	Raw material	Instruments fixing gradient	Technical method	Technological innovation/development	Comment
by open channels	bricks stone	clay, straw, sand	(1) *chorobates*	(a) plumb (b) water level horizontal tube with sighters	Vitr. oddly states that the *chorobates* is the most efficient, (2) and (3) being subject to error	Hero mentions *five* uses for the *dioptra*: (1) calculating rises and falls in a water-supply route; (2) locating starting points for a water-tunnel through a mountain; (3) siting air-shaft for such tunnels; (4) determining distance between points not visible to each other; (5) measure range by triangulation (Hero's *Dioptra*)
by enclosed piped systems*	lead or earthenware pipes	lead: solder clay, sand, bitumen	(2) *libra aquaria*			
			(3) *dioptra*	plane-table with worm gear adjustment	the *dioptra*, though much more sophisticated, could have been less reliable for this job, since a small sighting error would produce greater deviation	
water towers (*castella*)	bricks lead	clay, straw, sand				
settling tanks (*cisternae*)	lead	gravels				

* carried in tunnels, on solid supports, or arcading as terrain required.

Comparative costs of (a) open channelling and (b) enclosed piping systems. The consistent gradient needed by (a) required very heavy outlay in labour and materials, including inspection shafts for tunnels and double arcading in dressed masonry (Segovia) or even treble (Pont du Gard). Closed pipes much less costly but heavy on maintenance and impossible to inspect for leaks and blockages.

TABLE 10

CLOTHING (1) Woollen

Raw material	Operation	Instruments and equipment	Product	Technological innovation/development	Comment
wool	sheep-farming				
	shearing	shears	tunics cloaks		
	sorting				
	washing in hot water (dirt) scouring with detergents (grease)	vats or pits, natron, potash, soapwort, soapstone	togas	improved detergents; papyri of Hellenistic period mention use of natron + olive oil = soap	
	beating to detach fibres; plucking	mallets			
	carding or combing; roving (twisting)	thistlehead (carduus) or teazle comb			combing leaves the fibres parallel. Carding leaves them criss-crossed, giving a soft texture
	spinning dyeing	distaff and spindle vats, dyestuffs		Forbes (*Studies* IV, 23–4) associates better methods of spinning and weaving* (see Col. 6) with growth of centres at Mutina, Parma and Altinum, but no evidence.	on home spinning and weaving by slave women, see Colum. 12.3.6.
	weaving finishing (check flaws, mend, wash, bleach, felt, raise pile, shear) laundering	loom fuller's earth (= natural fine hydrated aluminum silicate = *creta fullonica* (Pliny 17.46)	improved type of vertical loom by Romans replaces a primitive Greek method of hanging weights on the warp. Improved Roman frame-loom	On Padua as manufacturing centre, Strabo V. 1.7.	Forbes associates this 'advance towards better processing of homespun outside the home' with the guilds of professional fullers mentioned in inscriptions (Walzing, *Corp. Prof.* IV. 24). Italy had no guilds of spinners and weavers: but evidence from Pompeii of improved organization of clothing manufacture, Moeller (1976)
	fulling of used garments	fuller's earth, sulphur (for fumigation)		improved bleaching agents (Pliny 35. 196–98) esp. *saxum* which increases in bulk when soaked	

TABLE 10 (continued)

CLOTHING (2) Linen

Raw material	Operation required	Instruments, equipment required	Products	Comment
flax (also hemp)	very complicated, with 6 or 7 separate treatments as follows:		flax	Pliny 19. 16–18 (detailed account of the various stages)
	(1) pull plants by hand		linen textiles (a) for garments (b) for sailcloth	Scheuermeier (1956), 231 ff. CIBA review 49 (1945)
	(2) make bundles and hang up to dry		hemp	(b) defined by Pliny as 'carminare' (NH 19.18)
	(3) shake out seeds		(a) rope	See Wild in Mus. Helv.
	(4) 'retting' for 2–3 weeks in water to soften in order to		(b) netting (fishing, hunting)	25 (1968), 139 ff.
	(5) remove bark by 'scutching' = beating with scutching knife	wooden scutching knife		
	(6) 'hackling' = straightening the fibres with the flax-hackle (aena) a special type of iron comb	hackle		

TABLE 11

TERRACOTTA

Terracotta	Operation	Instrument	Product	Technological innovation/development	Further treatment
	digging clay, washing, 'ripening' with older clay, drying, kneading				
for pipes roof tiles	moulding		water pipes roof tiles flues for heating systems		
for household pots	throwing on wheel	potter's wheel		(a) Greek black-figure techniques for high-grade potterx c. 600 BC, three firing (oxidizing, reducing and re-oxidizing).	Unglazed porous containers for storage of wine oil, etc. treated with pitch
	cutting and polishing, adding handles etc. (decorating)			(b) For red-figure reserve painting method.	
	firing	kiln		(c) Roman red-glazes technique developed from Greek, using 'peptization' technique (analogous to digestion). Roman moulded urn (c. 25 BC), using hemispherical boat-shaped moulds	
	glazing	glazes		(d) Greek moulded decoration (Megaron bowls) in late Hellenistic period leading to Roman relief decoration using hemispherical moulds	
	re-firing	kiln			

TABLE 12

LEATHER GOODS (1) Natural

Raw material	*Source*	*Principal product*
	Greek	Hides and
oxhide	Black Sea: export from Tanais-on-Don	slaves exchanged
	(Strabo 11.2.3 493).	for wine and textiles
cowhide	Crimea: Theodosia (Demosth. *Lakritos*)	
goatskin	Pontus: imported to Byzantium for processing	ropes (Cato's
	(Polyb. IV. 38)	spliced pressrope
sheepskin		*Agr.* 12)
		buckets
gazelle	North Africa, Egypt	thonging
	luxury leathers: from Babylonia, Syria,	harness and horse-
	Egypt, ? North Africa	trappings, saddles,
		boots
	Roman	
	Asia Minor: oxhides from Cilicia (Cic. *Pis.*	army: breastplates,
	36. 87)	shields & covers,
	leather from Tralles (probably luxury trade)	tents
	shoes from Caria and Colophon (probably	
	luxury trade	luxury leathers:
	hide from Italy and Gaul for shoemakers in Rome	shoes, water-skins,
		wineskins

Note: 1) guild of dealers at Ostia handling imports

2) Pompeii shows very different organization of leather manufacture

Britain an exporter of hides (Strabo IV. 5. 2.)

Illyria exporting to N. Italian manufacture (Strabo V. 1. 8. 214)

Army requirements very great; big exports from Gaul, Spain and Alpine provinces

LEATHER GOODS (2) Tanning

Terminology   *Processing methods*

GREEK

byrsodepsein (*byrsa* = skin or leather hide or wineskin)
skytodepsein (*depsein* = 'knead')
derma (= skin-leather)
dipthera (= prepared hide, leather)
skutos (= hide, especially dressed hide)

LATIN

corium (= hide, vegetable-dressed hide)
dorium perficere (= dress and hide),
aluta (= alum-dressed leather?)*
pellis alutacea (rare)
* Pliny, *NH* 23.125; Caesar, *BG* 3.13 (leather sails of Veneti).

Processing methods

1) *Curing* by salting on the flesh side by smoking
2) *Depilation* using urine or dove-dung
a) steeping in alkaline lye or other liquid infusion
b) stretching and scraping over a beam
no bating (liming followed by macerating with dung and other softening processes)
scudding of hairy side to remove remains of hair and other impurities
3) *Tanning:* modern methods are: vegetable (e.g. wattle-bark); mineral (e.g. chromium salts); oil (fish-oil); aldehyde (formalin)

Tanning: classical methods included:
a) smoking: very ancient (from northern hunters and fisherman)
b) oil tannage with brains, marrow, etc. plus smoking which came very early (caves of prehistoric period in Switzerland)
Earliest classical reference is to oil-tanning (Homer, *Iliad* 17.436 ff.)
c) mineral tannage is Roman (from Near East). No proven classical Greek references (Aristoph. *Clouds* 1237 refers to curing by salting)
d) vegetable tanning with very wide range (list is Dioscorides I, 142–56)

LEATHER GOODS(3) Organization of production

Organization	Profits	Operation	Item

Organization

Tanning a separate craft

One of the first guilds at Rome (Plut. *Numa* 17).
? 'No Greek tannery has been excavated'
(Forbes, *Studies* V. 51).

Roman tanneries have been found at Pompeii,
Bonn, Wiesbaden and Mainz (Roman Germany);
and the camp at Vindonissa (Windisch, near
Basle) has provided plenty of evidence.

Pompeii tannery (Mau, *Pompeii²*, 416).

Former town house converted. The *atrium*
(8.50×9.00 m = 28 ft × 30 ft) used for
depilation, scraping, scudding, etc. of hides which
arrived in bundles of ten.

In the back room, divided from the *atrium* by
a low wall, were 15 round pits varying
in diameter, from 1,25 to 1.60 m (= 4 ft—5¼ ft)
and *c.* 150 m (= 5 ft) deep, arranged in groups
of four for vegetable tanning. There were also
rectangular pits, *c.* 0.50 m deep (= 1.6 ft)
probably originally lined with timber, used
for mineral tanning. Pottery containers for
tanning liquors were placed on either side
of the rectangular pits.

An open verandah at the back was used for
preparation of tanning infusions. It contained
a masonry tank, in which the infusions were
made, and from which they flowed along an open
conduit into three large pottery containers
(dolia), partly buried in the floor.

Evidence of particular methods:
Leather scraps at Bonn had high fat content—
probably treated with vitriol and then with
melantaria. Those at Mainz had been tanned with
oak-bark.
'The finds at Vindonissa have been very carefully
investigated. They date from the first century
AD and show that mostly goat-skins were
worked. The tanning agent was the bark of the
spruce-fir.' Forbes, *Studies* V, 52–53.

Profits

Kleon inherited a
tannery and a shoe
factory from his father,
Kleainetos, in
Kydathenaion,
district of Athens

Anytos, accuser of
Socrates, was a
tanner, and presum-
ably wealthy.

Lysias and Polem-
archos (resident
aliens) ran a shield
factory at Piraeus,
with 120 slave
workers.

Pasion, also a
resident alien, owned
one which made
him one silver talent
p.a.

Timarchos inherited
a leather factory;
employed 9–10
slaves, who made
2 obols p.d.; foreman
3 obols.

Operation

hide-production
tanning
leather-making
shoe/boot
making

Item

army clothing
breastplates
shields
leggings
buckets
tents
army boots
civilian shoes
saddles
harness

TABLE 13

SOURCE AND MAIN USES OF THE MORE IMPORTANT MINERALS

Mineral	Deposit	Main use
Gold	GREECE	
		coins
	Rare in mainland Greece before Macedonian period (Philip II got 1000 T p.a.)	jewellery
		foil for plating and gilding
	Macedonia/Thrace (Archaic period), especially Mt Pangaeum (in 6th cent.)	
	Asyla, Hebrus, Skapte Hyle (4th cent.	
	Islands: especially Siphnos till 516 BC	
	Imported from Egypt (?) (Healy, 45). Danube Valley (Healy, 45, noting presence of tellurium) Asia Minor; especially Tmolus. Exploited by Seleucids with Caucasian and Black Sea deposits, later taken over by Armenia (Caballa), Arabia and Nubia (Ptolemaic period) Kings of Pergamum (W) and Pontus (E)	
	Note Persian gold came through Bactria from Siberia	
	ROME	
	Spain (esp. NW (Austrias) and SE (Sierra Nevada))	
	Portugal (Tagus Valley)	
	Gaul (Moselle/Rhine; Pyrenees, Massif Central)	
	Italy (Piedmont, Val d'Aosta, Aquileia, Salassi (expelled by Romans 25 BC)	
	Dalmatia (? till end of 1st century AD)	
	Transylvania (Dacia (Trajan settles miners from Albania))	
	Britain (Dolaucothi in Carmarthen)	
Silver	GREECE	
	Spain (Tartessos) Archaic/Class. Siphnos (till 516 BC)	coins
	Maced./Thrace till overrun by Persia (512 BC)—thus development of Attica (Laurion), Cappadocia (by Persians); Stageira (4th cent.) Melos, Lesbos	jewellery
	Sardinia (5th cent.); Chalcidice (Hellenistic period)	
	ROME	
	Spain (Pyrenees, New Carthage, Turdetania, Ilipa, Castulo, Rio Tinto)	
	Portugal (Minas de Mouros)	
	Gaul (Provence, Savoy, Aquitaine)	
	Sardinia. Attica (Laurion) (till 103 BC); Macedonia (from 167 BC— but soon exhausted)	
	Asia Minor (Cappadocia, Pontus,	
	Britain (as product of lead-refining)	

Mineral	Deposit	Main use
Copper	GREECE	

'Ancient Greece was not a copper-producing' (Forbes, *Studies* IX, 15) Copper was mined from an early date at Chalkis in Euboea ...in a mine where copper and iron were found together (Healy 1978, 57, citing Strabo X. 1. 9).

nails for shipwrights; wire (bronze)
rivets for building; mirror (bronze)

Macedonia (Forbes, *Studies* IX, 15)
Cyprus (frequent refs. in Homer)
Spain, little before Romans (excluding Rio Tinto)
Sinai; Sea of Marmara. Cyprus heavily exploited by Ptolemies

armour (bronze)
surgical instruments (bronze)
domestic utensils
sculpture (bronze)
doors, grilles (for temples and other public buildings)

ROME

Spain (much increased, especially NW after 25 BC).
Asia Minor (north of Pergamum, Cappadocia, Ephesus area).
Rare in Italy.
Britain (Cornwall, Shropshire, and especially North Wales).
Gaul (Provence, Savoy (expecially 'Sallustian' mine—Pliny 34.4))

Tin GREECE

None in Greece. Britain (Cornwall) from *c*. 500 BC (Forbes, *Studies* IX, 135)
Brittany from *c*. 500 BC Cassiterides = ('Tin Islands' probably = Cornwall plus other islands + Gaul).
On tin trade: Pytheas visits Land's End and St Michael's Mount; cf. Cary in *JHS* XLIII (1923), 166

Alloyed with copper = bronze
solder

ROME

Spain not on large scale before Roman take-over in 197 BC.
Massive increase after 25 BC Lusitania, Gallaecia (1st–2nd cent. AD), Galicia, Zamora.
Britain (Cornwall): abandoned after 43 BC in favour of Spanish sources; resumed in 3rd century.

Lead GREECE

Sardinia, Attica (Laurion), Ceos.

as waterproof 'mortar' in stonework (Pharos at Alexandria)
Large requirements for cisterns, vats, water pipes, bath linings, protective sheathing for ships, wooden part of anchors, seals for packages (Grenier, *Manuel,* vol. 2, 643 ff.

ROME

Spain (Baetica, Cantabria, New Carthage); most important deposits at Linares, Sardinia—largest at Iglesias, Sicily.*
Britain (Mendip area by AD 49; Flint by AD 70–74: Derbyshire)

Mercury GREECE

Spain (Almaden) Asia Minor (Ephesus); also Chalkis (Pliny 33.118)

ROME

Same sources as above exploited by Romans
Africa (Ethiopia)
Near East

* Italy imports lead for factory at Puteoli (Pliny, 33. 106)

TABLE 13 (continued)

Mineral	Deposit	Main use
Zinc	GREECE	
	Attica (Laurion) zinc carbonate	alloyed with copper = brass
	ROME	
	Gaul (Aix-La-Chapelle) Italy (Bergamo; Campania (early period); Transpadana Asia Minor Cyprus	
Iron	GREECE	
	Elba; Greek Islands (Siphnos, Seriphos, Cythnos, Gyaros, etc.) Euboea, Boeotia Asia Minor (Pontus, Amasia, Cappadocia) Cyprus (Soli, Paphos, Tamasos) (Amasia), Cappadocia, Armenia *Note:* on Chalybes tradition see Forbes, *Studies* IX, 268 f.	agricultural implements tools (carpenters, builders, masons, blacksmiths and miners, clamps, dowels, girders) (building) nails (the stockpile at Inchtulhill, Scotland, consisted of about a million iron nails, 6–12 inches long, and weighing a total of seven tons: Sprague de Camp, 1963, 188.
	ROME	
	Spain (Turdetani = Rio Tinto); central (between Douro, Ebro & Guadiana) Spain replaces Elba after Etruscan mines closed by Sulla Cantabria (Asturias), Bilbilis Gaul (E. Pyrenees, C. Aquitaine, Massif Centrale; Loire (Bituriges) Elba (vast resources of ore worked from Etruscan period to late Roman Empire—Virgil, *Aeneid* X. 174) Sardinia (Ferraria) C. Europe? Exploitation of Hallstatt ores Noricum (N. Carinthia) most famous in Empire; very high quality steel (Pliny 34. 145, etc.) Britain—widespread, all main types of ore exploited (Forest of Dean *(Ariconium),* Kentish Weald (main phase of devel. in Britain between AD 240 and AD 360 (Forbes, *Studies* IX, 267)	weapons (research now in progress on penetrating capacity of Greek weapons).

Bibliography

ABBREVIATIONS

Classical

ANN   *Annales* (Tacitus)
Corpus Inscriptionum Latinarum
HN   Pliny, *Historia Naturalis*
RR   *De Re Rustica* (Columella)
WD   *Works and Days* (Hesiod)

Modern

AA   *Archaeologischer Anzeiger*
AJA   *American Journal of Archaeology*
AJP   *American Journal of Philology*
BIA   British Institute of Archaeology
BSA   *Annals of the British School at Athens*
BSAF   *Bulletin de la Société Nationale des Antiquaires de France*
CIBA   Council of the Institute of British Archaeology
CR   *Classical Review*
CRAI   *Comptes rendus de l'Académie des Inscriptions et Belles Lettres,*
CQ   *Classical Quarterly*
EHR   *English Historical Review*
HT   *History of Technology* (journal)
IJNA   *International Journal of Nautical Archaeology*

JAS   *Journal of the Archaeological Society*
JHI   *Journal of the History of Ideas*
JHS   *Journal of Hellenic Studies*
JISI   *Journal of the Iron and Steel Research Institute*
JRAI   *Journal of the Royal Archaeological Institute*
JRS   *Journal of Roman Studies*
JSAH   *Journal of the Society of Architectural History*
MAAR   *Memoirs of the American Academy at Rome*
MEFR   *Mémoires de l'Ecole Française de Rome*
PBA   *Proceedings of the British Academy*
PBSR   *Papers of the British School at Rome*
PPS   *Proceedings of the Prehistoric Society*
RE   Pauly–Wissowa–Kroll, *Real-Encyclopädie d. klassischen Altumswissenschaft*
REA   *Revue des études anciennes*
TAPhA   *Transactions of the American Philosophical Association*
TAPA   *Transactions of the American Philological Association*
TAPS   *Transactions of the American Philosophical Society*
TC   *Technology and Culture* (journal)

General works

BERNAL, J. D. *Science in History,* 2 vols, Harmondsworth 1969

BLÜMNER, H. *Technologie und Terminologie der Gewerbe und Künste bei Griechen und Römer,* 4 vols, Leipzig 1875–87

CHILDE V. G. *Man Makes Himself,* London 1948

DAUMAS, M. 'The history of technology; its aims, its limits, its methods', trans. and introd. by A. R. Hall, *HT* I (1976), 85, 112

—— (ed.) *Histoire générale des techniques* I, 230–54, Paris 1961

DESHAYES, J. 'Greek Technology', *A History of Technology and Invention* I, ed. M. Daumas and trans. E. B. Hennessy, New York 1969

Dictionnaire archéologique des techniques, 2 vols, Paris 1963–64

DIEHLS, H. *Antike Technik,* 3rd ed. Leipzig, 1924

EDELSTEIN, L. 'Recent interpretations of ancient science'. Review of M. R. Cohen and I. E. Drabkin, *A Source Book in Greek Science,* and B. Farrington, *Greek Science* in *JHI* XIII, 4 Oct. 1952, 573–604

FARRINGTON, B. *Greek Science,* 2 vols, Harmondsworth 1944 and 1949

FELDHAUS, F. M. *Die Technik der Antike und des Mittelalters,* 2nd ed., Munich 1965

FINLEY, M. I. *The Ancient Economy,* London 1973

FORBES, R. J. *Studies in Ancient Technology,* 9 vols, Leiden 1955–64 (= *Studies*)

—— and DIJKSTERHUIS, E. J. *A History of Science and Technology,* 2 vols, Harmondsworth 1963

GILLE, B. 'Machines', *A History of Technology,* vol. II, ed. Singer et al., Oxford 1957, 629–38

HODGES, H. *Artifacts,* London 1964

—— *Technology in the Ancient World,* Harmondsworth 1971

KLEMM, F. *A History of Western Technology,* trans. D. W. Signer, London 1959

KLUGELHOFER, H. *Römische Technik,* Artemis, Zürich/-Stuttgart 1961

KRANZBERG, M. and PURSELL, C. W. (jr.) *Technology in Western Civilization,* 2 vols New York 1967

KRETSCHMER, F. *Bilddokumente römische Technik,* Düsseldorf 1958 (158 plates with 16 topics, no 4-wheeled vehicles, but good on cranes)

—— *Technik und Handwerk im Imperium Romanum,* Düsseldorf 1958

LANDELS, J. G. *Engineering in the Ancient World,* London 1978

LILLEY, S. *Men, Machines and History,* London 1948
—— 'Trading', *The Muses at Work,* ed. C. Roebuck. Cambridge, Mass. 1969

LLOYD, G. E. R. *Greek Science after Aristotle: Applied Mechanics and Technology,* London 1973, 91–112

NEUBURGER, A. *The Technical Arts and Sciences of the Ancients,* trans. H. L. Brose, London 1930

PLEKET, H. W. 'Technology in the Greco-Roman World', *Talanta* 5 (1973), 6–47

PRICE, D. J. DE SOLLA. 'Automata and the origins of mechanism', *TC* 5 (1964), 9–23

ROEBUCK, C. (ed.). *The Muses at Work. Arts, Crafts and Professions in ancient Greece and Rome.* Cambridge, Mass. 1969

SINGER, C., HOLMYARD, E. J. and HALL, A. R. (eds). *A History of Technology* vols I and II, Oxford 1956–57 (= *Hist. Tech.*)

SMITH, C. S. '*Art, technology and science: notes on their historical interaction', *TC* 11.4 (1970), 493–549
'Materials and the development of civilization', *Science* 148, No. 3672, 14 May 1965, 908–17
—— 'Technology in History', *Minerva* 8 (1970), 469–70

SPRAGUE DE CAMP, L. *The Ancient Engineers* London 1960

STAHL, W. H. *Roman Science,* Madison, Wisc. 1962

STRONG, D. E. and BROWN, P. D. C. (eds). *Roman Crafts, London 1976* (specialist articles on all main crafts)

USHER, A. P. *A History of Mechanical Inventions,* 2nd ed., Cambridge, Mass. 1954

Criticism of current literature on ancient technology

BICKERMANN, E. J. and MATTINGLY, G. Review of Singer, et. al., *A History of Technology,* vol. II, in *AJP* 79 (1958), 307–21

FINLEY, M. I. 'Technology in the Ancient World'. Review of L. A. Moritz, *Grain-mills...* (1958), Forbes, *Studies in Ancient Technology* (vols published to date) and Singer et al., *A History of Technology, in Econ. Hist. Rev.,* 2nd series, 12 (1959), 120–25

—— Review of H. Hodges, *Technology in the Ancient World,* C. Roebuck (ed.), *The Muses at Work,* K. D. White, *Roman Farming,* in *N. Y. Review of Books,* 10 June 1971

WHITE, K. D. 'A new model for the economy of the classical world: a critique of M. I. Finley', *Museum Africum* 5 (1976), 83–87

Technology: innovation, development, limitations

ANDRESKY, S. *Military Organization and Society,* 2nd ed., London 1968

CHAPOT, V. Sentiments des anciens sur le machinisme' *REA* (1938), 158–62

CROMBIE, ALISTAIR (ed.) *Scientific Change:* studies in the intellectual, social and technical conditions for scientific discovery and technical invention. London/New York 1961

DAUMAS, M. 'The history of technology, its aims, its limits, its methods', trans. A. R. Hall, *HT* I, 85–112

DODDS, E. R. *The Ancient Concept of Progress,* Oxford 1973

DRACHMANN, A. G. 'Hero's windmill'. *Centauros* 7 (1960), 145–51 (how working on a machine led Hero to think out a new invention)

—— *The Mechanical Technology of Greek and Roman Antiquity,* Copenhagen, 1963 (on the literary sources, especially Hero) (reprint of 1948 edition)

—— *Grosse griechische Erfinder,* Zürich 1967 (Greek weakness in technical development not due to slavery and Plato!)

EDELSTEIN, L. 'New interpretations of ancient science', *JHI* 13 (1952), 579–85

FINLEY, M. I. 'Technical innovation and economic progress in the ancient world', *Econ. Hist. Rev.* 2nd series, 18 (1965), 29–45
—— *The Ancient Economy,* London 1973, esp. 106 f.; 113 f.; 145–49.

HARRIS, J. R. 'The transfer of British metallurgical technology to France in the eighteenth century'. Seminar paper read to the Anglo-French Conference of Historians, London 1978

HODGEN, M. T. *Change and History,* New York 1952 (on the distribution and background of innovations, and the processes by which inventions are transmitted)

KOLENDO, J. 'Le travail à bras et le progrès technique dans l'agriculture de l'Italie antique', *Acta Poliniae Historica* XVIII (1968), 51–62

LANDELS, J. G., *Engineering in the Ancient World,* London 1978, ch. 8, 'The Origin of Theoretical Knowledge'

PARAIN, CH. 'Les anciennes techniques agricoles', *Rev. de synthèse* LXVIII (1957), 327 ff.

PLEKET, H. W. 'Technology and Society in the Greco-Roman World', *Acta Historica Neerlandica* II (1967), 16 ff.

PRICE, D. J. DE SOLLA, *Gears from the Greeks. The Antikythera System, a Calendar Computer System from c. 80 BC*, n.e. Yale 1975

REHM, A. 'Zur rolle der Technik in der Griechisch-römischen Antike', *Archiv f. Kulturgeschichte* 38 (1938), 135–62

SALANT, W. 'Science and Society in ancient Rome', *Science Monthly* 47 (1938), 525–35

SCHLEBEKER, J. T. 'Farmers and bureaucrats: reflections on technological innovation in agriculture', *Agric. Hist.* 51 (1977), 641–55

SMITH, C. S. 'Materials and the development of civilization', *Science* 148, no. 3672, 14 May 1965

—— Review of Ping-ti-Ho. *The Cradle of the East*. Hong Kong 1975, in *TC* 18.1 (1977), 80–86

STAHL, W. H., *Roman Science*, Madison, Wisc. 1962

TORREY, H. B. 'The evolution of mechanical ideas in Greek thought', *American Naturalist* LXXII (1938), 293–303

WHITE, K. D. 'Technology and industry in the Roman Empire'. *Acta Classica* II (1960), 78–89

—— 'Technology in North-West Europe before the Romans', *Museum Africum* 4 (1975), 43–46

—— 'Organization of work and productivity of labour in the Roman Empire', *Museum Africum* (forthcoming)

WHITE, L. jr. *Mediaeval Technology and Social Change*, Oxford 1962

WILSDORF, H. 'Technische Neuerungen in der Phase des Niedergang der Polis', *Hellenische Poleis*, Berlin 1974, 1781–821

Communications, trade and industry

General

AUSTIN, M. and VIDAL-NAQUET, P. *Economic and Social History of Ancient Greece*, Berkeley, Calif. 1977

BRAUDEL, F. *The Mediterranean and the Mediterranean World in the Age of Philip II.*, trans. Sian Reynolds, 2 vols, London 1973

CASSON, L. *Travel in the Ancient World*, London 1974

FINLEY, M. I. *The Ancient Economy*, London, 1973, especially 123–49

FREDERIKSEN, M. W. 'Theory, evidence and the ancient economy'. Review of M. I. Finley, *The Ancient Economy*, in *JRS* 65 (1975), 164–71

FRANK, T. (ed.). *An Economic Survey of Ancient Rome*. 6 vols, Baltimore 1933–40

HASEBROEK, J. *Trade and Politics in Ancient Greece*, trans. L. M. Fraser and D. C. McGregor, London 1933

HEICHELHEIM, F. M. *An Ancient Economic History*, 2nd ed., trans. J. Stevens, vol. I, ch. V, 193–295, Leiden 1958

HOPKINS, K. (ed.), *Trade and Commerce in the Greco-Roman World*, London 1983.

JONES, A. H. M. *The Later Roman Empire, 284–602,* 3 vols, Oxford 1964

LOANE, H. J. *Industry and Commerce of the City of Rome (50 B.C.–A.D. 200)*, Baltimore 1938

MICHELL, H. *The Economics of Ancient Greece*, 2nd ed., Cambridge 1957, 210–310

PERSSON, A. W., *Staat und Manufaktur in römischen Reich: eine Wirtschaftgeschichtliche Studie*, Lund 1923

Proceedings ... Aix. Proceedings of the 2nd International Conference of Economic History, Aix-en-Provence, 1962, Vol. I: *Trade and Politics in the Ancient World*. Paris/The Hague, 1965

ROSTOVTZEFF, M. I. *The Social and Economic History of the Hellenistic World*, 3 vols, Oxford 1941

—— *The Social and Economic History of the Roman Empire*, 2nd ed., trans. P. M. Fraser, 2 vols, Oxford 1957

ROUGÉ, J. *Recherches sur l'organisation du commerce maritime en méditerranée sous l'empire romain*, Paris 1966

SEMPLE, E. C. *The Geography of the Mediterranean Region in relation to Ancient History*, London 1932

Special topics

AUBIN, H. 'Der Rheinhandel in römische Zeit', *Bonn. Jb.* 130 (1925), 1 ff.

AURIGEMMA, S. 'L'elefanti di Leptis Magna e il commercio dell' avorio e delle *ferae Libycae* negli emporii tripolitani', *Rev. Arch. Africa Italiana* 7 (1940), 67

BADIAN, E. *Publicans and Sinners*, Oxford 1972

CARY, M. 'The sources of silver for the Greek world', *Mélanges Glotz*, vol. I. 133–42

CASSON, L. 'The grain trade of the Hellenistic world', *TAPA* 85 (1954), 168–87

CHARLESWORTH, M. P. *Trade-routes and Commerce of the Roman Empire*, 2nd (rev.) ed., New York 1970

COLLS, D. *et al.* 'L'épave Port-Vendres II et le commerce de la Bétique à l'époque de Claude', *Archaeonautica* I (1977)

COLINI, G. 'Portus Tiberinus' (on the river port in the Forum Boarium), *MAAR* 36 (1980), 43–53

GIANFROTTA, P. A. 'Ancore romane'; nuove materiali per lo studio dei traffici marittimi', *MAAR* 36 (1980), 103–16

GOMME, A. W. 'Traders and manufacturers in Greece', *Essays in Greek History and Literature*, Oxford 1937, 42–66

HARMAND, A. *L'Occident Romain*, 2nd ed., Paris 1969

HOPKINS, K. Review of F. Millar, *The Emperor in the Roman World*, London 1977, in *JRS* 68 (1978), 178–86

HOPPER, R. J. *Trade and Industry in Classical Greece*, London 1979

JONES, A. H. M. 'The economic life of the towns of

the Roman Empire', *Rec. Soc. Jean Bodin* 7 (1955), 161–94

—— 'The cloth industry under the Roman Empire', *Econ. Hist. Rev.* 13 (1960), 183–92

LE GALL, J. *Le Tibre: fleuve de Rome dans l'antiquité,* Paris 1953

MANACORDA, D. 'Questioni Cosane. Produzione agricola . . . etc. nell' Ager Cosanus nell' secolo a.C.' in *Instituto Gramsci, Società romana . . .* vol. 2, Rome 1981

MEIGGS, R. *Roman Ostia,* 2nd ed., Oxford 1973

—— *Trees and Timber in the Ancient Mediterrean World,* especially Ch. 5, 'Forests and fleets', 116–53; Ch. 12, 'The Timber Trade', 325–70

MOELLER, W. O. *The Wool Trade of Ancient Pompeii,* Leiden 1976

MOSSÉ, C. *The Ancient World at Work,* trans J. Lloyd, London 1969

PANCIERA, S. 'Olearii' (on the olive-oil trade). *MAAR* 36 (1980), 235–56

PANELLA, C. 'Retroterra, porti e mercati l'esempio dell ager Falernus', *MAAR* 36 (1980), 251–60

PARKER, A. J. 'The evidence provided by underwater archaeology for Roman trade in the western Mediterranean'. Paper read to XXIII Colston Research Symposium on Marine Archaeology, Bristol, 4–8 April 1971 (mimeographed report).

PEACOCK, D. P. S. *Pottery in the Roman World: an ethno-archaeological approach,* London 1982, Ch. 10, 'Pottery and the Roman Economy', especially 158–59.

RICKMAN, G. E. 'Articles of trade, and problems of production, transport and diffusion', *MAAR* 36 (1980), 261–76

—— *The Corn Supply of Ancient Rome,* Oxford 1980

ROEBUCK, C. *Ionian Trade and Colonization,* Arch. Inst. of America, monograph series, vol. IX, New York 1959

—— 'Trading', *The Muses at Work,* ed. C. Roebuck. Cambridge, Mass. 1969

TCHERNIA, A. 'Quelque problèmes concernant le commerce du vin et les amphores', *MAAR* 36 (1980), 305–12

TESTAGUZZA, O. 'The Port of Rome', *Archaeology* XVII (1964), 173–79

WARD PERKINS, J. B. 'On the marble trade'. Contrib. to the 1st Gramsci seminar. Rome 1979

—— 'The marble trade and its organization: evidence from Nicomedia', *MAAR* 36 (1980), 325–38

WHEELER, SIR M. *Rome Beyond the Imperial Frontiers,* London 1954

WILL, E. 'La Grèce archaïque, *2me Conf. inst. d'hist. écon.,* Aix-en-Provence 1962, Paris 1965

WILSON, F. H. 'Studies in the social and economic history of Ostia: Part I', *PBSR* XIII (1935), 41–68

ZEVI, F. and TCHERNIA, A. Amphores de Byzacène au Bas-Empire', *Antiq. afric.* III (1966), 171–214

Subsidiary bibliographies

For the relevant crafts which are not covered by the chapter bibliographies, the reader may refer to the bibliographies appended to the following chapters in Strong and Brown, *op. cit:* Ch. 6 *Pottery* by David Brown; Ch. 9 *Glass* by Jennifer Price; Ch. 12 *Woodworking* by Joan Liversidge; Ch. 13 *Leatherwork* by J. W. Waterer; Ch. 18 *Mosaic* (vaults and domes) by F. Sear; Ch. 19 *Mosaic* (floors) by D. S. Neal.

PART II

Chapter 6 Agriculture and food

Agricultural technology

ABEL, W. *Geschichte der Deutschen Landwirtschaft von Frühen Mittelalter bis zum 19 Jahrundert,* Stuttgart 1962

BILLIARD, R. *L'Agriculture dans l'antiquité d'après les Géorgiques de Virgile.* Paris 1928

BLOCH, M. *Land and Work in Mediaeval Europe,* trans J. L. Anderson, London 1967

CÜPPERS, H. 'Gallo-römische Mähmaschine auf einem Relief in Trier', *Trierer Zeitung* 27 (1964), 151–53.

FUSSELL, G. E. and KENNY, A. L'équipement d'une ferme romaine', *Annales (ESC)* (1966), 306–23

HAUDRICOURT, A. G. and DELAMARRE, M. J.-B. *L'Homme et la charrue à travers le monde,* Paris 1955

HEINZE, H. 'Die wirtschaftliche Entwicklung des Moselraumes zűr Römerzeit', *Trierer Zeitung* 39 (1976), 89–91 (on the recently discovered relief of a *vallus* at Koblenz).

KOLENDO, J. 'Avènement et propagation de la herse en Italie antique' *Archeologia* 22 (1971), 104–20

—— *L'Agricultura nell' Italia romana* Rome 1980 (with full bibliographical notices to date, especially on implements and techniques of cultivation)

—— 'Sur la roue dans l'agriculture des romains', *Ethnologie et histoire. Forces productives et problèmes de transition,* Paris 1975, 53–62

LESER, P. *Entstehung und Verbreitung des Pfluges,* Münster-i-Westfal., 1931 (repr. 1971)

MICHELL, H. *The Economics of Ancient Greece,* 2nd ed., Cambridge 1957, 38–88

PARAIN, CH. 'Das Problem der tatsächlichen Verbreitung der technischen Fortschritte in der römischen Landwirtschaft'. *Zeitschr. für Geschichtswissenschaft,* Heft II, VIII Jahrg. 1960, 357 ff.

—— 'Les anciennes techniques agricoles'. *Revue de Synthèse* LXVIII (1957), 327 ff

SCHLEBECKER, J. Farmers and Bureaucrats: Reflections on technological innovations in agriculture, *Agri. Hist.* 51 (1977), 641–55

WHITE, K. D. *Agricultural Implements of the Roman World,* Cambridge 1967 (with bibliography to date)

—— *Roman Farming,* London 1970

—— *Farm Equipment of the Roman World,* Cambridge 1975

—— 'The economics of the Gallo-Roman harvesting machine', 'Hommâges à Marcel Renard', *Coll. Latomus* 102 (1969), 804–9

—— 'The Great Chesterford scythes', *Proc. Hung. Agric. Mus.* 1971/2, 77–82.

Wine and oil production

BILLARD, R. *La Vigne,* Paris 1913

CAMPS-FABRER, H. *L'Olivier et l'huile dans l'Afrique romaine,* Algiers 1953

DRACHMANN, A. G. *Ancient Oil-mills and Presses.* Copenhagen 1932 (discusses, with trans. all relevant texts, together with all important monumental evidence)

FORBES, R. J. 'Fermented beverages 500 BC–1500 AD', *Studies* III, 106–30

JUNGST, E. and THIELSCHER, P. 'Cato's Kelter und Kollergänger; ein Beitrag zür Geschichte von Öl und Wein. *BJ* CLIV (1954), 32–93; CLVII (1957), 53–126 (full account of Cato's mills and presses, covering all aspects of design and operation)

MANZI, L. *La viticoltura e l'enologia Presso i Romani.* Rome 1883 (contains a full account of Roman methods of wine-making)

WHITE, K. D. *Farm Equipment of the Roman World,* Cambridge 1975, 112–17 (wine production), and Appendix A, pp. 225–33 (oil production)

Food technology

General

BENNETT, R. and ELTON, J. *A History of Corn-Milling,* London and Liverpool 1898–1904

BLÜMNER, H. *Technologie und Terminologie...* vol. I, ch. I

BROTHWELL, D. and P. *Food in Antiquity: a Survey of the Diet of Early People* London 1969

FORBES, R. J. *Studies* III, 91–96 (milling); 138–51 (grinding and pounding)

FRANCIS, C. A. *A History of Food and its Preservation,* Princeton 1937

JACOB, H. E. *Six Thousand Years of Bread,* New York 1944

JASNY, N. 'The daily bread of the ancient Greeks and Romans', *Osiris* IX (1950), 228–53

KRENKEL, W. 'Von Korn zum Brot', *Das Altertum* XI (1965), 209–33

LAVILLE, D. 'Découverte d'une carrière gallo-romaine specialisée dans la fabrication de meules à grain domestiques', *RAC Ant. Nat.* 2 (1963), 126–37

MAURIZIO, A. *Histoire de l'alimentation végétale depuis le préhistoire jusqu'à nos jours,* trans. from the Italian by F. Gidon, Paris 1958

MORITZ, L. A. *Grain-mills and Flour in Classical Antiquity,* Oxford 1958

RICKMAN, G. E. *Roman Granaries and Store Buildings,* Cambridge 1971

SPARKES, B. A. 'The Greek kitchen', *JHS* 82 (1962), 121–37

STORCK, J. and TEAGUE, W. D. *Flour for Man's Bread: A History of Milling,* St Paul, Minn. 1952

TENGSTROM, E. *Bread for the People. Studies in the Corn-supply of Rome During the Late Empire,* Stockholm 1974

VICKERY, K. F. *Food in Early Greece,* Urbana, Ill 1936

WHITE, D. A. 'A survey of millstones from Morgantina', *AJA* 67 (1963), 199–206

WHITE, K. D. 'Food requirements and food supplies in classical times in relation to the diet of the various classes', *Prog. Fd. Nutr. Sci.* 2 (1976), 144–91

ZNATCHKO-JAVORSKY, I. L. *The History of Alimentary Materials from Antiquity to the Middle of the 19th Century* (in Russian with English summary), Moscow 1963

Water-mills

BENOIT, F. 'L'usine de meunerie hydraulique de Barbegal', *Rev. arch. Africa* (1940), 19–80

BLOCH, M. 'Avènement et conquêtes du moulin à eau'. *Ann. d'hist. écon. et soc.* 7 (1935), 539 ff. Trans. J. E. Anderson: *Land and Work in Mediaeval Europe* (selected papers by M. B.), London 1966, 136 ff.

BRETT, S. G. 'A Byzantine water-mill', *Antiquity* 13 (1939), 354–56

CURWEN, E. C. 'The problem of early water-mills', *Antiquity* (1944), 130–46

FORBES, R. J. *Studies* II, 2nd ed. 1965, 86–104

GILLE, B. 'Le moulin à eau. Une révolution medievale', *Tech. et Civ.* 3 (1954), 2–3

KIECHLE, F. *Sklavenarbeit und technischer Fortschritt im römischen Reich,* Wiesbaden 1969, 117 ff.

KOEHNE, C. 'Die Mühle im Recht der Völker. Beitr. z Gesch. d. Technik u. Industrie', *Jhb. d. vereins Dt. Ingenieure* 6 (1915), 34

LAUFFER, S. *Diokletians Preisedikt* (Text und Kommentäre, Bd V), Berlin 1971, 14, 147; 257

MAROTI, E. 'Über die Verbreitung der Wassermühlen in Europa', *Act. Acad. Scient. Hung.* t. XXIII, fasc. 3–4, Budapest (1975), 255–80

MORITZ, L. A. *Grain-Mills and Flour in Classical Antiquity,* Oxford 1958

—— 'Vitruvius' Water-mill', *CR* 70 (1956), 193 ff.

NEUBURGER, A. *Die Technik des Altertums,* 3rd ed., 1930, 97, 232 (on the survival of a 'Vitruvian' mill in Switzerland until very recently)

PARSONS, A. W. 'A Roman water-mill in the Athenian Agora', *Hesperia* V (1936), 70–90

RADKE, G. art. 'Venafrum'. *RE* VIII A 669 (1955)

RICHMOND, Sir IAN. *Roman Britain,* 2nd ed. London 1963, 134

SAGUI, C. L. 'La Meunerie de Barbegal et les roues hydrauliques chez les anciens et au moyen age', *Isis* 38 (1948), 225–31

SIMPSON, F. G. *Water-Mills and Military Works on Hadrian's Wall. Excavations in Northumberland, 1907–1913,* ed. G. Simpson. Kendal 1976

USHER, A. P. *A History of Mechanical Inventions,* London 1954

VAN BUREN, A. W. and STEVENS, G. P. 'The Aqua Traiana and the mills on the Janiculum', *MAAR* I (1915), 59–61

WHITE, K. D. *Farm Equipment of the Roman World.* Cambridge 1975, 15 ff.

WIKANDER, O. 'Water-mills in ancient Rome'. *Opuscula Romana* XII Svensk. *Inst. Rom.* XXXVI, 4, Stockholm 1979, 13–36

Chapter 7 Building

Sources: Greek

BUNDGAARD, J. A. *Mnesicles: A Greek architect at work,* Copenhagen 1957 (contains texts, translations and commentaries on the more important surviving building inscriptions)

Sources: Roman

Vitruvius

BOETHIUS, A., 'Vitruvius and the Roman Architecture of his Age', *Acta Inst. Rom. Reg. Suec. Ser⁶* 1929, I, 114–43

CALLEBAT, L. *Vitruv. De l'Architecture.* Livre VIII, Paris 1973

CHOISY, A. *Vitruv.* 4 vols, Paris 1909

FENSTERBUSCH, C. *Vitruv. Zehn Bücher über Architektur.* Berlin, 1964. (German translation with minimal annotation.)

GRANGER, F. *Vitruvius on Architecture* (Loeb Classical Library), 2 vols, London/Cambridge, Mass. 1931–34

JÜNGST, E. and THIELSCHER, P. *Vitruv, über Baugrube, Baugrund und Grundbau:* ein Interpretationversuch. *Römische Mitteil,* 51 (1936), 145–80

McKAY, A. G. *Vitruvius, Architect and Engineer,* London 1978

MORGAN, M. H. *Vitruvius, the Ten Books on Architecture,* Cambridge, Mass. 1914; repr. pb. 1960

—— Addresses and Essays, New York 1910, 224–72

NOHL, H. *Index Vitruvianus,* Leipzig 1876

PLOMMER, W. H. *Vitruvius and Later Roman Building Manuals,* (Cambridge Classical Studies), Cambridge 1973

SOUBIRAN, J. *Vitruv. De l'Architecture.* Paris, ed. *Les Belles Lettres* (Budé series) in progress

TABARRONI, G. 'Vitruvio nella storia della scienza e della tecnica', *Atti d. Acad. d. Scienze dell' Istituto di Bologna; cl. d. sci. mor. mem.* 66 (1976), Bologna

THIELSCHER, P. 'Vitruvius', *RE* XVII col. 458, f. (1961)

Faventinus

FAVENTINUS, Marcus Cetius. *Compendium De diversis Fabricis Architectonicae.* Printed in the edition of Vitruvius by Valentine Rose, 1867

Palladius

PALLADIUS, Rutilius Taurus Aemilianus. *Opus agriculturae* in J. M. Gesner (ed.), *Scriptores Rei Rusticae Veteres Latini,* Leipzig 1774; also in J. G. Schneider's *SRRVL* of 1794–97

PLOMMER, W. H. *Vitruvius ... op. cit.,* 39–84

RODGERS, R. H. ed.). Palladius, *Opus Agriculturae,* etc. Leipzig (Teubner), 1975 (superseding the earlier edition by Schmitt, J. C., 1898)

Modern Works

General

ATKINSON, R. and BAGENAL, H. *Theory and Elements of Architecture,* vol. I, pt 1, London 1926 (outstanding on materials and structural techniques; profusely illustrated)

BOETHIUS, A. *Roman and Greek Town Architecture* (Acta Universitatis Gotonburgensis 54), Gothenburg 1948

BURFORD, A. M. *Craftsmen in Greek and Roman Society.* London 1972. (Index s.v. 'architects', 'carpenters', 'masons .)

DAVEY, N. *A History of Building Materials,* London 1961

DINSMOOR, W. B. *The Architecture of Ancient Greece.* 3rd (revised) ed., London 1950 (with comprehensive bibliographies to date)

PLOMMER, W. H. *Ancient and Classical Architecture,* vol. I of Simpson's *History of Architectural Development.* London 1957

ROBERTSON, D. S. *A Handbook of Greek and Roman Architecture.* 2nd ed., Cambridge 1945 (with extensive bibliographical appendices)

Greek building construction

BELL, M. 'Stylobate and roof in the Olympieion at Akragas'. *AJA* 84.3 (July, 1980), 359–72, (Proves from cisterns in the central court of archaic Temple of Artemis at Ephesus that its roof was not continuous and argues for the same at Akragas.)

BLÜMEL, C. *Griechische Bildbauerarbeit,* Berlin/Leipzig 1927

BOYD, T. D. 'The arch and the vault in Greek architecture', *AJA* 82 (1978), 83–100

BURFORD, A., *The Greek Temple Builders at Epidauros,* Liverpool 1969

BUNDGAARD, J. A., *Mnesicles: A Greek architect at work,* Copenhagen 1957

COULTON, J. J. 'Lifting in early Greek architecture', *JHS* 94 (1974), 1–19

—— *Greek Architects at Work,* London 1977

DINSMOOR, W. B. 'An archaeological earthquake at Olympia', *AJA* 45 (1941), 399–427

—— 'Structural iron in Greek architecture', *AJA* 26 (1922), 154–56

DORN, H. 'A note on the structural antecedents of the I-beam. The use of iron beams in Greek and Roman masonry construction', *TC* 9 (1968), 415–18. Reply by R. A. Jewett. 'The response', *Ibid.* 419–26; 'the rejoinder' by H. Dorn, *ibid.* 427–29

DRACHMANN, A. G. 'A note on ancient cranes', *HT* II, 658–62.

GARDNER, E. A. et. al. *Excavations at Megalopolis, 1890–1891. Suppl. Papers Soc. Prom. Hellenic Studies* I, London 1892

HEYMAN, J. '"Gothic" construction in ancient Greece', *JSAH* XXXI.1 (1972), 3–9

HODGE, A. T. *The Woodwork of Greek Roofs,* Cambridge 1960

JEWETT, R. A. 'Structural antecedents of the I-beam, 1800–1850', *TC* 9.3 (1967), 346–62

LAWRENCE, A. W. *Greek Architecture,* 3rd ed., Cambridge 1973

MAIER, F. G. *Griechische Mauerbauinschriften,* Heidelberg 1961

MARQUAND, A. *A Handbook of Greek Architecture,* London/New York 1909 (particularly strong on materials and technique)

MARTIN, R. *Manuel d'architecture grecque* I. *Materiaux et techniques,* Paris 1965

ORLANDOS, A. K. *The Construction Materials of the Ancient Greeks.* trans. fr. the Greek by V. Hadjimichali and K. Laummier. 2 parts, Paris 1966, 1968.

SCRANTON, R. 'Greek Building', *The Muses at Work. Arts, crafts and professions in ancient Greece and Rome,* ed. C. Roebuck, Cambridge, Mass. 1969

SHAW, J. W. 'A double-sheaved pulley-block from Kenchreae', *Hesperia* 36/4 (1967), 389–401

Roman building construction

BLAGG, T. F. C. 'Tools and techniques of the Roman stonemason in Britain', *Britannia* 7 (1976), 152–72

BLAKE, M. E. *Ancient Roman Construction in Italy from the Prehistoric Period to Augustus,* Washington 1947

—— *Roman Construction in Italy from Tiberius through the Flavians,* Washington 1959

BLANCHET, A. 'Recherches sur les tuiles et les briques de construction de la Gaule romaine', *Rev. arch.* II (1920), 189–210

BOETHIUS, A. *Roman Republican Architecture,* 2nd ed., Harmondsworth 1978, *Pelican History of Art*

—— *The Golden House of Nero* (Jerome Lectures ser. V), Ann Arbor, Mich. 1960

BROWN, F. E. *Roman Architecture,* New York 1961

CHOISY, A. *L'art de bâtir chez les romains,* Paris 1875

COZZO, G. *Ingegneria romana,* Rome (repr.), 1970 (1928) (important chapters on construction methods, especially referring to Colosseum, Pantheon, Fucino tunnel)

CREMA, L. *Architettura Romana, Enc. Classica,* ser. 3, vol. 13.1, Turin, etc. 1959

GUERRA, G. *Statica e tecnica costruttiva delle cupole antiche e moderne,* Naples 1958

LEA, F. M. and DESCH, C. H. *The Chemistry of Cement and Concrete,* New York/London 1935

LEACROFT, Maud R. *The Buildings of Ancient Rome,* London 1969

LUGLI, G. *La technica edilizia romana,* 2 vols, Rome 1957

MACDONALD, W. L. *The Architecture of the Roman Empire,* I: an introductory study. New Haven, Conn. 1965 (good discussion of techniques, and standards of workmanship, especially carpentry, 147 ff. and 158–59 ff.)

McDONALD, W. L. and BOYLE, B. M. 'The small baths at Hadrian's Villa', *JSAH* 39.1 (1980)

MEIGGS, R. 'Sea-borne timber supplies to Rome'. *MAAR* 36 (1980), 185–96 (on the use of long beams)

MIDDLETON, J. H. *Remains of Ancient Rome,* 2 vols, London/Edinburgh, 1892, vol. I, 36–103 (on the use of timber, especially for centering, and of concrete)

PACKER, J. E. Review of J. J. Coulton (1977), in *TC.* 19.3 (1978), 528–30

—— 'Roman Imperial Building' *The Muses at Work,* ed. C. Roebuck, Cambridge, Mass. 1969, 36–59

—— 'Structure and design in ancient Ostia', *TC* 9.3 (1968), 357–88

ROBERTSON, D. S. *A Handbook of Greek and Roman Architecture,* 2nd ed., Cambridge 1943, ch. 15. (on construction techniques for arches, vaults and domes)

SJÖQUIST, E. Review of Lugli, G. *La technica edilizia romana,* Rome 1957, in *AJA* 63 (1959), 104–6

TURIZIANI, R. in *The Chemistry of Cement,* ed. H. F. W Taylor, London 1964, 233–86

WARD PERKINS, J. B. *Roman Architecture,* New York 1977

Chapter 8
Civil engineering and surveying

Surveying: instruments and methods of land measurement

BLUME, F. *et al. Die Schriften der römischen Feldmesser (Gromatici Veteres),* 2 vols, 1848, 1852. Vol. I contains the text, ed. Lachmann, and the diagrams illustrating various topics and problems on land surveying: vol. II contains a number of essays and indices. Reprinted in 1962

THULIN, C. (ed.) *Corpus agrimensorum Romanorum,* Leipzig (Teubner). Contains the principal writings of the Roman surveyors, with black and white photographs of the MS miniatures

Herons Dioptra, *Herons von Alexandria Vermessungslehre und Dioptra,* ed. H. Schöne. Leipzig 1902–3

DELLA CORTE, M. 'Groma', *Monum. Ant.* 28 (1922), 5–181

DILKE, O. A. W. *The Roman Land Surveyors. An introduction to the Agrimensores,* Newton Abbot (David and Charles), 1971. Contains full bibliography to date; includes centuriation and cadastral material

HINRICHS, F. T. *Die Geschichte der Gromatischen Institutionen,* Wiesbaden 1974

KELSEY, F. W. Review of Della Corte, 'Groma' (1922), in *Class. Phil.* 21 (1926), 259–62

Civil engineering

Bridges
(see also Ch. 10, s.v. 'roads'; Ch. 12, s.v. 'aqueducts')

BALLANCE, M. H. 'The Roman bridges of the *Via Flaminia', PBSR* XIX (1951), 78–117

BRIEGLEB, J. *Die Vorrömischen Steinbrücken des Altertums,* Düsseldorf 1971

BROOKES, A. C. 'Minturnae: the Via Appia bridge', *AJA* 78 (1974), 41–48

BUNDGAARD, J. A. 'Caesar's bridges over the Rhine', *AA Copenhagen* 36 (1965), 87–104 (on pile-driving technique)

CÜPPERS, H. *Die Trierer Römerbrücken,* Mainz 1969

—— 'Vorrömische and römische Brücken über die Mosel', *Germania* 45 (1967), 1–2; 60–69

GAZZOLA, P. *Ponti romani,* Contributo ad un indice sistematico con studio critico bibliografico, Florence 1963

GERMAIN DE MONTAUZAN, C. DE. *La science et l'art de l'ingénieur aux premiers siècles de l'empire romain,* Paris 1909

JACKSON, D. A. and AMBROSE, T. M. 'A Roman Timber bridge at Aldwincle, Northamptonshire', *Britannia* 7 (1976), 39–72

STEINMAN, D. B. and WATSON, S. R. *Bridges and their Builders,* New York 1941–57

VAN BUREN, A. W. art. 'pons' in *RE* XXI.2 (1952), 2428–37

Canals

ALLEN, G. H. 'A problem of internal navigation in Roman Gaul', *Class. Weekly* 27 (11 Dec. 1933), 65–69 (on the Saône-Moselle project)

CLARK, J. G. D. 'Report on excavations on the Cambridgeshire Car Dyke', *Ant. Journ.* 29 (1949), 145–63

DE LALANDE, J. J. le F. *Des canaux de navigation.* Paris 1778, 158 (on the Aude-Narbonne canal)

MOORE, F. G. 'Three canal projects, Roman and Byzantine'. *AJA* 54 (April 1950), 94–111 (on the Marmara-Propontis project: the Saône-Moselle project, and the revival, by Justinian, of Pliny's revised version of the first-named)

SMITH, N. A. F. 'Roman canals', *Trans. Newcomen Soc.* 49 (1977/8), 75–86

SUMMERS, D. *The Great Level,* Newton Abbot 1976 (on the Car Dyke)

Dams

BENOÎT, F. 'Le barrage et l'aqueduc romain de St-Remy-de-Provence', *REA* 37 (1935), 331–39

BRUNHES, J. *L'Irrigation dans la péninsule ibérique et dans l'Afrique du Nord,* Paris 1902

CASADO, C. F. 'Las presas romanas en Espana', *Revista de Obras publicas,* VII Congrès Intern. de Grandes Presas, 1961, 357–63

CASTRO, GIL, J. de. 'El Pantano de Proserpina', *Rivista de Obras Publicas* (1933), 449–54

DUSSAUD, R. 'La Digue du Lac de Homs et le "Mur Egyptien" de Strabon', *Mon. et Mem.,* vol. 25 (1921–22), 133–41

GARCIA-DIEGO, J. A. 'Old dams in Estremadura', *HT* II (1977) 95–124

GOBLOT, H. 'Sur quelques barrages anciens et la genèse des barrages voûtés', *Rev. Hist. Sci.,* vol. XX, No. 2 (1967), 109–40

SALADIN, H. 'Description des antiquités de la Régence de Tunis', *Archives des Miss. Sci. et Litt.* 3ᵉ ser., vol. XIII (1886), 162–64

SMITH, N. A. F. *A History of Dams,* London 1971, 25–56

——, 'The Roman dams of Subiaco', *TC* 11, No. 1 (1970), 58–68

VITA-FINZI, C. 'Roman Dams in Tripolitania', *Antiquity* 35 (1961), 14–20

—— and BROGAN, O. 'Roman dams on the Wadi Megenin', *Libya Antiqua* II (1965), 65–71 and plates XXI–XXVII

Harbours

BARTOCCINI, R. 'Il porto romano di Lepcis Magna',

Bull. Cent. Stud. p.1. Storia dell' Architectura, Suppl. n. 13 (1958)

BESNIER, M. 'Pharus', Daremberg-Saglio, *Dict. Ant.* IV.2.427–32

BROWN, F. E. 'Cosa I: History and topography', *MAAR* 20 (1951), 89–96

—— 'The ports and fisheries of Cosa', *MAAR 32* (1978)

CASSON, L. *Ships and Seamanship in the Ancient World,* Princeton 1971, 361–70

D'ARMS, J. H. 'Puteoli in the 2nd century of the Roman Empire', *JRS* 64 (1974), 104–24

EUZENNAT, M. and SALVIAR, F. 'Marseille retrouve ses murs et son port Grecque', *Archeologia* 21 (1968), 5–17

GOODCHILD, R. G. art. 'Roads and land travel' *Hist. Tech,* vol. II, Oxford 1957, sect. VII. 516–24 (harbours, docks and lighthouses)

HAWTHORNE, J. G. 'Cenchreae, port of Corinth', *Archaeology* 18.3 (1965), 191–200

HÜTTER, S. 'Der Römische Leuchtturm von La Coruña', *Madrider Beiträge* 3 (1973)

LEHMANN-HARTLEBEN, K. 'Die Antiken Hafenanlagen des Mittelmeeres', *Klio,* Beih. 14, Leipzig, 1923; *idem* 'Limen', *RE* (1926)

LEWIS, J. D. 'Cosa: an early Roman harbour'. *Marine Archaeology,* ed. D. J. Blackman, 23rd Colston Symposium, April 4–8, 1971, London 1973, 233–58.

ROUGÉ, J. *Recherches sur l'organisation du commerce maritime...* Paris 1966, part I (sections 5–7); part 2 (section 1).

SAVILE, Sir L. H. 'Ancient harbours'. *Antiquity* 15 (1941), 209–32 (includes Alexandria, Tyre, Piraeus, Puteoli, Ostia, Centumcellae)

SCRANTON, R. and RAMAGE, E. 'Investigations at Corinthian Kenchreae', *Hesperia* 36 (1967), 124–86

SCRANTON, R., SHAW, J. W. and IBRAHIM, L. *Kenchreae: Eastern port of Corinth,* vol. I, *Topography and Architecture,* Leiden 1978, 13–52

SEMPLE, E. C. *The Geography of the Mediterranean Region in relation to Ancient History,* London 1932, 119–29

SHAW, J. W. 'Greek and Roman harbour installations', G. F. Bass (ed.), *A History of Seafaring,* London 1972, 87–112

STAGER, L. E. 'Carthage, 1977: the Punic and Roman harbours', *Archaeology* 30 (1977), 198–200

TAYLOR, J. au PLAT. 'Ports, harbours and other submerged sites', *Marine Archaeology,* London and New York 1966, 18–189

TESTAGUZZA, O. *Portus: Illustrazione dei porti di Claudio e Traiano e della città di Porto a Fiumicino,* Rome 1970

—— 'The Port of Rome', *Archaeology* 17 (1964), 173 f.

WHEELER, R. E. M. 'The Roman lighthouses at Dover', *Arch. Journ.* 36 (1929), 29–46

YORK, R. A. and DAVIDSON, B. P. *Roman Harbours of Algeria,* London 1968

Roads
See the bibliography to Ch. 10 *(Land transport)*

Chapter 9 Mining and metallurgy

Mining

ALFÖLDI, G. *Noricum* (trans. E. A. Birley), Oxford 1974

ALLAN, J. C. 'Considerations on the antiquity of mining in the Iberian peninsula', *JRAI* occasional paper No. 27

ARDAILLON, E. *Les mines du Laurion dans l'antiquité,* Paris 1897

BIRD, D. G. 'Roman gold-mines of North-west Spain', *BJ* (1972), 36–64

BLASQUEZ, J. M. in *Mineria Hispana* I (on literary sources for mining in Roman Spain)

BOON, G. C. and WILLIAMS, C. 'The Dolaucothi drainage wheel', *JRS* 56 (1966), 122 f.

BROMEHEAD, C. E. N. 'Ancient mining processes', *Antiquity* (1942), 193 ff.

—— 'Mining and quarrying', *HT* II, 1–10

COGHLAN, H. H. *Prehistoric and Early Iron in the Old World,* Oxford 1956

DAVIES, O. *Roman Mines in Europe,* Oxford 1935

DOUMERGUE, C. (on lead ingots as evidence of organization of lead production), *Arch. Esp. Aug.* XXI.10 (1966), 41 ff.; XXXII (1969), 159 ff. (see review by P. A. Brunt, *JRS* LXV (1975), 167)

FORBES, R. J. Studies VIII, 141 ff. (on mining methods and living conditions)

GOSSÉ, G. 'Las minas y el arte minero d'España en la antiguedad', *Ampurias* 4 (1942), 43–68

HALLEUX, R. 'Le problème des métaux dans la science antique', *Bull. Fac. phil. lett., Univ. de Liège,* Fasc. 209 (1974)

HEALY, J. F. *Mining and Metallurgy in the Greek and Roman World,* London, 1978

HOPPER, R. J. 'Mines and miners of ancient Athens', *Greece and Rome,* 2nd series 8 (1961), 138 ff.

—— 'The Attic silver mines in the fourth century BC', *BSA* 48 (1953), 200–54

—— 'The Laurion mines: a reconsideration', *BSA* 63 (1968), 293 ff.

JONES, G. D. B. and LEWIS, P. R. 'The Dolaucothi gold-mines'. *BJ* 171 (1971), 288–300

JONES, R. F. J. and BIRD, D. G. 'Roman goldmining in northwest Spain II', *JRS* 62 (1972), 59–74

JONES, J. E. 'Laveries (ergasteria) sur la partie nord de la haute Agrileza', *L'Antiquité Classique* XLV (1976), 149–72

LAUFFER, S. J. *Die Bergwerkssklaven von Laurieon,* Wiesbaden 1955

LEWIS, P. R. and JONES, G. D. B. 'Roman gold-mining in north-west Spain I, *JRS* 60 (1970), 169–85

LUZON, J. M. 'Los sistemas de desague en minas romanas del sudeste peninsular'. *Arch. Esp. Aug.* XLI (1968), 101–20 (On mine drainage)

ORTH, F., '*Bergbau*', *RE* Supplement, Bd. 4 (1924), 108–55

RAMIN, J. 'La Technique minière et métallurgique des anciens, *Coll. Latomus* 153, Brussels 1977

REIHE, D. 'Kultur und Technik, Heft I', *Mitt. Staatl. Mus. f. Mineralogie und Geologie zu Dresden,* Dresden

RICKARD, T. A. 'The mining of the Romans in Spain', *JRS* XVIII (1928), 129–43

—— *Man and Metals,* London 1932, vol. I, 495 ff.

SALKIELD, 'Ancient slags in the south west of the Iberian peninsula', *Mineria Hispana* I, 85 ff.

WHITTICK, G. C. 'Roman Mining in Britain'. *Proc. Newcomen Soc.,* March 1932

WILSDORF, H. *Bergleute und Hüttenmännern in Altertum bis zum Ausgang der römischen Republik,* Berlin 1952

Quarrying

BROMEHEAD, C. E. N. 'Mining and Quarrying . . . *HT* II, 23–32

DUBOIS, C. *Étude sur l'administration et l'exploitation des carrieres dans le monde romain,* Paris 1908

FORBES, R. J. *Studies* II, 2–5, 22–30

GNOLI, R. *Marmora Romana,* Rome 1971

PENSABENE, P. 'Considerazioni sul transporto di manufatti marmorei in étà imperiale a Roma e in altri centri occidentali', *Dialoghi d'Architettura* VI (1972), 358 ff. (with list of quarries and varieties of marble quarried)

WARD PERKINS, J. B. 'Quarrying in Antiquity: Technology, Tradition and Social Change', *PBA* 1972

Metallurgy

AIANO, R. *The Roman Iron and Steel Industry at the Time of the Empire,* M. A. diss. Aberystwyth 1975

AITCHISON, L. A. *History of Metals,* London 1960

ATKINSON, R. J. C. 'Wayland's Smithy', *Antiquity* 39 (1965), 126–33

BENOÎT, F. 'Soufflets de forge antiques', *REA* 50 (1948), 305–8 (on bellows).

BLÜMNER, H. *Technologie und Terminologie . . . ,* vol. IV, 1–378

BOULAKIA, J. D. C. 'Lead in the Roman world', *AJA* 76 (1972), 139–44

BROTHWELL, D. and HIGGS, E. (eds). *Science in Archaeology: a survey of progress and research,* London 1969

CALEY, E. R. *An Analysis of Ancient Metals,* London 1964

CAVE, J. F. 'A note on Roman metal-turning', *HT* II, (1977), 77–94

CLEERE, H. 'Ironmaking in a Roman furnace', *Britannia* 2 203–17

—— 'Ironmaking', in *Roman Crafts,* ed. D. E. Strong and D. Brown. London 1976, 127–41

—— 'On furnace types', *Ant. Journ.* 52 (1972)

—— 'Some operating parameters for Roman ironworks', *BIA* London 13 (1976), 233–46 (detailed study of the organization of a Roman mining complex in the Kentish Weald)

CRADDOCK, P. T. T. 'The composition of the copper alloys used by the Greek, Etruscan and Roman civilizations: I. Greeks before the archaic period', *JAS* 3 (1976), 93–113

ELIADE, C. *The Forge and the Crucible,* trans. S. Corrin, London 1962

FORBES, R. J. *Metallurgy in Antiquity,* Leiden 1950

—— Studies VII–IX, 2nd ed. 1972

GRAY, D. H. F. 'Metal-working in Homer', *JHS* LXIV (1954), 1 ff.

HALLEUX, R. *op.cit.* under *Mining*

HOOVER, L. H. and H. C. (eds). G. Agricola. *De Re Metallica,* London 1912

LUCAS, A. *Ancient Egyptian Materials and Industries,* 3rd rev. ed. London 1948

MANNING, W. H. 'Blacksmithing', *Roman Crafts,* ed. D. E. Strong and D. Brown, London 1976, 143–53

MUHLY, J. D. 'Copper and Tin: the distribution of mineral resources and the nature of the metals trade in the Bronze Age', *Trans. Conn. Acad. Arts and Sciences,* vol. 43 (1973), 155–535

MUTZ, A. *Die Kunst des Metalldrehens bei den Römern.* Basle/Stuttgart, 1972. (Lathe-turning of metals a common craft in the Roman world.)

RAMIN, J. *op.cit.* under *Mining*

ROSENFELD, A. *The Inorganic Raw Materials of Antiquity,* London 1965

SMITH, C. S. *A History of Metallography,* Chicago 1960

SMITH, C. S. 'The discovery of carbonic steel', *TC* 5.2 (1964), 149–75

TYLECOTE, R. F. *Metallurgy in Archaeology: a prehistory of metallurgy in the British Isles,* London 1962

WHITE, K. D. 'The Great Chesterford scythes'. *Proc. Hung. Agric. Mus.* (1971–72), 77–82

WYNNE, E. J. and TYLECOTE, R. F. 'An experimental investigation into primitive iron-smelting', *JISI* 190 (1958), 339–48

Chapter 10 Land transport

General history of transport

BAUDREY DE SAUNIER. *Histoire de la locomotion terrestre,* Paris 1936

BLUM, O. *Die Entwicklung des Verkehrs,* Berlin 1941

BOUMPHREY, G. M. *The Story of the Wheel,* London 1932

FABRE, M. *A History of Land Transportation,* vol. VII of *The New Illustrated Library of Science and Invention,* ed. Dr C. Canby, New York 1963

FINESTONE, H. S. *Man on the Move; the story of transportation,* New York 1967

LANDELS, J. G. *Engineering in the Ancient World,* London 1978, 170–85

LEIGHTON, A. C. *Transport and Communications in Early Mediaeval Europe,* Newton Abbot 1972 (discusses capacity of Roman freight vehicles)

PATERSON, J. *The History and Development of Road Transport,* London 1927

Roads

BESNIER, M. 'Via', *Dict. Ant.* V 785 ff.

BIRK, A. *Die Strasse,* Karlsbad 1954

CASTAGNOLI, F. *Via Appia,* Milan 1956

CHEVALLIER, R. 'Cadastres antiques et photographie aerienne: essai de methode', *Rev. arch. est.* (1957), 254–85

—— *Roman Roads,* trans. N. H. Field, London 1976

DUVAL, P.-M. 'La construction d'une voie romaine d'après les textes antiques', *BSAF* (1959), 176–86

FORBES, R. J. *Notes on the History of Ancient Roads and their Construction,* 2nd ed., Amsterdam, 1964

—— 'A bibliography of road-building, AD 300–1840', *Roads and Road Construction* (1938), 189–96

FREDERIKSEN, M. W. and WARD PERKINS, J. B. 'The ancient road-system of the central and western Ager Faliscus', *PBSR* n.s. 2 (1957), 67–208

FRIEDLÄNDER, L. *Darstellungen aus der Sittengeschichte Roms,* vol. I, 318–90

FUSTIER, P. *La Route: voies, antiques, chemins anciens, chaussées modernes,* Paris, 1968

—— 'Notes sur la construction des voies romaines en Italie', *REA* 62 (1960), 95–99; 63 (1961), 276–90

—— 'Etude technique sur un texte de l'empereur Julien relatif à la constitution des voies romaines', *REA* 65 (1963), 114–21

GOODCHILD, R. G. 'Roads and land travel, *Hist. Tech.* vol. II, Oxford 1956, sections III–IV (roads)

GREGORY, J. W. *The Story of the Road,* New York 1932

GRENIER, A. *Manuel d'archéologie gallo-romaine,* vol. II.1. *Les routes,* Paris 1969

HAGEN, V. W. VON. *Les Voies romaines,* Paris 1969

KAHANE, A. 'The paved Roman road east from Gabii', *PBSR* 41 (1973), 18–44

—— and Ward Perkins, J. B. 'The Via Gabina', *PBSR* 40 (1972), 106 ff.

LEE, C. E. *The Highways of Antiquity,* London 1947

MARGARY, I. D. *Roman Roads in Britain,* 2 vols, London, 1955–57

MATTY DE LA TOUR. *Les Voies romaines, système de construction et d'entretien,* 1865. Ms Bibl. de l'Institut, 1866, Part 2, vol. 3 (contains results of 300+ sections of roads)

MITCHELL, S. 'Requisitioned transport in the Roman Empire: a new inscription from Pisidia', *JRS* 66 (1976), 106–31

PEKARY, TH. *Untersuchungen zu der römischen Reichstrassen* Bonn 1968 (with exhaustive analysis of relevant texts and inscriptions)

PENSABENE, P. 'Considerazioni sul trasporto di manufatti marmorei in età imperiale a Roma e in altri centri cocidentali', *Dialoghi d'Architettura* VI (1972), 358 ff. (with list of quarries and varieties of marble quarried)

REBUFFORT, R., 'Deux ans de recherches dans le sud de la Tripolitaine', *CRAI,* April–June 1969, 189 ff.

SION, J. 'Quelques problèmes de transport dans l'antiquité, à point de vue d'un géographe mediterranéen', *Annales (ESC)* VII (1935), 62

STERPOS, D. *The Roman Road in Italy,* trans. F. Sear, *Quaderni di Autostrada* 17, ed. PRO, Autostrada S.p.A. Rome 1970

WISEMAN, T. P. 'Roman Republican road-building', *PBSR* 38 (1970), 122–35

Transport animals, and their role in land transport

BRUNEL, I. K. 'A treatise on draught', Youatt, W. *The Horse,* new ed., London 1857, 403–52

BURFORD, A. M. 'Heavy transport in classical antiquity', *Econ. Hist. Rev.* XIII.1 (August 1960), 1 ff, (on importance of oxen in road transport)

SAVORY, TH. H. 'The mule'. *Scientific American* 223, no. 6, (Dec. 1970), 102–9

VIGNERON, P. *Le Cheval dans l'antiquité gréco-romaine,* 2 vols, Nancy 1968

WHITE, K. D. *Farm Equipment of the Roman World,* Cambridge 1975, 51 f.

The harnessing of transport animals

BRUNT, P. A. Review of K. D. White, *Roman Farming.* London 1970, in *JRS* 62 (1972), 156 (supports the thesis of Lefebure des Noettes on equine harness)

GRAND, R. 'Utilisation de la force animale: vues sur les origines de l'attelage moderne', *Bull. de l'Acad. d'agric. de France,* vol. 33 (1947), 702–10

HAUDRICOURT, A. G. 'De l'origine de l'attelage moderne', *Ann. d'hist. econ. et soc.* (1936), 515–22

—— and DELAMARRE, J.-B. *L'homme et la charrue à travers le monde,* Paris 1955, ch. II, 155–87, 'Géographie et ethnologie de la voiture'

JOPE, E. M. 'Vehicles and harness', *Hist. Tech.,* vol. II Oxford 1956, 493–536

LEFEBURE DES NOETTES, Cdt R. J. C. E. *L'attelage. Le cheval à selle à travers les âges,* 2 vols, Paris 1931.

MEGNIN, P. *Histoire de la harnachement et de la ferrure du cheval,* Vincennes 1904

POLGE, H. 'L'amélioration de l'attelage a-t-elle réellement fait reculer le servage?', *Journ. des Savants,* Jan.-Mars, 1967, 4–42 (rebuts the thesis of Lefebure)

VIGNERON, P. *Le cheval dans l'antiquité greco-romaine,* vol. I, Nancy 1968, 108–37

WHITE, K. D. *Farm Equipment of the Roman World,* Cambridge 1975, 218–20

Wheels

CURLE, J. *A Roman Frontier Post and its People,* Glasgow 1911

HARRIS, H. A. 'Lubrication in Antiquity'. *Greece and Rome* 21/1 (April 1974), 33–36

JACOBI, L. (ed.). *Saalburgjahrbuch* III (1912), 68–70; figs 27–29, and pl. XVI

MACDONALD, G. and PARK, A. *The Roman Forts on the Bar Hill.* Glasgow 1906. (Details of wheels from Bar Hill and Glastonbury, Somerset.)

PIGGOTT, S. 'The earliest wheeled vehicles and the Caucasian evidence', *PPS* n.s. 34 (1968), 266–318
—— *The Earliest Wheeled Transport,* London 1983

WHITE, K. D. 'Notes on goods of agricultural use' in K. T. Erim and J. Reynolds, 'The Aphrodisias copy of Diocletian's Edict ...', *JRS* 63 (1973), 99–110

Chapter 11 Ships and water transport

General

ASSMAN, A. 'Seewesen' in A. Baumeister. *Denkmäler des klassischen Altertums* III, Munich and Leipzig 1889, 1953–69

BASS, G. F. *Archaeology under Water,* London and New York 1966

BASS, G. F. (ed.). *A History of Seafaring,* London 1972

BLACKMAN, D. J. (ed.). *Marine Archaeology,* vol. 23 of the Colston Papers, Colston Research Society, Bristol 1973

CASSON, L. *The Ancient Mariners,* New York 1959
—— *Ships and Seamanship in the Ancient World,* Princeton, N.J. 1971

CLOWES, G. S. L. *Sailing Ships: Their History and Development.* 2 parts, London (Science Museum), Pt I, 1932; Pt II, 1952

FROST, H. *Under the Mediterranean,* London 1963

HORNELL, J. *Water Transport,* Cambridge 1946

KÖSTER, A. *Das Antike Seewesen,* Berlin 1923

MILTNER, 'Seewesen', *RE Supplementband* V. 906–62 (1931)

MORRISON, J. S. and WILLIAMS, R. T. *Greek Oared Ships,* Cambridge 1968

PARKER, A. J. *Ancient Shipwrecks of the Mediterranean and the Roman Provinces* (forthcoming). A computerized catalogue
—— and PAINTER, J. M. 'A computer-based index of ancient shipwrecks', *IJNA* 8 (1979), 69–87

ROUGÉ, J. *Recherches sur l'organisation du commerce maritime en Méditerranée sous l'empire romain,* École pratique des hautes études, VIᶜ section, Centre de Recherches Historiques, Ports, Routes, Trafics 21, Paris 1966

ROUSSEAU, P. *Histoire des transports,* Paris 1963

RUPP, D. W. *A Catalogue of Ship Representations in Roman Mosaics and Monumental Painting,* M.A. Diss. Univ. of Pennsylvania, n.d.

TAYLOR, L. du PLAT (ed.). *Marine Archaeology,* London 1965

THROCKMORTON, P. *Shipwrecks and Archaeology,* Boston 1970

TORR, C. *Ancient Ships,* Cambridge 1895; 2nd ed., Chicago 1964

Special

BASS, G. 'Yassi Ada'; Underwater excavations at Yassi Ada: A Byzantine shipwreck', *AA* 1962, 537–63

BENOIT, P. L'épare du "Grand Congloué", *Gallia,* Suppl. XIV, Paris 1961

CARLINI, CDT. 'Le gouvernail dans l'antiquité', *Bulletin de Techniques, Assoc, maritime et aeronautique,* Paris 1935

CASSON, L. 'Harbour and river boats of ancient Rome', *JRS* 55 (1965), 31–39
—— 'The Isis and her voyage', *TAPA* 81 (1950), 43–56
—— 'Speed under sail of ancient ships', *TAPA* 82 (1951), 136–48
—— 'Studies in ancient sails and rigging', Essays in honor of C. B. Welles (American Studies in Papyrology I), New Haven, Conn. 1966, 43–58
—— 'The size of ancient merchant ships' in *Stud. in onore di Aristide Calderini e Roberto Paribeni,* I, Milan, 1956, 231–38

de SAINT-DENIS, E. *Le vocabulaire des manoeuvres nautiques en latin,* Macon 1935
—— 'La vitesse des navires antiques' *Rev. arch.* (1941), II, 121–38

DE VRIES, K. and KATZEV, M. L. 'Greek, Etruscan and Phoenician Ships and Shipping' in Bass 1972, 38–64

DUNCAN–JONES, R. P. 'Giant cargo-ships in antiquity', *CQ* 27 (1977), 331–32
—— *The Economy of the Roman Empire,* Cambridge 1974, App. 17, 366 ff. (on the evidence of the Price-Edict).

DUVAL, P. 'La forme des navires romains', *MEFR* 61 (1949), 119–49

EMANUELE, P. D. 'Ancient square-rigging, with or

without lifts', *IJNA* 6.3 (1977), 181–85

FOUCHER, L. 'Navires et barques figurés sur les mosaïques découvertes à Sousse et aux environs', *Musée Alaoui, Notes et documents* 15. Tunis 1957

GÖTTLICHER, A. '*Naves longae:* Bau, Bewaffung und Einsatz römische Kriegsschiffe', *Antike Welt* 7.2 (1976), 49–55

—— '*Naves onerariae:* Bau und Einsatz römischer Handelsschiffe', *Antike Welt* 8.3 (1977), 47–54

KATZEV, M. L. 'Resurrecting the oldest known Greek ship'. *Nat. Geogr. Mag.* 157 (June 1970), 840–57 (on the Kyrenia freighter)

LANE, F. 'Tonnages, mediaeval and modern', *Econ. Hist. Rev*, 2nd ser., 17 (1964), 213–33

LAMBOGLIA, N. 'Il rilevamento totale della nave romana di Alberga', *Riv. Stud. Lig.* 27 (1961), 213–20 (on estimates of cargo tonnage)

LEFEBURE DES NOETTES, Cdt. R. J. C. E. *De la marine antique à la marine moderne: la révolution du gouvernail,* Paris 1935

LE GALL, J. 'Un modèle reduit de navire romain'. Mélanges Charles-Picard II. *Rev. arch.* 29 (1949), 613–16

—— *Le Tibre: fleuve de Rome dans l'antiquité,* Paris 1953

LEHMANN-HARTLEBEN, K. *Die antiken Hafenanlagen des Mittelmeeres. Klio,* Beiheft 14, Leipzig 1923

MANACORDA, D. *Relitti sottomarini di étà repubblicana* stato della ricerca (1979) (with full bibliographical appendix of wrecked ships).

MARSDEN, P. 'A boat of the Roman period . . .' *Trans. Lond. and Middx. Arch. Soc.* 21 (1965), 118–31

—— *A Roman Ship from Blackfriars,* London 1967

MOLL, F. *Das Schiff in der bildenden Kunst vom Altertum zum Ausgang des Mittelalters,* Bonn 1929. (with over 4,000 illustrations.)

PENSABENE, P. 'A cargo of marble shipwrecked at Porto Scito near Crotone', *IJNA* 7.2 (1978), 105–18

POMEY, P. and TCHERNIA, A. 'Les bateaux de commerce romains: capacités maritimes', *Gallia,* Paris

POUJADE, J. *La route des Indes et ses navires.* Paris 1946

ROUGÉ, J. *La marine dans l'antiquité,* Paris 1975

RUPP, D. *op. cit.* under *General*

SMITH, J. *The Voyage and Shipwreck of St Paul,* London 1848

TARN, W. W. *Hellenistic Military and Naval Developments.* Cambridge 1930

TCHERNIA, A., POMEY, P. and HESNARD, 'L'Épave romaine de la Mandrague de Gien (Var)'. XXXIVC Suppl. a *Gallia,* Paris 1978

THROCKMORTON, P. 'The Antikythera ship', *TAPS* 55.3 (1965), 40–47

—— 'The Romans on the Sea' in Bass 1972 66 ff.

UCELLI, G. *Le nave di Nemi,* Rome 1950

WALLINGA, H. T. 'Nautika I: the units of capacity for ancient ships', *Mnemosyne* XVII (1964), 1–40

WARD PERKINS, J. B. and THROCKMORTON, P. 'The San Pietro Wreck', *Archaeology* 18.3 (1965), 201–9

WEINBERG, G. D. 'The Antikythera shipwreck reconsidered', *TAPA* N. S. 55.3 (1965)

Anchors

BENOIT, P. In *Riv. Stud. Lig.* 62 (1952), 269 ff.

BOON, G. C. 'A Greco-Roman anchor-stock from North Wales', *Ant. Journ.* 57 (1977), 10–30

GARGALLO, F. 'Anchors of antiquity', *Archaeology* 14 (1951), 31–35.

GIANFROTTA, P. A. 'First elements for the dating of stone anchor-stocks', *IJNA* 6.4 (1977), 285–92

KAPITAN, G. in *Sicilia Archeologica* 4 (1971), 13–22

LLORES, M. B. *Las anforas romanas en España.* Zaragoza, 1970. (On the lead parts of anchors.)

MERCANTI, M. P. *Ancorae Antiquae,* Rome 1979

Chapter 12 Hydraulic engineering

Water supply—general

BISWAS, A. K. *History of Hydrology,* Amsterdam/London 1970, 37–77 (Greek period); 79–101 (Roman period)

BROMEHEAD, C. E. N. 'The early history of water-supply', *Geogr. Journ.* 90 (1924), 183–96

BUFFET, B. and EVRARD, R. *L'eau potable à travers les âges,* Liège 1950

CALDERINI, A. 'Machine idrofore secondo i papiri greci', *Rend. Ist. Lombardo, Sci. e Lettr.,* 2nd ser., 53 (1920), 620–31

CLARK, J. G. D. 'Water in antiquity', *Antiquity* 18 (1944), 1–15

DROWER, M. S. 'Water-supply, irrigation and agriculture', *Hist. Tech.,* vol. I, 525–40

DUPONT, G. *L'eau dans l'antiquité,* Paris 1938

FELDHAUS, F. M. *Die Maschine im Leben der Völker,* ed. Birkhauser, Stuttgart/Basle 1955

FORBES, R. J. Hydraulic engineering and sanitation', *Hist. Tech.,* vol. II, 663–89

—— *Studies* I (water supply); II (irrigation and drainage: power)

KRETCHMER, F. 'La robinetterie romaine'. *Rev. arch. est. C-est* 11 (1960), 89–113

MAHUL, J. 'Les tuyaux de plomb: histoire et progrès de leur fabrication', *Nature,* Paris, 65, no. 3014 (1037), pp. 503–10

ROUSE, HUNTER, and INCE, SIMON. *A History of Hydraulics,* Iowa 1957

SCHIOLER, T. *Roman and Islamic Water-Lifting Machines,* Odense 1973

SMITH, N. A. F. *A History of Dams,* London 1971, pp. 1–56, (Ch. I, Antiquity; Ch. II, The Romans.)

—— 'Attitudes to Roman engineering and the ques-

SPRAGUE DE CAMP, L; *The Ancient Engineers,* London 1960, 194²-202

VAN BUREN, A. W. art. 'Wasserleitungen', *RE* VIII A 1 (1955), cols 453–85

Water supply: regional see also bibliography, Ch. 3, s.v. 'Dams')

BEAN, G. E. *Turkey's Southern Shore.* London 1968, 75 ff, (on the water-supply system at Aspendos)

BIREBENT, *Aquae romanae: recerches d'hydraulique dans l'est algérien,* Algiers 1962

BLANCHÈRE, M. R. de la. 'Un chapitre d'histoire pontine', *Mem. Acad. Inscr.* X (1893), 18 ff.; 72 ff. (on the draining of the Pomptine marshes)
'L'Aménagement de l'eau et l'installation rural dans l'Afrique ancienne'. *Nouv. Arch. Miss Sci. Litt.* VII, pt 2 (1895), 1–109. (Comprehensive survey of the evidence to date on water-supply and water-control in the region.)

BRUNHES, J. B. 'L'irrigation, ses conditions géographiques, ses modes et son organisation dans la peninsule Iberique et dans l'Afrique du nord' Paris 1902.

BURNS, A. 'Ancient Greek water supply and city planning: a study of Syracuse and Akragas', *TC* 15.3 (1974), 289–412
'The tunnel of Eupalinus and the tunnel problem of Hero of Alexandria', *TC* 12 (1971), 50–65

CALDERINI, A. 'Macchine idrofore secondo i papiri greci'. *R.C. Ist. Lomb. Sci. Lett.,* sec. ser., 53 (1920), 620–31

DILKE, O. A. W. and M. K. 'Terracina and the Pomptine Marshes', *Greece and Rome,* 2nd ser., VIII (1961), 172–78

FABRICIUS, ERNST. 'Altertümer auf der Insel Samos', *Mitt. d. Deutschen Archäologischen Institutes,* Ath, (1884), 165–92

GAUCKLER, P. *Enquête sur les installations hydrauliques romaines en Tunisie,* Tunis 1897–1901. 5 vols

GEFFROY, A. 'L'archéologie du lac Fucin', *Rev. arch.* n.s. 36.2 (1878), 1–11, pl. XIIIa. (On the bas-reliefs relating to the *emissarium* of the Fucine Lake.)

GOODFIELD, J. and TOULMIN, S. 'How was the tunnel of Eupalinus aligned?', *Isis* 56 (1965), 45–56

GRABER, F. 'Die Wasserleitungen', *Die Alterthümer von Pergamon.* Bd I. Stadt und Landschaft, 3, 365–424. Berlin 1912. (Pl 86–104 cover the pipeline and surviving material *in situ*.)

GSELL, S. *Enquête administratif sur les travaux hydrauliques anciens en Algérie,* Paris 1902

HANSON, J. *Municipal and military water supply and drainage in Roman Britain.* Diss (unpubl.) University of London, Inst. of Arch.

JUDSON, S. and KAHANE, A. 'Underground drainageways in southern Etruria and South Latium', *PBSR* n.s. 18 (1963), 34–99

KASTENBEIN, W. 'Untersuchungen am Stollen des Eupalinos', *Arch. Anzeiger* 75 (1960), 178–98

LEVEAU, P. and PAILLET, J.-L. *L'alimentation en eau de Caesarea de Mauretanie et l'aqueduc de Cherchel,* Paris 1976

PACHTÈRE, F. G. de, 'Le Réglement de l'irrigation de Lamasba, *MEFR* 28 (1908), 373–405 (on *CIL* VIII, 4440), 663–79

SAMESRAUTHER, E. 'Römische Wasserleitungen in den Rheinlanden', *Ber. Römisch-Germanischen Kommission,* 26 (1936), 24–157

SCHUCHHARDT, KARL. 'Die Arbeiten zu Pergamon, 1886–1898', *Mitt. d Deutschen Archäologischen Institut,* Ath., Abt. XXIV (1899), 103–12

SHAW, Brent D. 'Lamesba: An Ancient Irrigation Community', *Antiquités Africaines 18* (1982), 61–101

SMITH, N. A. F. 'The Roman dams of Subiaco', *TC* 11 (1970), 58–68

VITA-FINZI, C. 'Roman dams in Tripolitania', *Antiquity* 35 (1961), 15 ff.

Aqueducts

ASHBY, T. *The Aqueducts of Ancient Rome,* ed. Ian Richmond, Oxford 1935

BENOIT, F. 'L'usine hydraulique de meunerie de Barbegal', *Rev. Arch.* 22 (1940) (on the feeder aqueduct supplying the multiple water-mill system)

BLACKMAN, D. R. 'The volume of water delivered by the four major aqueducts of Rome', *PBSR* n.s. 33 (1978), 52–72

CASADO, F. *Acueductos romanos en España.* Madrid 1972 (complete survey, including details of construction and engineering problems; fully illustrated) *Enciclopedia Italiana,* s.v. 'Acquedotto'.

GERMAIN DE MONTAUZAN, C. *Les aqueducs antiques de Lyon,* Paris 1908.

GRAEBER, F. 'Wasserleitungen in Pergamon', *Altertümer von Pergamon,* I (text) pp. 365–412; II, (ps) Beibl. 86–104

GRENIER, A. *Manuel d'Archéologie Gallo-Romaine,* vol. IV, pt. 1 'aqueducs', Paris 1960

GRIMAL, P. 'Vitruv et la technique des aqueducs', *Rev. phil.* XIX (1943), 162–74

HERSCHEL, C. *Frontinus and the Water-Supply of Rome,* London 1913

LEVEAU, P. and PAILLET, J.-L. *Op. cit.* under *water supply: regional*

MATTHEWS, K. D. 'Roman aqueducts: technical aspects of their construction', *Expedition* 13.1 (1970), 2–16

PRAGER, F. D. 'Vitruvius and the elevated aqueducts' *HT* III (1978), 105–22

THOMPSON, F. H. 'The Roman aqueduct at Lincoln', *Arch. Journ.* CXI (1955), 106 ff.

VAN DEMAN, E. B. *The Building of the Roman Aqueducts,* Washington (Carnegie Inst. Publ. 423), 1934

VAN DEMAN, E. B. *The Building of the Roman Aqueducts,* Washington (Carnegie Inst. Publ. 423), 1934

Pumps (incl. screws, bucket-chain hoists, wheels and force-pumps)

BATTERSBY, T. W. *Roman Force Pumps: a Preliminary Survey,* Diss. University of London, 1981 (full discussion of evidence on 20 pumps)

BOON, G. C. *Roman Silchester,* London 1957, 159–61 and WILLIAMS, C. 'The Dolaucothi drainage-wheel', *JRS* LVI (1966), 122–27

DAVIES, F. 'Notes on a Roman force-pump found at Bolsena'. *Archaeologia* 55.1 (1896), 254 ff.

FORBES, R. J. *Studies* II, s.v. *Water supply-general; Studies* VII, ch. 3 (on mine drainage equipment)

LANDELS, J. G. *Engineering in the Ancient World,* London 1978, 58–84 (on water pumps)

NEYSER, A. 'Eine römische doppelkolbendruckpumpe aus dem Vicus Belginum', *Trirerer Zeitung* XXXIV (1972), 109–21

PELMET, R. E. in *Trans. Inst. Min. Met.* 36 (1926/27), 299–310 (on mine drainage equipment)

SANQUER, R. in *Gallia* 31 (1973), fasc. 2, 355–60 (8-cylinder force-pump from St Malo, 1971)

SHAPIRO, SHELDON. 'The origin of the suction-pump', *TC* 5.4 (1964), 566–74

Notes

Introduction

1 M. I. Finley, in *The New York Review of Books,* 10 June 1971.

2 J. D. Bernal (1969, vol. 1, 292).

3 J. Goodfield and S. Toulmin, 'How was the tunnel of Eupalinos aligned?', *Isis* 56 (1965), 45–56.

4 A. Burns, 'The tunnel of Eupalinus and the tunnel problem of Hero of Alexandria', *TC* 12 (1971), 50–65.

5 M. R. Cohen and I. E. Drabkin, *A Source Book in Greek Science,* New York/London 1948, 9.

6 Singer *et al., Hist. Tech.*

7 Forbes, *Studies.*

8 'Our chief sources of information for most purposes are the actual archaeological remains', Cohen and Drabkin, *op. cit.* 314.

9 Hodges (1971).

10 Finley, art. cit.

11 E. W. Marsden, *Greek and Roman Artillery,* vol. 1, *Historical development,* Oxford 1969; vol. 2, *The technical treatises,* Oxford 1971; Healy (1978), Vigneron (1968), White (1967), (1975).

12 Landels (1978). It should be noted that Landel's book is not a comprehensive technical treatise, being strictly limited to 'the application of power for engineering or similar purposes' (Author's *Preface,* p.7.). Moreover, the title is somewhat misleading, the treatment being virtually confined to Greece and Rome.

Chapter 1

13 See Childe (1948).

14 For some pertinent remarks on man's interaction with his physical environment, and its bearing on the history of technology, see the introductory chapter of Usher (1954, 123).

15 On the economics of the heading machine, and the factors likely to have hindered its spread see my article, 'The economics of the Gallo-Roman harvesting machine', *Hommages à Marcel Renard, Coll. Latomus* 102, 1969, 804–9.

16 See H. Cleere, 'Iron Making', *Roman Crafts,* ed. D. Strong and D. Brown, London 1976; Healy (1978), 232 f.).

17 See J. R. Harris, 'The transfer of British metallurgical technology to France in the eighteenth century', seminar paper read to the Anglo-French Conference of Historians, London 1978.

18 Lloyd (1973, 99).

19 On the range of the legends see Eliade (1962, 27–33; 87–108; Forbes (1950).

20 See Forbes, *Studies* IX, 198 ff., and esp. 218 ff.

21 E. R. Dodds, *The Ancient Concept of Progress,* Oxford 1973.

22 2 Kings 24.14, 'And he carried away all the mighty men of valour, . . . and all the craftmen and smiths: none remained, save the poorest sort of people of the land.'

23 K. D. White, 'Technology in nort-west Europe before the Romans', *Museum Africum* 4 (1975), 43–46.

24 S. Andreski, *Military Organization and Society,* 2nd ed., London 1968, 162; the converse, as Andreski points out (*op. cit.,* 181 ff.) may also be ture: while war has provided a powerful stimulus to technical innovation and development, universal peace might in the future destroy the incentive for technical development.

25 Foreign correspondents: as Cicero's correspondence shows, provincial governors and business friends abroad frequently provided him with information on a great variety of topics, apart from matters connected with their work.

26 See Stahl (1962). Archimedes is said to have written no technical manuals, but Plutarch's statement can-

not be used to infer that no one else wrote any; Varro's flockmaster *De Re Rustica* could certainly read; and the format and style of the instructions that make up the bulk of Cato, *De Agri Cultura* strongly suggest a close connection with this class of technical work. See further Finley (1973, 145), who accepts that there were experts and expertise in all fields of activity contributing to manufacture, engineering, food processing and navigation.

27 R. J. Forbes, in Singer et al., *Hist. Tech.* II, 603 (rather too sweeping); for a considered, better documented, assessment see the review article by L. Edelstein (cited below).

28 The assertion that the Greeks were mere speculators has not passed without challenge; for a reasoned analysis, showing that both quantitative assessment and repetitive experiments were carried out by some scientists in certain fields, see. Edelstein', 'Recent interpretations of ancient science', *JHI* XIII. 4 October 1952, reviewing Cohen and Drabkin (1948), and Farrington (1944 and 1949).

29 See the pertinent comment on this passage by Lloyd, *op. cit.*, (n. 18).

Chapter 2

30 N. Lewis and M. L. Reinhold, *Roman Civilization*, 2 vols, New York 1951 and 1955; vol. II, 298.

31 Usher, *loc. cit.* (n. 14).

32 See Drachmann (1932, 77 ff.); White (1975, 231 f.).

33 Collected and discussed in an unpublished thesis (n.d.) by D. Rupp, formerly of the University of Pennsylvania, now of Brock University, Ontario. The text contains drawings showing the distribution of masts and sails, but no photographs.

34 There in no commentary in any language on the works of Hero, and the only available text (in the Teubner series) was edited between the years 1899 and 1914. A French edition of Vitruvius, *De Architectura,* with translation and commentary, is in course of publication in the Budé series.

35 Landels (1978, 19); Hero was not the only technologist to amuse himself in his spare time by making ingenious toys; the German scientist and engineer Mercklein (fl. *c.* 1740) produced a number of 'toys' and 'contrivances' (from a paper (not yet published) by Harris, 1978).

36 The developments include improvements in valve-design, and the attachment to the outlet pipe of a swivelling device, so that wheen fitted on a chassis the pump could serve as a fire-engine. The four force-pumps recently recovered from the wreck of a Roman merchantman (details is Landels 1978, 81 ff.) show and advanced standard of workmanship on the pistons and cylinders. See n.70.

37 Pre-classical work: J. M. Coles, *Archaeology by Exper-*

iment, London 1973; Classical period: H. Cleere in *Britannia* 2 (1971), 203–51 (experiments in iron-making in the Roman fashion); Landels (1978, 107 ff.) (design and construction of a Roman catapult).

Chapter 3

38 See Austin and Vidal-Naquet (1977, 120 ff. and 310 ff.).

39 These include the recently discovered portions of the Aphrodisias copy of the *Edictum de pretiis* of Diocletian, containing the Latin version of the section on vehicles and their components (detailed comments in Ch. 10, and the new inscription from Pisidia (see the article by S. Mitchell, *JRS* 66, 1976, 106–31) giving valuable information on the maintenance of public roads and the requisitioning of transport.

40 Braudel (1973, vol. 1, *passim*).

41 Working for oneself: Aristotle, *Rhetoric* 1.9.1367 a 32; hirelings: Joh. 10:13.

42 Austin and Vidal-Naquet (1977, 15).

43 This is the thesis propounded by Finley (1973), Chs II, IV, V, and especially Ch. II, 'Orders and Status'; criticized at length by M. W. Frederiksen in *JRS* 65 1975, 164–71; for some interesting *contra* evidence from Pompeii see Moeller (1976).

44 Plutarch, *Cato maior* 21. 5–6: 'As he applied himself more vigorously to the pursuit of gain, he came to regard agriculture as more of an entertainment than a source of profit, and invested his capital in business that was safe and secure. He bought ponds, hot springs, districts given over to laundering, pitch factories, etc.'

45 Moeller (1976).

46 On private contracting: E. Badian, *Publicans and Sinners,* Otago 1972.

47 Livy 44.16.4. Badian *(op. cit.)* calculated the value of the clothing at 1,150,000 *denarii,* the wealth of 'about a dozen Roman knights'. Moeller (1976, 110, n. 19), says this amount 'could completely outfit a modern army corps'.

48 Moeller (1976, 40) cites evidence for classifying the house of M. Terentius Eudoxus (Reg. VI.XIII.6) as a 'true spinning and weaving factory, producing for consumption outside the household'; the work force included at least seven male weavers and eleven female spinners.

49 J. B. Ward-Perkins, 'Quarrying in Antiquity: Technology, Tradition and Social Change', *PBA* (1972). This excellent study is a model of what should be done for other areas of industrial activity.

50 Parallels to this type of situation can be found today in many parts of the developing countries of Africa and Asia.

51 For a good summary of the controversy over the nature of the ancient Greek economy see Austin and

Vidal-Naquet (1977, 3 ff.).

52 'Farmers and Bureaucrats: reflections on technological innovation in agriculture', *Agric. Hist.* 51 (1977), 641–55.

53 A recent example is the long ban (recently lifted) on public television in the Republic of South Africa, inspired by fear for the survival of the Afrikaans language and its culture.

54 Maudslay's crucial contribution was his invention of the slide-rest for the turret lathe, which made the lathe self-operating; on his association with Bentham (1757–1831) and Brunel (1769–1847) see the article by D. F. Galloway in Singer et al. *Hist. Tech,* vol. v, 636 f.

55 Ping-ti-Ho, *The Cradle of the East:* an enquiry into the indigenous origins of techniques and ideas of neolithic and early historic China, 5000–1000 BC, Hong Kong 1975; see the review by C. S. Smith, in *TC* 80–86.

56 E.g. E. C. Semple, *The Geography of the Mediterranean Region in relation to Ancient History,* London 1932; J. M. Houston, *The Western Mediterranean world,* London 1964.

57 White (1970, Ch. VI).

58 *Drainage in southern Etruria:* S. Judson and A. Kahane, 'Underground drainage ways in southern Etruria and south Latium', *PBSR* XXXI (1963), 74–99; in North Africa, J. Birebent, *Aquae romanae:* recherche d'hydraulique romaine dans l'est algérien', Algiers, 1962.

59 Rome's food-supply problems: K. D. White, *Prog. Fd. Nutr. Sci.,* vol. II (1976), 171 f. (early famines); 172 f. (late Republic and early Empire).

60 Finley, (1973, 126).

61 Xenophon, *Ways and Means,* II.2; on Piraeus as a clearing-house: Isocrates, *Panegyricus* 42: 'She set up Piraeus as a market where there is such an abundance of goods that things which are found elsewhere only with difficulty are all easy to procure here.' On Athenian food-supply policy see Austin and Vidal-Naquet (1977, 115 f. and refs cited); for Xenophon's remarkable proposal to exploit the silver in a big way, *Ways and Means,* IV. 1–17 (Austin and Vida-Naquet, no. 96 with notes).

62 'It is one of the paradoxes of the history of Athens that, although she was primarily a naval power, she did not have in Attica adequate resources for her navy.' (Austin and Vidal-Naquet (1977, 117), who go on to explain that in order to obtain these vital commodities the Athenians sometimes used political muscle, at other times were forced back on to diplomacy!

63 Circumference of the earth: the claim of 'remarkable accuracy' commonly made rests on insecure foundations, since we do not know the value E. attached to his basic measuring unit; if it was the Olympic *stadion,* his degree of longitude would be more than 10

per cent too large. The new method, however, was destined for a long innings; details in Lloyd (1973, 49 f.).

Chapter 4

64 See White (1970, 452) on the persistence of traditional techniques.

65 Loss of evidence: wooden implements and, more importantly, the wooden portions of machines, have a very poor chance of survival, save in the dry atmosphere of Egypt; this often results in failure to recognize evidence, or to identify the surviving metal parts; absence of dating; without an accurate chronological sequence, development becomes a matter of sheer conjecture; the eighteen or so literary references known to me concerning the water-mill can be closely dated (Appendix 6), but the equally numerous physical remains of Ctesibian pumps cannot! (See below, n.70).

66 R. Needham, quoted by K. Hopkins, reviewing F. Millar, *The Emperor in the Roman World,* London 1977, in *JRS* 68 (1978), 180.

67 The cradle-scythe persisted in many parts of Europe for generations after the mechanical reapers (see E. J. Collins, *From Sickle to Combine,* Reading 1970).

68 C. S. Smith, 'Technology in History', *Minerva* 8 (1970), 470.

69 See bibliography to Ch. 12. s.v. 'pumps'. Texts: Vitruv. 10. 7; Hero, *Pneumatike* I. 28; Pliny (*HN* 19.60) calls them *organa pneumatika.* Landels' recent discussion (1978, 75–83), is brief, but up-to-date.

70 I am very much indebted here to T. W. Battersby, whose unpublished dissertation has been made available to me, and whose researches mark an important step forward in the study of these machines. Of the 22 pumps identified and discussed by him 9 are certainly of bronze, 6 certainly of oak, and the remaining 7 possibly of oak. The bronze types are of much smaller capacity, with outputs ranging from 278 l.p.h. to *c.* 1000 l.p.h., while the smallest of the oak types has an output of 1,788 l.p.h., and the largest 5,737 l.p.h. Tests on one of a set of four, discovered in a wrecked Roman freighter of the mid-first century AD, showed the remarkably high level of efficiency of 95.45 per cent.

71 See Forbes, *Studies,* VII, 192–218: 'Ancient Mining Techniques'.

72 Healy (1978, 278), n. 30: the output of metal was 63 per cent, which was regarded as a 'very positive result'.

73 Miners' lamps: Diodorus 3.12.2: modern commentators have been naturally sceptical, e.g. Healy (1978, 83):' 'there is no independent proof of this practive nor would the lamps available to the Greeks have been suitable or effective for such a use'.

74 Noxious fumes: Strabo, 3.2.8; Pliny *HN* 33.98, etc.

Indifference to the health of categories of workers regarded as expendable (viz. common criminals and the like) was common enough; cf. the comment by Tacitus on the Jews banished to Sardinia to put down brigands: 'should they succumb to the pestilential climate, the loss would be insignificant' (*Ann.* 2.86).

75 The *Lex metalli Vipascensis:* the contents of this document of Hadrianic date are summarized by Healy (1978, 130 ff.); the amenities which the contractors were obliged to provide included hot baths all the year round, and provisions for cobbling, shaving and the cleaning of clothes; the work-force was no longer expendable!

76 H. Cleere, 'Iron making' in Strong and Brown (1976).

77 Ausonius, *Mosella* 361–64.

78 Healy (1976, 199): in the course of an excellent chapter on alloys, Healy notes that the ancient world had only a limited number of alloys in common use.

79 See below, p. 42.

80 Harden (1968, 63); pl. IX(c)—a superb example of this technique.

81 The enthusiasm is immoderate and the language exaggerated; but there is no disputing the fact that the glass industry did undergo rapid development, with or without 'aggressive marketing techniques'! Compare the rather more sober assessment quoted below.

82 Sergius Orata: the literary references are numerous: Varro, *De Re Rustica,* 3.3.10; Columella 8.16.5 (on fishponds); Macrobius, *Sat.* 2.11 (oyster-beds); Cicero, *ap. Non.* 194.13 (underfloor heating, fishponds).

83 Statius, who as a young man will doubtless have seen the technical extravagance of Nero's masterpiece, the Golden House, was intrigued by the contemporary developments in bath establishments; see *Silvae* I.3 (The Pleasaunce of Vopiscus at Tivoli); I.5 (The Baths of Claudius Etruscus—'where not even the man who comes fresh from the Baths of Nero would be loath once more to sweat!'.)

84 The Exeter system: Bidwell (1980), 30, fig. 16, has a conjectural restoration of the baths and furnace house and full discussion of the technical development there.

85 Boscoreale no. 13: see Rostovtzeff, (1941, vol. 2, n. 26, p. 552); White (1970, 442).

86 See W. L. McDonald and B. M. Boyle, 'The small baths at Hadrian's Villa', *JSAH* 39.1 (1980, 1–27).

87 Shortly before the publication of *Ancient Engineering,* a number of startling discoveries were made, affecting the design of torsion-spring catapults; most, but not all, of the new evidence was covered by D. Baatz's article in *Britannia* 9 (1978), 1–17; see further Appendix 14.

Chapter 5

89 See Appendix 3. The high output figures commonly given for the undershot water-mill have no sound basis. Forbes, *Studies* II.88 assumes without discussion that Roman mills gave five revolutions of the millstones to one of the wheel! But Vitruvius' short and perfunctory account (10.4.2) states unequivocally that the horizontal gear wheel is smaller than the vertical one, giving a ratio of less than 1:1, making the process very slow, if sure!

90 Landels (1978, 66–68); an ingenious device for eliminating spillage, the hollow-rim wheel, for which the evidence is entirely archaeological is described and illustrated by Landels (68–70, fig. 16). Eight pairs were used to lift water 100 ft out of the Rio Tinto mines. See Ch. 9.

91 Lynn White, Jr. (1962).

92 Water-organ *(hydraulis)*: described by Vitruvius (10.8) and by Hero (*Pneumatike* I.43); the constant air pressure was maintained in a hydraulic reservoir, the air being pumped in by a Ctesibian pump, operated by a small windmill (see Landels 1978, 26 f. and fig. 6).

Chapter 6

93 Multi-purpose tools: White (1967, 59 ff.) (five varieties of pickaxe): *idem,* 66 ff. (six varieties of mattock).

94 White (1967, 71–103) (twelve varieties of *falx*).

95 White (1967, 157–73): and now Kolendo (1980, 162, n. 23), on the Koblenz relief.

96 H. Cüppers, *Trierer Zeitung,* 27 (1964, 151); White *Antiquity* 40 (1966, 49).

97 White (1967, 152 ff.) ('drags and threshing machines').

98 White (1975, 75–76) (winnowing basket, *vannus*).

99 Importance of relishes: a recent estimate puts the proportion of cereal products (bread or porridge) as high as 70 per cent. Hence the widespread use of piquant relishes, the best known of wich are the fish-sauce *garum* and its cheaper relatives.

100 The derivation of *pilum* = throwing spear, from the same root *pinsere—pistillum* (= 'little pounder')—as *pilum*—'pestle' is indisputable, but the priority of pounder over weapon is unproven.

101 On the alleged time-lag for the water-mill, Bloch, (1967, 136–68); Lynn White, jr. (1962). We now have a substantial number of finds as well as of literary references (eleven of the former have been claimed for Britain alone (Wikander 1979, 13–36). For a more balanced view of the problem of the spread of the water-mill in Europe, Maroti (1975, 255–280); for lists and commentary, Appendix 6.

102 So Forbes, *Studies* II. 88; See Appendix 3.

103 On the floating mills Wikander (1979) (part of a valuable and detailed study of the evidence for water-

104 Moritz (1958, 151 ff.); the recently constructed Pompeian mill at the Museum of London is unfortunately not a working model; and no results of any experimental milling have so far been published.

105 In the anonymous poem known as *Moretum (The Country Salad)*, which vividly portrays the early morning activities of a smallholding peasant; see White (1977, 1, 21 f.).

106 See White (1975, 112 ff.) (wine-making); (225 ff.) (oil-making)

107 The Villa Albani relief (Drachmann 1932, fig. 21, 151 description, 67); (the drum is clearly shown, but not how it works!); the Rondanini relief (see *ill. 65*), also Drachmann, 42 and 68, which shows both a mill and a press, which appears to represent Hero's lever-and-screw press, as described in our text (p. 69). See his diagram of the press, fig. 20, 151. The advantage of the weighted type is that pressure is sustained, whereas with the ordinary capstan press it is spasmodic.

Chapter 7

108 For an up-to-date survey of the main landmarks in the development of Greek monumental design, see Coulton (1977, 31–32).

109 Acquired stability of cracked architraves: J. Heyman, *JSAH* 31 (1972), 3–9.

110 On problems of scale arising from increases in spans, Coulton (1977, 74 ff., esp. 81 ff.); solutions included reducing clear spans by use of internal supporting colonnades; see list of spans, Appendix 9.

111 Hodge (1960); for the most recent discussion, Coulton (1977, 157–58). Wide spans need not imply a trussed roof; thery could be bridged by very large timber beams (for a beam of record size, Pliny *HN* 16.200). The 'ultimate solution', to leave the shrine unroofed, was done occasionally (Coulton 1977, 79 ff.). For an extreme position in this highly controversial argument, Plommer (1957, 303): 'we cannot say that Vitruvius knew it' (the tie-beam truss) 'for the roof of his basilica at Fano, as he describes it (5.1.6) could have either bearer-beams or tie-beams.

112 Text of the Arsenal inscription: *IG* II–III², 2.1.1668; for the portion relevant to actual building operations, Bundgaard (1957, 117–32), text with parallel English translation and commentary; restoration: Coulton (1977, pl. 14); Prostoon inscription: Bundgaard (1957, 100–10), text, etc. as above.

113 Thermum buildings: Megaron 'B': Dinsmoor (1950, 42 and fig. 14); Temple of Apollo: Dinsmoor (1950 51–53 and fig. 18); Coulton (1977, 36–37 and fig. 7).

114 Samos: first Temple of Hera: Coulton (1977, 31 and fig. 4).

115 Introduction of roofing tiles: Boardman (1973, 73).

116 Olympia: Heraion: Dinsmoor (1950, 53 f.); Coulton (1977, 43–44 and fig. 11).

117 Terracotta revetments, acroteria, etc. Martin, *Manuel* I (1965, 87–112).

118 *Thersilion*: E. A. Gardner *et al., Excavations at Megalopolis* 1890–91 (*JHS* Suppl. I (1892), 17–33; Dinsmoor (1950, 242 f. and fig. 89).

119 *Katagogion*: Dinsmoor 251 and fig. 91.

120 Preparation and centering of drums: Dinsmoor (1950, 171–72); use of clamps: Dinsmoor (1950, 174 f. and fig. 64).

121 'Bearer' beams: Dinsmoor (1950, 176 and fig. 66) (Propylaea at Athens); Coulton (1977, 148) (with full discussion of the structural problem in the central span of this structure).

122 Four compared with six.

123 See the Prostoon inscription from Eleusis (text and translation in Bundgaard (1968, 100–10); on cost of transport: R. S. Stainer, in *JHS* 72 (1953, 70–71; use of oxen: Burford (1960, 1–18); Burford (1969, 184–91).

124 Cranes: Coulton (1977, 144–45); Landels (1978, 84–94). The crane on the Haterii monument (see *ill. 3*) has 5 pulleys and 7 men on the 'square-cage' wheel, and a maximum loading of 21 tons. The hoisting power of a man on pulleys is estimated to be 2 tons for a 10 ft lift, with 5 pulleys and a 6-inch diameter axle; estimated time taken 30 minutes (Landels 1978, 89), who has much to report on the structural and operational limitations of these machines.

125 *Anathyrosis*: Coulton (1977, 46–48).

126 Treatment of Doric column: Dinsmoor (1950, 175–76 and fig. 65); the 'fusion' of drums first noticed on a corner column of the Parthenon by Stuart in 1760.

127 'Specifications': Bundgaard 1968, 111 ff. ('not drawings, but a comprehensive system of very detailed working descriptions' *idem*, 116); Coulton (1977, 54–58).

128 Greek use of arch and vault: Coulton (1977, 140; 143–44), stressing the powerful influence of temple-building tradition in retaining structural systems unmodified by new elements; see also Boyd (1978, 83–100).

129 On the influence of fine local building materials on Roman building skills, see Appendix 9.

130 Chemical reaction of pozzolana cement: Atkinson and Bagenal (1926, 54, n. 3); on various uses of pozzolana: Davey (1961, 103).

131 See especially Boethius (1948, 114 ff.) (limitations of Vitruvius' opinions of concrete); Lugli (1957), who omits the important developments of the 2nd century completely!; on these see Boethius and Ward Perkins (1970), F. Rakob, 'Bautypus and Bautechnik', *Hellenismus in Mittelitalien* (1977, 366–78), stresses the

originality of Roman builders of this period.

132 Plommer (1973, 143 ff.).

133 Nero's Golden House: Boethius (1960, Ch. 3, 94–188).

134 Carpentry for frames: Macdonald (1965, 147–48, 158–59).

135 See the detailed account in Appendix 9.

136 Porticus Aemilia: Boethius (1978, 128–29; fig. 125).

137 Baths of Caracalla: Robertson (1945, 258–60; fig. 110); Macdonald (1965), reconstruction: Crema (1959, fig. 701) (the usual ground-plan illustrations reflect the complexity of the structure, but not the astonishing skill of the designer); Basilica of Maxentius: Robertson (1945, 261–22) ("perhaps, even in ruin, the most impressive of all Roman monuments"); reconstruction: *idem*, 262, fig 111.

138 In the provinces as well as in Italy: see the account of the aqueduct system at Aspendos in Pamphylia, the high standard of which matches that of its magnificent theatre, which was built entirely of masonry, and with workmanship of the highest quality, as we can still see! But much provincial building was of poor quality: see Plommer (1956, 284–85), on the contrast between the high standards maintained in imperial Ostia and the shoddy work that marred the last days of Pompeii.

139 See Daumas, (1961, vol. I, 237 f., and fig. 84, 238); the innovations and improvements in the design of tools are not unrelated to the high standards achieved by those who used them!

140 Ward Perkins (1977, 247).

Chapter 8

141 Vitruvius, *De Architectura* I. 1. See Burford (1972, 101 ff., esp. 102).

142 Herodotus, 3. 60.

143 See Appendix 11.

144 Sprague de Camp, (1960, 87–113) (The Greek Engineers); 114–63 (The Hellenistic Engineers).

145 Demosthenes, *Against Kallikles,* 11.

146 'rutted' roads: noticed long ago by the observant Col. Mure (*Journal of a Tour in Greece,* vol. II, 251 (1842), who aptly described them as 'stone railways'; see Forbes, *Studies* II, 499, who notes the consistent depth (7–15 cm range), and gauge (138–44 cm range) as evidence of planning! 'When carved in rock, as was often the case in Greece, such a road was practically wear-proof and weather-proof' Casson 1974, 69). Hence the rut roads from Athens to the great sanctuary at Eleusis, home of the Eleusinian Mysteries, from Sparta to the sanctuary of Amyclae, and other heavily used roads; 'macadamised' roads: Mitchell (1940, 250).

147 Pausanias on road conditions: 8.54.3 (good road for carriages); 10.32.8 (Tithorea-Delphi: unsuitable for vehicles); 10.5.2 (Daulis-Delphi: increasingly difficult even for foot passengers); 2.11.3 (Sicyon–Titane: too narrow for wagons). Livy on Macedonian roads: 7.37.

148 Animals not shod: except with leather or straw shoes *(soleae)* for use in slippery conditions: Casson (1974, 181).

149 Sterpos *(p. 29.)* reproduces an early nineteenth-century drawing of a section through the Via Appia made in 1813; the report of the engineer who made it showed how the results 'contradicted the most authoritative theories on the foundations and superstructure of Roman roads'.

150 Statius, *Silvae* I. 3: the marvels of Vopiscus' villa at Tivoli (water in every bedroom v. 37; steaming baths, vv. 63 ff.; an aqueduct, vv. 66–67).

151 Development in road-construction: roads laid out *de novo* on a dead straight line, involving cuttings and viaducts and tunnels where the terrain is broken, come rather late in the story. The differences between earlier and later systems may be seen by comparing the old Via Gabina with the later Via Praenestina (A. Kahane and J. B. Ward Perkins, *PBSR* n.s. 27 (1972), 106 f.) and noting the difference between the humble low-level bridge of the former with its 'grandiose, still-surviving, high-level successor' *(loc. cit.).* The earlier routes were not laid out—they 'passed imperceptibly from footpath to pack trail, from pack trail to farm track, and from farm track to waggon road' *(ibid.).*

152 A. C. Brookes, 'Minturnae: The *Via Appia* bridge'. *AJA* 78.1 (1974), 41–48.

153 E. M. Wightman, review of H. Cüppers, *Die Trierer Romerbrücken,* Mainz (1969), in *JRS* 62 (1972), 209.

154 Open channel system: the old view, that this method was preferred by Roman engineers is no longer tenable, in view of the large and increasing number of piped systems attested: see Appendix 12. And for full discussion of the technical questions, N. A. F. Smith, 'Attitudes to Roman engineering and the queston of the inverted siphon', *HT* (1976), 45–71.

155 On this important aspect of civil engineering Forbes *Studies* II. 41 ff. is now quite inadequate; there is no good treatment of dams in the classical period; see the short, but very informative, chapter on Roman dams in Smith (1971, 25–56).

156 For the Lake of Homs scheme: Smith (1971, 39–42; fig. 4).

157 See C. Vita-Finzi, 'Roman dams in Tripolitania', *Antiquity* 35 (1961, 14–20); C. Vita-Finzi and O. Brogan, 'Roman dams on the Wadi Megenin', *Libya Antiqua* II (1965), 65–71.

158 See Smith (1971, 49), who points out that the standard achueved in the best Roman dams was not reached again until the nineteenth century.

159 Harbours: the comprehensive study by K. Lehmann-

Hartleben, *Die Antiken Hafenanlagen: Klio Beiheft*
14, Leipzig (1923), is still indispensable, but much
new information is being made available through underwater surveys; see now J. W. Shaw in Bass (1972,
Ch. 4, 87–112); on the second harbour of Corinth, at
Kenchreae, see Hawthorne, (1965), 191–200, Scranton *et al* 1978.

160 Harbour works at Alexandria: see Shaw in Bass,
(1972, 94–95) who notes that the massive underwater
structures have been variously attributed to all ages
from early Egyptian to Hellenistic and even Roman
times. Only underwater excavation can provide a
chronology.

161 Lighthouses: F. Besnier's art. 'Pharos' in *Dict. Ant.*
is excellent, though now outdated; on the lighthouse
at Coruña, still wonderfully preserved *(ill. 107)* see
the recent study by Hütter (1973).

162 See the full account in Testaguzza (1964) with full
photographic coverage of recent excavations.

163 Carthage: the commercial harbour: see the interim
report on the recent excavations by L. E. Stager in
Archaeology 30 (1977), 198–200.

164 Cosa: see Lewis, (1973); Brown (1951).

165 Details of the scheme: Brown (1951, 89–96).

166 Sir M. Wheeler, *Roman Art and Architecture,* London
(1964), pl. 185, 199; Singer, *Hist. Tech.* II, 520, fig.
473.

167 Canals: an important, but strangely neglected topic.
There is a brief, but stimulating discussion of the
technical problems in N. A. Smith, 'Roman canals',
Trans. Newcomen. Soc. 49 (1977–78), 75–86.

168 Canals are briefly touched on by Forbes (*Studies* II, 19
f.; 44 ff.), but only with reference to irrigation; the
few Roman references (44 ff.) are purely descriptive,
and there is no mention of construction techniques!

169 Canals with locks?: see the full discussion, using all
the evidence, by F. G. Moore, 'Three canal projects,
Roman and Byzantine', *AJA* 54 (1950), 97–111: lock
gates aré supposedly an invention of Leonardo, but
Strabo's account (17.1.25.804) of Ptolemy II's completion of a Suez canal is unequivocally in favour of
an artifical barrier.

170 E.g. Cicero, *De Oratore* I.62 (on Hermodoros of
Salamis (2nd cent. BC) designer and builder of dry-
docks); Athenaeus, *Doctors at Dinner* 5.204 c–d (on
the dry-dock designed by a Phoenician for Ptolemy
IV of Egypt).

Chapter 9

171 'They were mainly interested in the origin and
sources of metals and ores, a subject which would
now belong to economic geology and petrology.'
(Healy (1978, 15.). In his excellent introduction on
the geological background (*op. cit.,* 15–29), Healy,
having noted the vast practical knowledge of

minerals and metals revealed by Theophrastus' *De
Lapidibus,* expresses surprise that the Greeks and
Romans never developed any geological theories.
The reason for this failure is surely to be found in a
difference of attitude. The motivation which caused
the fact-accumulating, cataloguing methods of the
earlier naturalists to give way to the systematic
methods of modern botanists and zoologists was lacking.

172 Forbes, *Studies* VII, 165. Were the Egyptian
engineers more advanced in this area than their
Greek and Roman successors, or is the Turin papyrus,
as one more example among many of a document
which owes its uniqueness to the highly favourable
conditions of the Egyptian climate to preservation of
documents actually written on papyrus?

173 Forbes, *Studies* VII, 165, denies the Romans any
knowledge of faulting or other geological complica-
tions: *contra* Healy (1978, 87), with specific examples.

174 Forbes, *Studies* VII, 145 ff., fig. 13. The chronology of
the Stageira workings not secure: 'Mines at Stageira,
known to have been exploited in the time of Alexander the Great, may well have been in operation
several centuries earlier' (Healy 1978, 78).

175 Laurion mines: Ardaillon (1897); the most recent
studies are by R. J. Hopper, *BSA* 63 (1968), 293 ff.,
'The Laurion mines: a reconsideration'; and *BSA* 48
(1953), 200 ff., 'The Attic silver mines in the fourth
century B.C.'. Comprehensive account in Healy
(1978, 78 ff.), emphasizing development from early
opencast to deep vein mining.

176 See S. Judson and A. Kahane, 'Underground drainageways in southern Etruria', *PBSR* n.s. 18 (1963),
34–99.

177 Healy (1978, 102, n. 242), citing Allan, *Iberian Mining,*
15 ff.; and Vitruvius, *De Architectura* 10.2.1 ff.

178 See D. G. Bird, 'The Roman gold-mines of north-
west Spain', *Bonn. Jb.* 172 (1972), 36–64 (with useful
survey of work to date); P. R. Lewis and G. D. B.
Jones, 'Roman gold-mining in north-west Spain,
JRS 60 (1970), 169–85 (Las Medulas mine, 174–78;
Puerto del Palo mine, 178–81).

179 G. D. B. Jones and P. R. Lewis, 'The Dolaucothi
gold-mines', *Bonn. Jb.* 171 (1971), 288–300.

180 One of many cases of innovation in the Roman period (see Ch. 4).

181 Jones and Lewis (above, n. 179), 178–81.

182 The latter resembled 'Pompeian' grain-mills (see Ch.
6); Healy (1978, 142 and fig. 21).

183 This process, technically know as 'jigging', is still
used (Healy, 144).

184 For a summary of research findings, Forbes, *Studies*
VIII, 26 ff.

185 Healy (1978, 234). Pausanias (2.3.3) reports the process as hearsay only. Plutarch (*Moralia* 5.262 f.)
thinks it may be a lost art, which would account for

the disappearance of bronze weapons!

186 For the details of this highly skilled operation, see Forbes, *Studies* IX. 20 ff. and for recent experiments, Healy, 158 ff.

187 Healy (1978, 190 ff.); not a mechanical method of removing one element from a natural mixture, but a true chemical process, according to Caley, *Theophrastus on Stones*, 188 (cited by Healy *(loc. cit.)* who reports that experiment has proved the method first described by Theophrastus *(op. cit.* 60) to be viable.

188 Forbes, *Studies* VIII. 196 ff. esp. 109 ff.; Healy (1978, 179–81); Tylecote (1962, 75 ff.). Healy, discussing Pliny's confused account *(HN* 34.159) of the process of lead refining, notes that in desilvering lead by the cupellation process Greek and Roman craftsmen achieved remarkable success; 'the Greeks were able to desilver lead to 0.02 silver, while the Romans were even more successful (0.01 or even 0.002 per cent in some cases)', 180.

189 Mendip lead mines: Tylecote (1962, 82); Healy (1978, 181 ff.).

190 On recent experiments with Roman-type furnaces, Wynne and Tylecote, *JISI* 190 (1958, 339–48), summarized by Healy (1978, 186 ff.); more recently, H. Cleere, *Britannia* 2 (1971), 203–17; *idem,* 'Ironmaking', in Strong and Brown (1976, 127–42).

191 Possible use of fluxes: rests on the interpretation of a not very lucid passage in Aristotle's *Meteorologica* (4.383 a–b); *pro:* Caley and Richards (eds), *Theophrastus on Stones* (1956), 76 f., who identify the references in the two writers to the melting of 'fire-resisting stones' during smelting as evidence for the addition of acidic and limestone fluxes; *contra:* Tylecote, *op. cit.* 186 f.; see Healy (1978, 185–86), translation of both passages with full discussion of the problem.

192 See above.

193 Report by Dr Brown, *ISRI,* on the scythe-blades from Great Chesterford: see White, *Proc. Hung. Agric. Mus.* (1971–72), Budapest, 77 ff.

194 *Illustrated London News,* October 8th, 1955, 614 and fig. 2.

Chapter 10

195 Vigneron (1968, 140); White (1975, 51 ff.; 218 ff.).

196 Vigneron, *loc. cit.* (n. 195); Sion (1935, 62).

197 White (1977, pl. 20, top), in many parts of Africa the commonest way of transporting water from wells or standpipes in urban areas is in 4-gallon paraffin drums, a pair being suspended from a stick across the shoulder; the milkmaid's yoke, which gives both comfort and easier balance, is known both in Europe and the USA.

198 Frontinus, *Strategemata,* 4.7.

199 White (1975, pl. 12b) (amphora on pole); pl. 5b (manure on hurdle).

200 Landels (1978, 170).

201 White (1975, pl. 16) (sack of grain); (pl. 9a) (off-loading wine).

202 Rickman (1970, 11), who notes (p. 86) that the design of the granaries at Ostia rules out the use of wheeled transport.

203 Traffic in the main streets of many third-world cities is often reduced to walking pace or less by the intrusion into the stream of manually operated vehicles, such as the 2- or 3-man handcarts used in Nigerian town for local handling of grain, timber, etc.

204 *Experiences,* Oxford (1969, 28).

205 See Hodges (1971, 402).

206 *Salagassus inscription:* S. Mitchell, 'Requisitioned transport in the Roman Empire', *JRS* 66 (1976), 106–31.

207 White (1975, 108–10); used extensively in Spain and other mountainous regions in the Mediterranean area.

208 Cato, *De Agri Cultura,* 22 (the text unfortunately gives no mileage!).

209 There is still much investigation to be done on the complex problems of harnessing in relation to the anatomy of bovines and equines: see the penetrating analysis by I. K. Brunel, *A Treatise on Draught* (see bibliography to this chapter). The latest detailed treatment of the problems is that of H. Polge (1967, 28–54), who argues convincingly that factors other than harnessing difficulties must be taken into account: see also White (1975, 219–20); Chevallier (1976, 180).

210 Landels (1978, 177 f.).

211 By P. A. Brunt, reviewing White, *Farm Equipment of the Roman World,* in *JRS* 66 (1976).

212 *Finley* (1973, 126–28).

213 On long-distance trade in items other than luxuries, see the review of Finley's *The Ancient Economy* by M. W. Frederiksen in *JRS* 65 (1975), 166 ff.

214 It is usually assumed that neither Greeks nor Romans knew how to harness animals in tandem ('in line ahead'); the Gallo-Roman relief from Langres showing a 'four-in-hand' vehicle 'casts doubt on the assertion that the Romans could only harness animals side by side', Leighton (1972, 81). This careful, well-documented study provides much food for thought on many questions relating to animal transport, and is by no means restricted to the Middle Ages. Langres 'four-in-hand': Appendix 10, ill. 132. See also R. W. Bulliet, *The Camel and the Wheel,* Cambridge, Mass. 1975.

215 Private communication from a veterinarian; see also Savory (1970).

216 The record of pre-classical archaeology is much more satisfactory; restorations have been made of many of the important remains, e.g. of the Dejbjerg wagon and the Anglesey chariot (illustrated in Singer et al., *Hist. Tech.* II, figs 481 and 482).

217 Vigneron (1968); see also G. Raepsaet, *Attelages antiques dans le Nord de la Gaule... Trierer/Zeitschrift* 45 (1982), 215 ff.

218 For a full discussion of this topic see Haudricourt-Delamarre (1955, 155–87).

219 S. Piggott, 'The earliest wheeled vehicles and the Caucasian evidence', *PPS* n.s. 34 (1968), 266–318.

220 Cato, *De Agri Cultura*, 20.2, with Brehaut's (1933) commentary *ad loc.*

221 Virgil, *Georgic* 3, 585–86.

222 The suggestion was made by E. M. Jope in Singer et al *Hist. Tech.* vol. II, 551; queried by H. A. Harris, 'Lubrication in antiquity', *Greece and Rome*, 21/1, (April 1974), 33–36.

223 Anglesey chariot: Singer et al., *Hist. Tech.*, vol. II, 539, fig. 482; tyres in the new fragment of the *Edictum de pretiis* White, *JRS* 63 (1973), 105 f.

224 9.2.52.2; for the translated text, Appendix 10, p. 209.

225 *axungia:* large quantities of pork-fat appear in the quartermaster's stores list at the Roman fort of Vindolanda, close to Hadrian's Wall in Britain. Could this have been needed for the regular lubrication of the sinew-ropes of the catapults? Animal sinew loses its elasticity without lubrication.

226 The latter is of course the modern horse-collar, which first appears in Europe around the 10th–12th centuries (Forbes, *Studies* II, 85).

227 For a review of the evidence on this controversial topic, see above p. 134.

Chapter 11

228 In some of the better known areas, such as the Riviera coast, looting has increased in recent years; some known and recorded wrecks no longer exist, thanks to the predatory activities of 'sporting' divers.

229 See the reports in *IJNA* (1972–); in *Riv di Studi Lig.* (Italian Mediterranean coast); *Gallia* (French Mediterranean coast).

230 The cannon of Nelson's day were only slightly more effective than the most advanced torsion catapults of classical times. Accurate gunnery came only with the modern naval gun.

231 Personal communication from Professor J. E. Gordon, of the Department of Materials Technology, University of Reading.

232 By J. S. Morrison; see his *Greek Oared Ships* (1968, 268 ff.), citing all the evidence; for a succinct account, with diagrams, Landels (1978, 143–45); see also *Scientific American*, Oct. 1981.

233 See Morrison (1968, 285–86); for development of 'fivers', Casson (1971, 100–3).

234 The subject is treated in full by Wallings (1956), and briefly by Casson (1971, 121). The main function of the 'boarding bridge' was to hold the enemy vessel fast.

235 Ptolemy IV's monstrous 'fortier' with its 4,000 oarsmen had space for several batteries of arrow-shooters and stone-throwers; but, as Tarn explains (1930, 121), a catapult mounted in the bows of a quinquereme had to be compensated by ballast, which in turn 'submerged the vessel above its most efficient waterline'. Heavier siege-engines required a pair of them to be lashed together (Plut. *Marcellus* 17).

236 Lefebure des Noettes (1935). The same writer argued (*op. cit.* 49 and 69–70) that sea-going merchantmen did not exceed 60 tons burden. A major argument in this book, unsupported by solid evidence, is that the ancient steering oars were inferior to the later stern-post rudder (see Casson 1971, 224).

237 Casson (1971, 281–95 and Tables 1–6); *idem*, (273–80 and notes 13–28).

238 *Catalogue of Ship Representations in Roman Mosaics and Monumental Painting,* diss., Univ. of Pennsylvania, n.d. The author's aim was exclusively artistic, not technical, and sculptured representations are excluded; but the work contains valuable materials for the study of types of rig.

239 Information on cargoes is very scattered: Casson, *Index,* s.v. (no register of known cargoes, but most of the 32 wrecks he reports are locatable; see Appendix 10). For more recent finds, see *IJNA* and the other references in n. 229.

240 Kyrenia wreck: *Nat. Geogr. Mag.* (1970, 841–57).

241 A notable exception, is the wonderfully preserved wreck of Anse Gerbal, discovered off Port Vendres (Pyrenees coast) in 1973. The keel and the garboard strakes were recovered almost intact (ill. ...). Equally impressive are the remains of a much larger wreck, the Madrague de Giens, both timbers and cargo of amphorae beautifully preserved (B. Liou, *Gallia* 23 (1975), 585–89).

242 Wrecks show great variety of woods: Grand Congloué had keel and ribs of pine but her floors were of oak, while the Mahdia wreck was of elm (Casson 1971, 213, n. 52, with refs).

243 Casson (1971, 224 ff.); *contra* Morrison, who argued (*GOS* 291–92) for lateral as well as pivotal movement, The monumental evidence (e.g. *ill. 152* appears to support Casson).

244 See the observations of B. Liou in *Gallia* 23 (1975), 578) (on the Pointe de la Luque wreck); *idem*, art. cit., 585–89 (on the much larger Madrague de Giens wreck, where the same structural feature occurs, but the ribs are fastened with pegs, not tree-nails, and copper nails).

245 Mercanti (1979) who notes that the anchor with moveable stock makes its first appearance in the second century BC.

246 Rougé (1966, 67) using a very inaccurate calculation of amphora capacity reached the absurdly low figure

for a cargo of 3,000 of just over 30 tons; for lists of cargoes, see Parker and Taylor (1965). Up-to-date lists are badly needed!

247 P. Pensabene, 'A cargo of marble shipwrecked at Porto Scito near Crotone' in *IJNA* 7.2 (1978), 105–18 notes that (1) Capo Scito must be rounded by ships making for the Italian side of the Adriatic, (2) her cargo was mixed marble from several different quarries in Anatolia, (3) the cargo included 8 columns in two sizes, 5 blocks varying in size from 15 to 60 Roman cu. ft and valued at between 3,000 and 12,000 *denarii*. Date, *c.* AD 200.

Chapter 12

248 Wells: Forbes (*Studies* I, 146 ff.); water-raising devices: White (1975, 44–48); Landels (1978), s.v. 'water pumps', pp. 58–83.

249 Qanats: Forbes (*Studies* I, 152 ff.); as 'horizontal wells': N. A. Smith, *Man and Water,* London 1975, 70: qanats represent 'an extension of the concept of wells to a novel technique for locating and utilizing undergound water' *(loc. cit.).*

250 Gradients: see the recent study of Rome's aqueducts by D. R. Blackman in *PBSR* n.s. 33 (1978), 52–72. B. emphasizes the extreme difficulty involved in estimating the behaviour of water running in open channels: 'even if somebody of observational experience were available there must have been unexpected and embarrassing deviations' (p. 70).

251 'Hezekiah (727–669 BC) made a pool and a conduit, and brought water into the city': 2 Kings 20: 20; 2 Chron. 32: 30. Forbes *(Studies* I. 152).

252 Dioptra: see Dilke (1971, 76–8). *Dioptra* reconstructed, with some portions of Hero's description conjecturally supplied, to fill the gaps in the ms.

253 Siphons: Forbes' account (*Studies* I. 161) is very perfunctory, and completely ignores the difficulties arising in Vitruvius' account; see now Landels (1978, 42 f.); for a full study, set against the general background of Roman civil engineering, see Smith (1976, 45–73) with up-to-date bibliography.

254 So Forbes *(Studies* I. 161): 'In Roman times the siphon was used only in some few cases, probably because of leakages and the relatively poor materials available for high pressures'; followed by Landels (1978, 42 ff.). Were Roman pipelines inaccessible for maintenance as he declares (p. 42)? The archaeological evidence does not support this view (see Appendix 12).

255 So Smith (1976, 58): 'It is conceivable that Vitruvius' imperfect reading of some Greek work on siphons had caused him to confuse the inverted siphon with the true siphon, that is to say a pipeline which at some point carries the liquid flow above the hydraulic gradient.' This would account for the reference to

the need of 'water-cushions' (Morgan's version of the otherwise unknown word *colliviaria*), or 'stand-pipes' (Granger's even more unlikely rendering!) or 'air-valves', as others have suggested. See Appendix 12 for further discussion of the problem.

256 See Thompson (1955, 106 ff.); in view of the extensive remains at Lincoln, which were partly known before the major excavations reported by Thompson (art. cit.), it is strange to see the report from Forbes (*Studies* I; 164) in the same year (1955) that 'there is nothing left of the Lincoln system'.

257 T. W. Battersby (diss. (unpubl.), Univ. of London, 1980) has given details of 21 known examples of the force-pump, including a set of four, used pumping out the bilges, having a combined lift of 25 20 l (500 galls) per hour, at the astonishing level of 95.45 per cent efficiency.

258 Lead poisoning: Vitruv. (8.6.11) notes that lead poisoning is an occupational disease of plumbers, and mentions the characteristic pallor which replaces the natural colour of the body. The main danger is not in the pipes, where the water is frequently in motion, but in the reservoir, where it is stagnant. Where neither springs nor wells are available, he recommends cisterns to catch rain water; these are to be made of the concrete known as 'signinum work', and fitted with compartments to ensure cleansing.

259 On putty: 'Their joints are to be coated with quicklime worked up with oil' (Vitruv. 8.6.8).

260 A slight emendation of the obscure passage in Vitruvius' account of closed pipe systems discussed (above, n. 255) would give a meaningful reference to his type of device (details in Landels 1978, 46 ff.).

261 *Contra* Landels (1978, 49) who ascribes the failure to measure rate of flow to 'neglect or ignorance of the dynamic features'. Frontinus is no scientist, but an administrator, and rule of thumb methods are good enough for him. Contrast the approach of Hero to the same problem (Appendix 12).

262 Blackman, (1978), the latest contributor to the subject, estimates the daily flow of four aqueducts at 7 m³/s or 600,000 m³/day; cf. Ashby's estimate of 8.8 m³/s, making Blackman's estimate at least credible.

263 Public latrines: Sprague de Camp (1960, 203, f.).

264 Water from Marcia: Pliny, *HN* 31.41; sterilizing by boiling: *Pliny HN* 31.40; Oribassius, *Collect. Med.* 5.1.11; *Digest* 33.7.18.3; *Geoponika* 2.47 fin.

265 The three instruments are discussed by Dilke (1971), 74–76 *(chorobates),* 75–79 *(dioptra),* 79 *(libra aquaria),* with reconstructions *(ills 172–74).*

266 Uses of the *dioptra:* Hero says it is to be used for all types of survey work, and for astronomy; Vitruvius is lukewarm; the instrument is not mentioned in any of the writings of the Land Surveyors, 'presumably regarded as too elaborate, expensive and unwieldy for regular use' (Dilke (1971, 79).

List of Illustrations

1 Greek black-figured vase depicting a blacksmith's forge. From Blümner 1887, vi, 53.
2 Bricklaying scene. Wall painting, tomb of Trebius Iustus, Via Latina, Rome.
3 Tread-whell crane. Monument of the Haterii, Musei Vaticani, Rome.
4 Ctesibian water-pump from Silchester.
5 Ctesibian water-pump in section.
6 Diagram of nozzle attached to the Ctesibian water-pump for fire-fighting. From Landels 1978, Drawn by Edgar Holloway.
7 Three types of water-lifting machine described by Vitruvius. From Ucelli, *Storia della technica,* Milan 1945.
8 Bas-relief from Linares, a mining centre in Spain. Society of Antiquaries, London.
9 Mosaic from Sousse, Tunisia, *c.*250 AD, depicting a ship unloading. Bardo Museum, Tunis.
10 A pair of multiple drainage wheels, Rio Tinto, Spain. From Healy 1978.
11 Roman freighter under full sail, 3rd century AD. Relief found at Portus. Museo, Torlonia, Rome.
12 Diagram of Archimedean screw from Centenillo, Spain. From Healy 1978.
13 Plan showing site of Athens and the harbours of the new port of Piraeus. From M. Grant, *Ancient History Atlas,* London 1971. Drawn by Arthur Banks.
14 Map showing supply areas of raw materials in the Roman World. From M. Grant, *Ancient History Atlas,* London 1971. Drawn by Arthur Banks.
15 'Balanced' Roman sickle *(falx messoria)* from Pompeii. From White 1967. Drawn by L. A. Thompson.
16 Two panels from the Porte de Mars, Reims. From White 1967.
17 *Carpentum,* restoration from Palladius' detailed description. From White 1967. Drawn by L. A. Thompson.
18 Greek black-figure amphora, 6th century BC. Hermitage Museum, Leningrad.
19 Catonian lever-and-drum press. From Drachmann 1932. Drawn by author.
20 Modern Catonian lever-press with counter-weight and capstan, northern Algeria. From Camps-Fabrer 1953.
21 Restored oil press from Pompeii.
22 Vitruvius' aqueducts for lead and clay pipes.
23 Vitruvius' bucket-chain. From Landels 1978. Drawn by Edgar Holloway.
24 Laurion, diagram of shaft furnace built by Tylecote on the basis of a 2nd-century AD furnace from Norfolk. From Healy 1978.
25 Washery B at Agrileza, Laurion. From Healy 1978.
26 Reconstruction of a primitive bloomery hearth. From Healy 1978.
27 Black-figure Corinthian votive plaque showing a potter at work. From Neuburger, A., *Die Technik des Altertums,* Leipzig 1919.
28 Greek black-figure *oenochoe* depicting a shaft furnace in operation. British Museum, London.
29 Interior of a Roman pottery mould. From Arezzo. From D. E. Strong, and D. Brown, *Roman Crafts,* London 1976.
30 Dyer (?) from a funerary moment at Arlon, Belgium. From J. Wild, *Textile Manufacture in the north-west Roman Provinces.*
31 Scenes from a fuller's tomstone from Sens, Yonne, France. Ditto 30.
32 Dyeing and weaving works at Isthmis, Greece. Ground plan. From C. Kardara, in *AJA* 65(1961).
33 Black-figure vase from Thebes, depicting Circe standing at a Greek vertical warp-weight loom. Redrawn from a vase at the Ashmolean Museum, Oxford. From A. Macleish, *Greek Exploration and Seafaring,* London 1977.
34 Roman glass bowl with pinched vertical ribbing. Ashmolean Museum, Oxford.
35 Wall mosaic, depicting a Hellenistic galley, 1st century BC. Palazzo Barberini, Palestrina.
36 Forum baths, Pompeii.
37 Reconstruction of Hero's *cheiroballistra.* From D. Beatz, *Britannia* ix, 1978.
38 Bas-relief from Arezzano showing two men turning a capstan. Museo Torlonia, Rome.
39 Relief from the Vigna delle tre Madonne showing a horse mill. Musei Vaticani, Rome.
40 Animal-operated *sakiyeh* near Puerto d'Andraix, Majorca.
41 *Sakiyeh* from Puerto d'Andraix showing vertical wheel and bucket-chain.
42 Undershot waterwheel (above) and overshot waterwheel (below). From Landels 1978. Drawn by Holloway. Holloway.
43 Large Roman freighter sailing full and bye to the breeze.
44 Vine-dresser's knife *(falx vinitoria).* From White 1967. Drawn by author.
45 Sole-ard (Algerian) closely resembling Roman design.
46 Socketed ploughshare. From White 1967. After H. J. Höpfen.
47 Reconstruction of harvesting machine type I *(vallus).* From White 1967. Drawn by H. Cüppers.
48 Buzenol harvesting machine.
49 Wooden pestle and mortar (African).
50 Terracotta from Thebes, late 6th century BC, depicting a man working a saddle quern *(mola trusatilis).*
51 Diagram to illustrate operation of Olynthian grain-mill. From Moritz 1958.
52 Diagram showing the parts of a hand-mill (Cato's 'Spanish mill'). From Moritz 1958.
53 Upper stone of small mill found at Saalburg, Germany. From Moritz 1958.
54 Section through the stones of a normal Pompeian mill. From Moritz 1958.
55 Mills from the bakery in the Vico Storto, Pompeii. From Moritz 1958.
56 Roman water-mill according to Vitruvius. From Moritz 1958.

57 Primitive Greek lever-press with stone-weight. From Forbes, *Studies* II.

58 Catonian press beam and capstan, from the model collection in the Museo Nazionale, Naples.

59 Hero's lever-and-screw press with hanging stone weight. From Drachmann 1932. Drawn by author.

60 Lever-and-screw press still in use in Morocco. From Camps-Fabrer 1953.

61 Single screw portable press as described by Hero. From Drachmann 1932. Drawn by author.

62 Wedge-and-beam press. Redrawn from a wall-painting at Pompeii.

63 Reconstruction of a Greek oil-mill *(trapetum)* from Olynthus. From Parsons, *Hesperia* 5, 1936.

64 Columella's *mola olearia*. From a relief in the Palazzo Rondanini, Rome.

65 The Rondanini relief, Palazzo Rondanini, Rome.

66 *a* Roof construction of Philon's arsenal at Piraeus. *b* Tie beam truss. *c* Transverse section of the arsenal. From Dinsmoor 1950. Drawn by author.

67 Ground plans of Greek temples: *a* Thermum, Temple of Apollo. *b* Samos, first Temple of Hera. *c* Olympia Heraion. *d* Paestum, Temple of Hera. *a, c* and *d* from Dinsmoor 1950, drawn by author. *b* from Coulton 1977 drawn by author.

68 *a* Eleusis, Hall of the Mysteries. Iktinos' original design, *c.* 340 BC. *b* Megapolis, the Thersilion. *c* Epidauros, the Hotel *(katagogion)*. *a* from Coulton 1977, drawn by author. *b* and *c* from Dinsmoor 1950, drawn by author.

69 Construction of columns with centering pin. From Dinsmoor 1950. Drawn by author.

70 Early Greek construction techniques. From Coulton 1977. Drawn by author.

71 *a* Metagenes' cradle for moving heavy beams. *b* Colossal stone transport. From Coulton 1977. Drawn by author.

72 Crane fitted with reduction gear. From Landels 1978. Drawn by Edgar Holloway.

73 Pulley and sheave from Zugmantel, West Germany.

74 Drum of crane replaced by tread-wheel operated by men from inside. From Landels 1978. Drawn by Edgar Holloway.

75 Weight-reducing techniques for heavy beams. From Coulton 1977. Drawn by Edgar Holloway.

76 Aids to lifting blocks showing Lewis bolt. From Landels 1978. Drawn by Edgar Holloway.

77 Section of drum to show the various processes and layers. From Dinsmoor 1950.

78 Tenons for lifting columns into position on the Parthenon. From Marquand 1909.

79 Triglyphal frieze of the Parthenon: axonometric view. From Marquand 1909.

80 Diagram for constructing stylobate curvature. From Dinsmoor 1950. Drawn by author.

81 Roman barrel-vaulting; *a* single, *b* intersecting. From Banister Fletcher, *A History of Architecture on the comparative method*. London, 1951.

82 Types of Roman wall construction.

83 Pantheon, Rome. Sketch showing construction. From Picard 1965.

84 Pantheon, Rome. Interior restored. Painting by Giovanni Paolo Panini, *c.* 1740. National Gallery of Art, Washington, D. C. Samuel H. Kress Collection 1939.

85 Vaulted roof, Nimes, the so-called 'Baths of Diana'.

86 Porticus Aemilia, Rome.

87 Basilica of Maxentius in section to show construction. From Picard 1965.

88 Colosseum, Rome. Auditorium in section to show construction. From Picard 1965.

89 Pantheon, Rome. Sketch showing the brick arches in the walls and the coffering of the dome.

90 Diversionary dams in the Wadi Meginin, Tripolitania. From Vita-Finzi 1961

91 Rutted road in Greece, From *Hist. Tech.* II.

92 The Via Appia showing paving blocks and kerb. From Sterpos 1970.

93 Section of the Via Flacca at Pisco Montano near Terracina. From Sterpos 1970.

94 Section through the Via Appia, nade in 1813.

95 Section of the Via Mansuerisca in the Haute-Fagnes, Belgium. From J. Mertens-M. Chevallier, *Roman Roads*, Berkeley, Calif. 1976

96 Roman road on Blackstone Edge, Yorkshire, section. From Richmond, in *Roman Engineering*, ed. R. Bosanquet.

97 Section of a Roman mountain road, Via delle Gallie, near Donnaz, Val d'Aosta.

98 Bridge over the Tiber at Narni, on the Via Flaminia.

99 Reconstruction of the Roman bridge at Trier. From H. Cüppers, *Die Trierer Römerbrücken*, Mainz, 1969.

100 Forlo tunnel on the Via Flaminia, dug 76–77 AD. From Sterpos 1970.

101 Section of tunnel on the Via Domitiana. From Sterpos 1970.

102 Cross-section of a *substructio* in a Roman aqueduct. From Landels 1978. Drawn by Edgar Holloway.

103 Cross-section of a typical Roman aqueduct. From Landels 1978. Drawn by Edgar Holloway.

104 The Cornalvo dam, Mérida, Spain, early 2nd century AD.

105 The Proserpina dam, Mérida, Spain, early 2nd century AD.

106 Harbour plan, Alexandria. From Sir L. H. Savile, 1941.

107 Roman lighthouse at Brigantium (La Coruña), Spain. From S. Hütter. *Der Römischen Leuchtturm*, 1973.

108 Lepcis Magna showing remains of the lighthouse.

109 Diagram of the harbour of Portus at Ostia.

110 Bronze sestertius of Nero, AD 64–65.

111 Carthage, commercial harbour.

112 Harbour plan Cosa. From Brown 1951.

113 Cosa Harbour, northern end of the later of the two-sluiced channels. From Brown 1951.

114 Wall-painting from Herculaneum depicting the harbour of Puteoli (Pozzuoli). Museo Nazionale, Naples.

115 Ostia, Piazzale delle Corporazione, Mosaic showing square-rigged Roman freighter about to pass the Claudian lighthouse.

116 Harbour basin of Cesarea.

117 Lepcis Magna, quayside.

118 Corinthian painted pinax, Pentaskovfi, Corinth. Berlin, Staatliche Museen.

119 Laurion, cross-section of silver-bearing strata. From Healy 1978.

120 Well-preserved miner's ore-bucket of esparto grass.

Urheberecht Bergsakademia, Freiberg.

121 Las Medulas mine complex. From Barri-Jones 1970

122 Puerto del Palo, mine complex. From Barri-Jones 1970

123 Ore-grinding mill in section. From Healy 1978.

124 Ore-washing table at agrileza, Laurion. Second half of 5th century BC.

125 Bowl furnace (experimental) constructed and used by Wynne & Tylecote 1958. From Healy 1978.

126 Reconstruction of a Roman shaft furnace. From Healy 1978.

127 Roman silver ingot. British Museum, London.

128 Cross-section of Romano-British scythe blade from Great Chesterford.

129 Six-oared tug *(navis codicaria)* for harbour work. Wall plaque from a mausoleum, Isola Sacra. 3rd century AD.

130 Sketch of breaststrap and girth harness. From Haudricourt–Delamarre 1955.

131 Bas-relief from Langres, northern France showing a four-wheeled carriage, pulled by four heavy horses. Musée de Langres.

132 Bas-relief from Langres, northern France, depicting a four-wheeler carrying a capacious wine barrel, drawn by mules. Musée St–Didier, Langres.

133 Diagram of a primitive two-wheeler from the Landes district, southern France. From Haudricourt-Delamarre, 1955.

134 Plan and elevation of the Dejbjerg wagon. From *Hist. Tech*-II.

135 *a* Three-piece solid wheel; *b* wheel with single-bar containing the hub and two cross-members; *c* multi-spoked wheel with eight spokes and six felloes.

136 One piece felloe wagon wheel from Saalburg. From J. Curle, *A Roman Frontier Post and its People,* Glasgow 1911.

137 Bas-relief from Klagenfurt, Austria, depicting covered travelling coach drawn by two horses.

138 Heavy Roman cartwheel from Newstead, Scotland. From J. Curle, *A Roman Frontier Post and its People,* Glasgow 1911.

139 Bas-relief showing a shafted two-wheeler with a basketry doby drawn by two mules. Musée de Montauban, Buzenol.

140 Bas-relief showing a pair of draught mules pulling a shafted vehicle. Musée d'Arlon, Belgium.

141 High-sided travelling carriage with light eight-spoked wheels drawn by a pair of mules. Musée Calvet, Avignon.

142 Diagram showing method of sailing against the wind with the sail brailed up. From Casson. 1971.

143 Diagram showing various types of rigging and sails.

144 Relief from tombstone of Alexander of Miletus, 2nd century AD, showing lateen–rigged vessel.

145 Detail from tombstone of Naevoleia Tyche, shipper of Pompeii, showing sailing vessel entering port, *c.* AD 50.

146 Diagram to show method of attaching garboard strakes to keel, floor timber and mast socket. From Casson 1971.

147 Shell-first technique of boat-building; A Swedish shipwright at work, 1929. From Casson 1971.

148 Model of Roman freighter under construction.

149 Funeral relief (late 2nd century AD) showing the shipwright Longidienus preparing a rib for insertion into the finished hull. Museo Archeologico, Ravenna.

150 Diagram to show edge-to-edge attachment of planking by mortises and tenons. From Casson 1971.

151 Hull assembly showing mortising. From *JNA* 1977.

152 Keel of Roman freighter, viewed from above.

153 Mosaic from the Piazzale delle Corporazione, Ostia, depicting a three-masted freighter with large foresail.

154 Mosaic from Sousse, Tunisia, depicting a Roman freighter with oars as well as a large foresail.

155 Relief, 2nd century AD, showing a freighter entering the harbour at Ostia and unloading. Museo Torlonia, Rome.

156 Housing and brackets for steering oar from ship of Odysseus, Museo Sperlonga, Italy.

157 Wooden anchor of Nemi barge with lead stock.

158 Reconstruction of an anchor with arms and shank secured by a lead collar.

159 Iron anchor with removable stock cased in wood. First half 1st century AD.

160 Mosaic from Ostia showing trans-shipment of cargo of amphorae from freighter to river craft.

161 Windlass and ropes for a domestic well, Casa a Graticcio, Pompeii.

162 Well bucket from Newstead, Scotland.

163 Egyptian wall painting showing a gardener watering from a *shaduf* with twin buckets. From a tomb in Thebes. From R. Remondon, *Histoire Générale de Travail,* Paris 1959.

164 Diagrammatic section of the aqueduct of Samos, probably 6th century BC.

165 Twin aqueducts carrying water across the River Louros, Epirus, Greece.

166 Aqueducts of ill. 165 used to carry water underground to Nikopolis.

167 Central arcaded portion of Rome aqueducts. Segovia, Spain.

168 The great aqueduct intersection outside Rome showing combined channels of Anio Novus and Claudia. Deutsches Museum, Mainz.

169 Map showing routes taken by the principal aqueducts serving Rome. From *Hist. Tech*. II.

170 Two channels crossing the Via Praenestina and the Via Labicane at Porta Maggiore, Rome.

171 One or many surviving water storage cisterns in North Africa.

172 Surveying instrument. Reconstruction of Hero's *dioptra* by O. A. W. Dilke, *Roman Land Surveyors,* Newton Abbot 1971.

173 Surveying instruments, front and side views. Hero's *dioptra* Reconstruction by O. A. W. Dilke op. cit.

174 Surveying instrument *(chorobates),* Reconstruction by O. A. W. Dilke, *op. cit.*

175 Surveying instrument, Hero's 'leveller'.

176 Surveying instrument *(groma)* from restored model in the Museo Nazionale, Naples.

177 Ktesibios' water organ, restored. From Sprague de Camp 1960.

267

Index

Figures in italic refer to the pages on which illustrations appear

Acropolis, Athens 78, *82*
acroteria 76, 204
agora 83
agricultural machinery 10, 29, 53, 60–1, *61*, 174–5, 183
agriculture 9–10, 12, 18, 21, 22–3, 28–30, 58, 195
air-locks 32, 161, 162
Alexander the Great 26
Alexandria 26, 105, *105*
Ammianus Marcellinus 218
amphitheatres 89
amphorae 147, 153, 154, 206, 207, 211, 212
Ampurias 217
amurca 62–3, 71
anchors 151, *152*, 153, 213
Ancient Economy, The (Finley) 18, 172
animal power 33–4, 51–5
antefixes 76
Antikythera machine 179
Aqua Claudia 163
Aqua Marcia 168
aqueducts 35, 87, 91, 100–2, *101*, 116, 119, 161, 162–4, *163, 164*, 165, *166, 167*, 168, 171, 192, 197, 206, 215
Aquileia 191
arcading 87, 102, 162, 163
archaeology, underwater 141, 145
arches 73, 83, 86, 97, 98, 99, 204, 207
Archimedean screw 15, 23, *23*, 32, 35, 154, 173, 178, 185, 192, 194
Archimedes 6, 7, 13, 15, 25, 47, 178–9
architects 73, 202–3, 205
Ardaillon, E. 116
Ariccia 99
Aristotle 18, 177
Arles 163
Arretine pottery 38
arrow-heads 217
Arsenal inscription 205
artillery 47
ashlar 205
Aspendos 162, 206
Assembly buildings 76–7
Assyria 11, 14
Athens 20, *24*, 24–5, 197, 200
Augustus, emperor 83
Ausonius 36, 56
automata 181
Automatopoietike (Hero) 180, 181
automation of catapult 47, 218

axe/adze 28
axle-grease 137
axle pivot 134
axle, pivoted 133, 134, 208, 209

Baatz, dr. D. 217, 218
Baden-Baden 209
ballista 46, 218, 219
balneum pensile 46
Barbegal 56, 194, 197, 200
Bardown 216–17
barley 63
Baronius Sura 19
Basilica of Maxentius *88*, 89
bath-houses 44–6, *45*
Baths of Caracalla 84, 89, 169
Baths of Diocletian 73
battering-ram 193, 220
beams, weight-reduced *80*
bearers, iron 78
bellows 36
Bernal, J. D. 6, 179–80
Biton 179
Blackstone Edge 96, *96*
Bloch, Marc 196
'bloom', iron 125
blue frit (Egyptian blue) 44
boiler, water 45
bonding courses 205, 206
Boscoreale 45
bosses, handling 81
bow 219
brace 50
brakes 136–7
brass 123
bread 63, 67
bricklaying *13*
bricks 74, 84, 86, 204
brick sizes 204
bridges 86, 97–9, *98, 99*, 164, 206
bronze-casting 11, 36, 122, 164–5
bucket *157*
bucket-chain hoist 17, 33, *33*, 175, 192–3, 194
bucket hoist 17, 175
bucket, miner's 115, *115*
bucket-wheel 32, 53–4, *54*
building materials 73, 74, 84, 86, 203, 204–5
burglars 74
Burns, Alfred 6

buttresses 206, 207
Buzenol 60, 61, *61*

caementa 205
Caesarea *111*
calix 166
Callixenos of Rhodes 48
camels 132
canals 110, 112, 132, 216, 227–9
capstan 50, *51*, 68, *68*, 79
carding 40
cargoes 18, 131, 145, 153–4, 211–12
carpentry 50
carpentum 29–30, *30*, 131
carriage, 4-wheeled *130*, 133, 136, *137*, *139*
carriage, 2-wheeled 133, *134*, 136, *138*
Carthage 107, *107*, 163
Casson, Lionel 26, 108, 211, 212
catapults 17, 47, *47*, 50, 91, 177, 178, 179, 187, 217–19, 220
'Catonian' press 31, *31*, 68, *68*, 70, 174
Cato the Elder 19, 30, 32, 52, 59, 60, 67, 130, 174, 183
cement 83, 85, 192, 245
centering, timber 87, 206
chain drive 218
charcoal 10, 12, 216, 217
chariots 133, 136, 139
cheiroballistra 47
Chersiphron 79, *79*
Chevallier, Robert 95
China 22
chisel 126
chorobates 171, *172*
cisterns 168, *169*
clamps, bonding 78
Claudius, emperor 106, 163, 212
Cleere, Henry 36, 216–17
climate 22–3
Cloaca Maxima 168
clothing 189–90, 233–4
coal 10
coach, travelling 136, *137*
coffer-dams 98, 108
coke 10
collar (harness) 137–8, *139*, 208
Colosseum 73, 84, *88*, 89, 207
Columella 32, 58, 60, 62, 71, 72
columns, construction of *78, 81, 82, 82*
columns, shipping of 154
columns, use of 76, 77

combing 40
compartment-wheel *17*
compressed air 57
'computer' 179
concrete 85, 86, 87, 204, 207
construction methods, Greek *78*, 78–9, 81–2, 204, 205–6, 207
construction methods, Roman 83–90 *84*, 204–5, 206
contracting, private 19
copper 113, 114, 121–3
Corinth 25, 104, 108
Corpus Agrimensorum (Frontinus) 188
Coruña, La 106, *106*
Cosa 107–8, *108, 109*
cow 51
'cradle', drum-moving 79, *79*
craftsmen's skills 90
cramps 78
crane 14, 15, *15*, 79, *80*, 81, 91, 104, 175, 182
crane, pivoting 48, 220
crank 50
Cremona 217, 218
cross-bow 217
Ctesibian water-pump *16*, 17, 34, 177
cupellation 123
cutting, road 99

dams *91*, 98, 102–4, *102, 103*, 168–9
Darby, Abraham 10
Daumas, Henri 192, 193
De Architectura (Vitruvius) 49, 65, 87, 91, 123, 161, 171, 176, 202
defensive weapons 47
deforestation 216, 217
Delos 104
Demosthenes 92
De Rebus Bellicis (anon) 54
diet, classical 63
Diodorus 115
Dionysius of Alexandria 47
dioptra 101, 171, 172
Dioptra (Hero) 101, 159, 180, 214
dockers 127–8
docks, dry 112
documentary evidence 7, 15, 17, 20, 46, 60–1, 67, 73, 75, 96, 145
Dolaucothi 117
domes 86, *89*, 192, 207
donkey *51*, 52, 128, 129, 130, 132
Donnaz 97
doorways 74
Drachmann, Dr. A. G. 31, 68, 69, 70, 181, 185
drainage canals 110, 112
drainage, mine 34, 35
drainage systems 23, 115
drainage wheels *21*
dyeing 39, *39*, 40–1, *41*, 169

earthenware pipes 165, 214
eclipse 180
economy, classical 18–21
Edictum de pretiis 44, 56, 66, 136, 189, 208
Egypt 23, 24, 41, 53, 86, 174, 190, 191
electricity 21
Eleusis *77*, 79, 132
engineering, civil 91–112, 226
engineering, hydraulic *16, 17*, 46, 102–3, 119, 157–72, 231–2
Engineering in the Ancient World (Landels) 8, 12
engineers 92, 99, 203
environment, alteration of 9
Epidaurus 77, *77*
Eratosthenes 26
Erech 42
Erechtheum 82
Eretria 104
Etruria 23
Eupalinus of Megara 6, 91, 159–60, 215
Exeter 45

factories 190, 191
falces 59
Faventinus 85
fertilizing 195
Finley, Professor M. I. 6, 18, 34, 172, 190, 192, 193
fire-fighting *16*
fishery 108
fish farm 46
flame throwers 141
flux 125
food processing 30, 46, 63, 222–3
food production 22, 221
food supply 24
food technology 63–4
forge *11*
Forlo tunnel 100, *100*
forum 83
foundations 85, 207
frame saw 50
Franklin, Benjamin 165
Frederiksen, Martin 189
Frontinus, Sextus Julius 165–8, 187–8, 214, 215
fulling 39, *39*, 169
freighters *see* ships, cargo
furnace, bowl *122*
furnace, domestic 45
furnace, shaft *35*, 36, *37*, *123*

Garigliano, river 98
gear-wheels 177, 182
geography, influence of physical 23
geological map 113–14
glass-blowing 42
glass decoration 42

glass-working *41*, 41–2
glass-works 42
glazes 37, 38–9
glosses 38–9
Golden House of Nero 86, 206
gold extraction 36–7, 123–4
gold-glass 42
gold mining 34, 36, 113, 114, 116–20, 121, 171, 173
gold refining 36, 123
Gornea 217
grabs 220
grain 63–7, 128, 153
grain harvester 10, 29, 53, 60–1, *61*, 174–5, 183
grain-mills 30
grain-ships 145
grain supply 24, 26, 212
granaries 62–3

Hadrian's Villa, Tivoli 45–6
Hadrian's Wall 196, 198, 200
haematite 114
Hagia Sophia 84
Haltern 217
harbours 92, 104–10, *105–9, 111*
harness 130, *130*, 133, 137–40
harnessing, tandem 209
harrow 58, 195
harvesting machine *see* grain-harvester
Hassall, Mark 173
Haterii, monument of *15*
Hatra 217, 218
heading machine 10, 195
Healy, J. F. 34, 35, 113
hearth, bloomery 36, *36*
heat, conservation of 45
heating, domestic 44, 45–6
heating, industrial 44
heating, underfloor 44, 45
heating, water 45
helepolis 220
Herculaneum 45, 71
Hero of Alexandria 7, 15, 17, 31, *47*, 57, 68, 69, 70, 101, 159–60, 171, 172, 180, 195, 214
Herodotus 157, 158–9, 215
Herschel, Clement 188
Hiero of Syracuse 212
Hipparchus of Nicaea 26
Hippocratic bench 46
Historia Naturalis (Pliny) 12, 106, 116, 119, 124, 125, 183–5
Hodges, Henry 7
Homer 211, 213
Homs, Lake of 102–3, 104
hopper-rubber 30, 65, 195
horses 51, 52, 53, 128, 130, *130*, 133, 138, 139, 140, 141
hotel *(katagogion)* 77–8, *77*

houses, private 73, 75–6, 187, 207
hub-cores, mill-wheel 198, 200
hushing 117, 171
hydrostatics 178–9
hypocaust 44, 45, 46
Hypsaeus, Lucius Veranius 19

Igel monument 208
industrial organization 190–1
innovation and development 27
innovation, technical 21, 27
inscriptions 36, 73, 79, 82, 97, 129, 132,
   153, 158, 160, 169, 175, 202, 203,
   205, 215
inventions 14, 192
inventions, transmission of 10–13
instruments, surgical 46
insulae 85
iron 10, 11, 17, 113, 121, 125
iron, bloomery 36
iron, cast 36
iron-working 36, 125–6, 216–17
iron, wrought 10, 11, 125, 174
irrigation 23, 168–9
Isis 144, 145, 155, 212

Janiculum hill, Rome 197, 199
Jerome, St 156
jet propulsion 14, 57
joints, fusion of 82
Juvenal 135

Kasserine dam 104
katagogion 77–8, *77*
keels 151, *151*
Kelsey, F. W. 175
knife, vine-dresser's 28, 58, *58*
krepidoma 207
Ktesibios 17, *173, 177*

laconicum 45
Laird Clowes, G. S. 210
Lamesba 169
Landels, dr. John 8, 12, 17, 33, 34, 47,
lamps, miners' 35
Landels, Dr. John 8, 12, 17, 33, 34, 47,
   48, 49, 50, 52, 53, 54, 57, 101, 165,
   175, 195, 214, 219
Langers *133*, 208, 209
latrines 168
Laurion 20, 35, *35*, 113, 114–15, *115*,
   116, 120, 124, 125
lead 34, 124–5, 153, 162, 165, 206, 214
lead poisoning 165
leather 174, 236–7
Leighton, Albert C. 208
Lepcis Magna *106*, 109, 110, *111*
lever 10
lever and capstan press 68, *68*
lever and screw press *69*

lever press *31*, 32, 67, *68*, 70
Lewis bolt 81
Lex Iulia Municipalis 209
libra aquaria 171–2
lifting devices 81, *82*
lighters 153
lighthouse 91, 105–6, *106, 110*, 150
limestone 84
Lincoln 164, 198, 200
linen 41
Lipsada 114
lithoboloi 220
Livy 93
locks 112
loess 22
looms 41, *41*
Louros, river *163*
lubrication 137
Lucian of Samosata 155, 212
Lyons 163

Madalianus, L. Cornelius 110
maintenance, tunnel 101
Maiuri, Amedeo 99
maize 30, *63*, 64
Majorca *54*
marble 132, 212
Marcellus (Plutarch) 13, 220
Marsden, Eric 47, 217
masonry, dry 78
masonry, mortared 84, *84*
mattock 28, 58
Maudslay, Henry 21
maza 63
mechane 23
measurement, system of 205
mechanics, principles of 7
Mechanike (Hero) 15, 68, 70, 181, 185
Mediaeval Technology and Social Change
   (White) 172
medicine 46
Megalopolis 77
Mendip Hills 124
merchantmen *see* ships, cargo
mercury 123–4
Mérida *102*, 103, *103*
metallurgy 12, 36, 120–26
metopes 82
milestones 97, 138
mills, animal operated 33, 52, 65, 66,
   67, 194, 195, 198
mills, floating 66, 197, 199
mill, grain 30, 51, *51*, 52, 56, 64, 193
mills, hand *64, 65,* 66, 67, 120, 194, 195
mills, horizontal 199
mill, oil 32, 71–2, *71, 72*, 134
mills, revolving *(trapetum)* 71, *71*
mills, saw 56
mills, water 53, 55, 56, 64, 65–6, *66*,
   169, 193, 194, 195, 196–201

mills, wind 56–7
mills, woollen 191
millstones 30, 64, *65*, 71, 198, 200
Milvian Bridge 86
mine drainage 113
mine regulations 36
mine ventilation 34–5, 116
minerals 113, 120, 238–40
mining 19, *20*, 34,6 113, 114–20, *114,
   115, 118*, 153, 169, 171, 215
Minturnae 98
missiles 47
Mnesicles 78
models, scale 181, 203
mola olearia 32, *72*
mola trusatilis 30
mole 104, 105, 108, *109*, 110, *111*
mortar 84, 205
mosaics *20*, 42–4, *43, 150*, 153, *154*, 196
mosaics, vault 43–4
mould, pottery *38*
mould-pressing glassmaking 42
mud-brick 74, 84, 204
mules 51, 52, 53, 127, 128, 129, 130,
   132, *133, 138, 139*, 208
muscle power, human 49–51, 54–5, 153
muscular movements 49, 50

Narni 98, *98*
Nemi barges 151, *152*
Nero, emperor 153
Neumagen 208
Nonius Datus 160–1, 215
nymphaeum 44, 46

obelisk 212
oil 63, 71, 153, 184, 225
oil-mill 32, 71–2, *71, 72*, 134
oil-press *32*, 67–8, 174, 184
olive-crusher 130
Olympia 76
onager 218
On Dyeing (Bolus) 41
open channel aqueducts 32, 100, 163,
   165
Opus Agriculturae (Palladius) 12
opus incertum 84, 86, 205
opus quadratum 84
opus reticulatum 84, 86, 205
opus signinum 104
opus testaceum 84, 86, 204, 205
Orata, C. Sergius 44, 46
ore-washing 34–5, *35*, 120–21, *121*
organa 220
organum pneumaticum 17
organ 46, 57, *173, 177*, 178, 179
Orsova 217
Ostia 106, *107*, 108, 109, 110, *110*, 128,
   150, 153, *154*, 204, 213
output, machine 193–4

ovens 44
oxen 51, 52, 53, 54, 79, 128, 130–1, 137, 139, 140

pack-animals 128, 129, 131, 132, 140
Paconius 79, *79*
Paestum 76
Palladius 12, 60, 61, 62, 85, 165
Pantheon *85*, 86, *89*, 90, 206–7
papyri 41, 54, 113, 176
Parthenon 74, 82, *82*, 83, 203, 205
pavimentum sectile 43
pavimentum tessellatum 43
pavimentum vermiculatum 43
pebble mosaic 42
pentekontor 142
Pergamum 162, 165
periscope, German 175
Persson, A. W. 190
pestle and mortar 31, *31*, 63, *63*, 64, 120
Pharos 92, 105
Philon of Byzantium 75, 175, 178, 192, 219
pick-axe 58
pick/mattock 28, 58
piles 207
Ping-ti-Ho, Dr. 22
piped water-supply 32
pipes, water 159, 160, 161, 162, 163–5, 166, 206, 214
Piraeus *24*, 25, 73, 75, 104, 155, 205
piston 34, 181
pivot, axle 134
planetarium 179
plastering 204
Pliny the Elder 11, 12, 13, 17, 30, 31, 34, 39, 43, 44, 58, 59, 60, 61, 63, 67, 68, 70, 106, 114, 115, 116, 117, 119, 124, 125, 147, 155, 165, 168, 183–5, 212
Pliny the Younger 112, 131
plostellum Punicum 30
plough 29, 53, 58, 59, *59*, 174, 176, 184, 195
plough, wheeled 12, 60, 184
ploughshares 59–60, *60*, 174, 184, 195
plumbing 45
Plutarch 13, 19, 48, 143, 220
Pneumatika (Hero) 180–1
pole, load-carrying 127
Polybius 47
Pompeii 19, *32*, 44, 45, *45*, 52, 53, *65*, 70, 96, 148, 174, 191
Pont du Gard 87, 163
population growth 24
porridge 31, 63
Porte de Mars, Reims *29*
porterage 127–8
Porticus Aemilia *88*, 89
Portus (Ostia) *107*, 108, 110, 153, 213

Porto Terres 99
post-and-lintel doorway 74
pottery 37–9, 174
pounding 31, *31*
power resources 49–57, 195
pozzolana 83, 84, 106, 205
Pozzuoli 106, *109*
Praxiteles 204
pressure pipes 206
pressure towers 166, 206
presses 31–2, *32*, 67–71
presses, beam 185
presses, cloth 41
presses, lever 67, 70, 184, 196
presses, lever and capstan 68, 184, 196
presses, lever and screw 68, 69, *69*, 184–5
presses, oil *32*, 67–8, 174, 184
presses, screw 15, 31, 32, 67, 70, 182, 184–5, 192, 196
presses, wedge 70, *71*, 184
presses, wine 67
Procopius 98, 99
Prometheus 10, 51
Propylaea 78
prospecting 34, 113, 114, 117
Puerto del Palo *118*, 119, 171
pulley 10, 14, *80*, 177, 178, 181
puls 63
pump, Ctesibian force- 16, 17, 34, 154, 158, 164, 169, 177–8, 179, 192, 194
Punic cart 30, 62
puppet theatre 181
Puteoli—see Pozzuoli

qanat (water tunnels) 158, 160
quarrying 19, 78
querns 30, *64*, 120, 194
quinquireme 142

rainfall 23
ramming 142
ramps 206
raw materials, supply of 22, 25–6, *28*, 34
reaping 29
reconstruction of processes 15, 17, 36, 174–5, 217
reliefs *15, 20, 22, 29, 39, 51, 57, 61, 72, 128, 130, 132, 133, 138, 139, 145, 150*, 193, 208
relieving arches *89*, 90
reservoirs 102, 197
ribbing, brick 90, 207
rigging 210
Rimini 98
Rio Tinto *21, 33*, 35, 51, 113
roads 23, 92–7, *92–97*
roads, construction of 94–6, 209
roads, Greek 92–3, *92*, 128

roads, pre-Greek 92, 93
roads, Roman 92, 93–7, *93–97*, 209, 215–16
roads, 'rutted' 92, *92*
roofs 73, 74, *74*
roofspan 207–8
ruts 92, *92*, 96, 97, *97*

Saalburg *65*, 217
saddle, pack 129
saddle-quern 30, *64*, 194
sails 143–4, *144*, 148, 149, 151, 210
sakiyeh 32, 53, *54*
Saldae 160–1, 215
sambuca 220
Samos 76, 158–60
sand-core glassmaking 41
saw, frame 50
saw-mill 56
scaling-ladder 179, 220
Schlebeker, J. T. 21
Schramm, Major 217
Schuman, Theodor 37
scoop-wheel *17*
screw 10, 15, 181
screw, Archimedean 15, 23, *23*, 32, 35, 185
screw-cutter 15, 21, 31, 70, 182, 185
screw and lever press 68, 69, *69*
screw-press 15, 31, 32, 67, 182, 192
screw-pump 23, *23*
scythe 29, *29*, 50, 60, 125, *126*, 174, 183, 195
Segesta 82
Segovia 163, *164*
Selinus 207
Seneca 100
serracum 135
settling tanks 166, 214
shaduf 17, 158, *158*
shafts 53, 101, 114, 116, 139, 173
shear-legs 79
ship-building 46, 211
ships 15, 18, 20, *22*, 48, 141–8, *145, 150*, 177, 179, 210–13
ships, cargo 141, 143, 145–8, 153, 154, 210–13
ships, construction of 146–8, *146, 147*
ships, fighting 48, 54, 141–3
ships, multiple-banked 142
ships, sailing capabilities of 143
ships, speed of 143–4, 154
shuttering 86, 87, 205
sickle 28–9, *29*, 50
siege-engines 179, 220
sieve 30
sifter 30, 67
Silchester *16*, 194n.
silting 103, 109, 110, 161, 165, 214, 215
silver, extraction of 124, 125

silver-mining 18, 20, 25, 34–5, 36, 113, 114–16, 216
silversmithing 36
siphons 161, 163, 164, 165, 178, 214, 215
slag 10, 122, 125, 216
slaves 20, 189
sludge-cocks 165, 214
smelting 44, 121–2
smith 10–11, *11*
sole-ards 59, *59*
Sostratos of Knidos 92, 105
Spain 20, 23, 33, 34, 35, 116, 215
Stabiae *109*
staircases 186
Statius 96, 209
steam power 57, 180, 195
steel 10, 17, 126, 136
steering 149, *149*, 156
stone, building 78, 81, 86, 154, 203–/205
storage, grain 62–3
Strabo 25, 125, 213
Strategemata (Frontinus) 188
Straton 177
structural engineering 73
structural problems 76–7
Studies in Ancient Technology (Forbes) 7
substructio 100, *101*, 102
Suessa Aurunca 99
surgery 46
surveying 101, 160, *170*, 171–2, *171*, 188
suspensurae 44
swing-beam 17
Synesius 155–6
Syracuse 24, 47

Tacitus 112
tacking 143, 144, *144*
technology, ancient 6–7, 9
technology, mediaeval 6
temples 74, 75, *75*, 82, 85, 187, 204, 207–8
terracotta 76, 235
textiles 19, 39
theatres 76, 83, 89
Theodosian Code 129, 209
Thermum 75
Thersilion 77, *77*
Thompson, E. A. 193
threshing machine 30, 62, 195
Thucydides 25
tie-beam truss 83
tiles 76, 89, 154, 203, 204, 206
tile, standard 203
timber used in building 75, 90, 204, 216
toilet basin 178
tools, agricultural 28, 29–30, 49–50, 57–8

tools, builders 205
tools, carpenters' 90, 126, 193
tools, miners' 115
tools, surveyors' 171–2
Tophet, sanctuary of 107
toys 57, 178
trade, Mediterranean 189–90, 191, 201–2
trade with Britain 216
traffic conditions 209
Trajan's Column 217, 218
transport, land 79, *79*, 97, 127–40, 208–9, 230
transport, mediaeval 140
transport, water 131, 141, 143, 230
trapetum 32, 71, *71*
travertine 84
tread-wheel 15, *15*, 50–1, 54, 79, *80*, 113
treatises, technical 176–88
tribulum 30
Trier 99, *99*
Tripolitania 102, 103, 168
trireme 142
tufa 84
tug *128*
tunnels, road 99–100, *100*
tunnels, water 6, 100, 101, *101*, *117*, 158–61, 215
tyres 136

Ulpian 132
'untrussed' roofs 74

Vaihingen 209
vallus 29, *29*, 30, *61*
Varro, Marcus 30, 52, 62, 129, 131, 183
vase-paintings *11*, *31*, *37*, *41*, *118*
vaults, barrel/tunnel 73, *83*, 83–4, 86, 87, *87*, 89, 206, 207
vaults, groined *88*, 89
vehicles, wheeled 12, 131, 132–6, *133*, *134*, *137*, *138*, *139*, 208–9
Venafro (Venafrum) 197, 200
venter 33, 161
ventilation, mine 34–5, 116
Via Appia *93*, 94, *94*, 98, 168
Via delle Gallie 97
Via Domitiana 96, *100*
Via Egnatia 23
Via Flacca *93*, 97
Via Flaminia 100, *100*
Via Mansuerisca 95, *95*
viaducts 97, 99
viticulture 58–9, *58*, 60
Vitruvius 14, 17, *17*, 32, *33*, 43, 44, 45, 49, 53, 54, 55, *65*, *66*, 79, 84, 85, 87, 94, 98, 101, 108, 123, 161, 165, 171, 172, *173*, 177, 185–7, 203, 207

voussoirs 86, 99, 206

wagons 131, 133, *133*, 134, 135, 208, 209
wales 147, 148
wall-painting *13*, *109*, *158*
Ward Perkins, John 19, 86, 206
warfare 7, 11, 47–8
warfare, naval 141–3
warships 48, 54, 141–2
water heating 45
water in metallurgy 120–1
water-mills 53, 55, 56, 64, 65–6, *66*, 91, 100–4, 116, 160, 162–9, 171, 193, 194, 195, 196–201, 206, 214–15
water organ 46, *173*, 177, 178, 179
water pipes 159, 160, 161, 162, 214
water power 21, 35, 49, 55–6
water pump 192–3
water purification 168
water-raising devices 17, 23, 32–4, *33*, 53, 55, 176, 194
water supply 6, 23, 32, 157–171, 187
water tower (*castellum*) 166, 206
water tunnels 158–61, 171, 215
waterwheels *17*, 21, 33, 34, 35, 51, 53, 55, 56, 169, 174, 176
weapons 46–8
wedge 10
wedge-press 67, *71*
wells 116, 157–8, 164
wheat 63, 67
wheel, bucket 32, 53–4, *54*
wheel, drainage *21*, 192
wheel, overshot 55, *55*, 197, 198
wheel of pots 32, 34, 53, 54
wheel, potter's 37
wheel, undershot 55, *55*, 197, 198
wheel, vehicle 135–6, *135*, *136*, 192, 208
wheel, water *17*, 21, 33, 34, 35, 51, 53, 55, 56, 174, 176, 192
whetstone 12
white gold 36
windlass 10, 115, *157*, 158, 181
windmill 56–7
window-glass 42, 192
wind power 56
wine 63, 67, 153
wine production 224
wine wagon *133*, 208
winnowing 62
wool industry 19, 39–41, 191
wrecks 211–13

Xenophon 139
Xenophon, pseud. 25–6

yokes 127, 130, 137, 138, 139, 209